AF572592

The Institute of Mathematics
and its Applications
Conference Series

Previous volumes in this series were published by
Academic Press to whom all enquiries should be addressed.
Forthcoming volumes will be published by
Oxford University Press throughout the world.

NEW SERIES

1. *Supercomputers and parallel computation* Edited by D. J. Paddon
2. *The mathematical basis of finite element methods*
   Edited by David F. Griffiths
3. *Multigrid methods for integral and differential equations*
   Edited by D. J. Paddon and H. Holstein
4. *Turbulence and diffusion in stable environments* Edited by J. C. R. Hunt
5. *Wave propagation and scattering* Edited by B. J. Uscinski
6. *The mathematics of surfaces* Edited by J. A. Gregory
7. *Numerical methods for fluid dynamics II*
   Edited by K. W. Morton and M. J. Baines
8. *Analysing conflict and its resolution* Edited by P. G. Bennett
9. *The state of the art in numerical analysis*
   Edited by A. Iserles and M. J. D. Powell
10. *Algorithms for approximation* Edited by J. C. Mason and M. G. Cox
11. *The mathematics of surfaces II* Edited by R. R. Martin
12. *Mathematics in signal processing*
    Edited by T. S. Durrani, J. B. Abbiss, J. E. Hudson, R. N. Madan,
    J. G. McWhirter, and T. A. Moore
13. *Simulation and optimization of large systems*
    Edited by Andrzej J. Osiadacz
14. *Computers in mathematical research*
    Edited by N. M. Stephens and M. P. Thorne
15. *Stably stratified flow and dense gas dispersion*
    Edited by J. S. Puttock
16. *Mathematical modelling in non-destructive testing*
    Edited by Michael Blakemore and George A. Georgiou
17. *Numerical methods for fluid dynamics III*
    Edited by K. W. Morton and M. J. Baines
18. *Mathematics in oil production*
    Edited by Sir Sam Edwards and P. R. King
19. *Mathematics in major accident risk assessment*
    Edited by R. A. Cox
20. *Cryptography and coding*
    Edited by Henry J. Beker and F. C. Piper
21. *Mathematics in remote sensing*
    Edited by S. R. Brooks
22. *Applications of matrix theory*
    Edited by M. J. C. Gover and S. Barnett

Continued overleaf

23. *The mathematics of surfaces III*
Edited by D. C. Handscomb
24. *The interface of mathematics and particle physics*
Edited by D. G. Quillen, G. B. Segal, and Tsou S. T.
25. *Computational methods in aeronautical fluid dynamics*
Edited by P. Stow
26. *Mathematics in signal processing II*
Edited by J. G. McWhirter
27. *Mathematical structures for software engineering*
Edited by Bernard de Neumann, Dan Simpson, and Gil Slater
28. *Computer modelling in the environmental sciences*
Edited by D. G. Farmer and M. J. Rycroft
29. *Statistics in medicine*
Edited by F. Dunstan and J. Pickles

Computer Modelling in the Environmental Sciences

Based on the proceedings of a conference organized by the Natural Environment Research Council in association with the Environmental Mathematics Group of The Institute of Mathematics and its Applications and held at the British Geological Survey headquarters at Keyworth in April 1990

Edited by
D. G. FARMER
Principal Systems Consultant
NERC Computer Services
Swindon

M. J. RYCROFT
Head of College of Aeronautics
Cranfield Institute of Technology
Bedford

CLARENDON PRESS · OXFORD · 1991

Oxford University Press, Walton Street, Oxford OX2 6DP

Oxford New York Toronto
Delhi Bombay Calcutta Madras Karachi
Petaling Jaya Singapore Hong Kong Tokyo
Nairobi Dar es Salaam Cape Town
Melbourne Auckland

and associated companies in
Berlin Ibadan

Oxford is a trade mark of Oxford University Press

Published in the United States
by Oxford University Press, New York

British Library Cataloguing in Publication Data
A catalogue record for this book is available from the British Library

Library of Congress Cataloging in Publication Data
Computer modelling in the environmental sciences: based on the proceedings of a conference organized by the National Environment Research Council in association with the Environmental Mathematics Group of the Institute of Mathematics and Its Applications and held at the British Geological Survey headquarters at Keyworth in April 1990/edited by D. G. Farmer, M. J. Rycroft.
(The Institute of Mathematics and Its Applications conference series; new ser., 28)
Papers from the Conference on Computer Modelling in the Environmental Sciences.
1. Earth sciences–Computer simulation–Congresses. I. Farmer. D. G. II. Rycroft, Michael J. III. National Environment Research Council (Great Britain) IV. Institute of Mathematics and Its Applications. Environmental Mathematics Group. V. Conference on Computer Modelling in the Environmental Sciences (1990: Keyworth, England) VI. Series.
QE48.8.C612 1991 550'.1'13—dc20 91–4440
ISBN 0–19–853394–2

Printed in Great Britain
by Bookcraft (Bath) Ltd
Midsomer Norton, Avon

Preface

The Conference on *Computer Modelling in the Environmental Sciences* was organised by the Natural Environment Research Council (NERC) in association with the Environmental Mathematics Group of the IMA (Institute of Mathematics and its Applications). It was held over two days in April 1990 at the headquarters of the British Geological Survey (a component body of NERC) at Keyworth, near Nottingham.

The meeting was prompted by recognition that in all branches of the environmental sciences there is a discernable trend towards attempting to model natural systems and predict their behaviour, rather than passive observation. Advancing scientific knowledge has made the design of models feasible, while developments in computer technology have made their implementation practicable.

Few meetings or books address the application of computer modelling *across* the full range of environmental sciences. In contrast, the Conference brought together workers in the component disciplines (oceanography, meteorology, hydrology, earth and life sciences) and those concerned with the enabling computer technology. The aim was to promote a mutual exchange and cross-fertilisation of ideas on the opportunities and pitfalls of such modelling, and on the methodologies and equipment which can be used.

The outcome is this book which provides a cross-section of the state-of-the-art in computer modelling spanning the environmental sciences. Many of the topics are relevant to matters of topical public concern such as global atmospheric effects and world climate, oceanographic phenomena and coastal flooding, pollution dispersal and acid rain, earthquake prediction etc. Indeed, computer modelling is now widely recognised as a *vital* tool in tackling these issues.

Perhaps one of the major themes emerging from the Conference was the sheer variety: of environmental problems being addressed by computer modelling; of model complexity, scale and sophistication; of numerical and computational methods involved; and in the computing equipment being used, ranging from advanced supercomputers to the now ubiquitous personal computer.

We would like to thank our colleagues who were involved in arranging the Conference. Thanks especially to Dr Peter White (UK Meteorological Office), who chaired the Programme Committee, and our co-members Dr Richard Haworth (British Geological Survey), Dr Paul Whitehead (Institute of Hydrology), Dr David Webb (Institute of Oceanographic Sciences) and Dr Roger Evans (Science and Engineering Research Council). Appreciation is also due to John Down, Head of NERC Computer Services (NCS), who initiated the idea for the Conference, and to other NCS staff involved. Most notably, thanks to George Darwall who chaired the Organising Committee and did so much other work to make the meeting a success, and to Norman Wiseman and Dr Fred Hopper, NCS staff at Keyworth. Also at Keyworth, we are most grateful to the Director of the British Geological Survey (BGS), Dr Peter Cook, for the use of his facilities, and to the capable and smooth administrative support from Val Burman and her staff.

In addition, thanks must go to all the delegates who took part in the Conference, and of course especially to those who have made the contributions published in this volume. Others to whom we are grateful are the manufacturers, suppliers, groups and individuals who provided exhibits and demonstrations at the meeting, and particularly to IBM (UK), Cray Research (UK), Sun Microsystems and Convex Computer who provided financial and other support.

The Proceedings are arranged essentially in the order of presentation of papers at the Conference. Although there are not clear divisions, this provides something of an excursion through the various environmental disciplines: oceanography, hydrology, geology and geophysics, atmospheric sciences and meteorology, life sciences and ecology, coastal and sedimentary processes.

All of the basic text was taken from the authors in computer-readable form. The contributions were standardised, with the addition of scientific notation, using a modern word processing program on a personal computer. Proofs, for checking by the authors, and the final camera-ready copy of the text were produced on a laser printer, with addition of the diagrams and final publication by the Oxford University Press (OUP). Appreciation is again due to the authors for their prompt review of the proofs, and Clara Farmer for her assistance in proof-reading all the papers.

Finally, our thanks must go to Catherine Richards of the IMA, and especially Martin Gilchrist at the OUP for guiding us through the overall publication process.

David Farmer and Michael Rycroft

The IMA and NERC are most grateful to Cray Research (UK) and IBM (UK) for generously contributing to the publication costs of these proceedings, thereby enabling inclusion of the colour plates.

Table of contents

List of contributors

Allen, C.M., *Institute of Environmental and Biological Sciences, University of Lancaster, Lancaster, England LA1 4YQ*

Aspinall, R., *Macaulay Land Use Research Institute, Craigiebuckler, Aberdeen, Scotland AB9 2QJ*

Baker, C., *Department of Theoretical Mechanics, University of Nottingham, University Park, Nottingham, England NG7 2RD*

Baxter, M.S., *Scottish Universities Research and Reactor Centre, East Kilbride, Glasgow G75, Scotland*

Begg, F., *Scottish Universities Research and Reactor Centre, East Kilbride, Glasgow G75, Scotland*

Bell, S.B.M., *NERC Unit for Thematic Information Systems, Department of Geography, University of Reading, Whiteknights, PO Box 227, Reading, England RG6 2AB*

Beven, K.J., *Institute of Environmental and Biological Sciences, University of Lancaster, Lancaster, England LA1 4YQ*

Binley, A.M., *Institute of Environmental and Biological Sciences, University of Lancaster, Lancaster, England LA1 4YQ*

Bradley, P., *Scottish Universities Research and Reactor Centre, East Kilbride, Glasgow G75, Scotland*

Britter, R.E., *Department of Engineering, University of Cambridge and Cambridge Environmental Research Consultants Ltd, Sheraton House, Castle Park, Cambridge, England CB3 0AX*

Cape, J.N., *Institute of Terrestrial Ecology, NERC, Bush Estate, Penicuik, Midlothian, Scotland EH26 0QB*

Carruthers, D.J., *Cambridge Environmental Research Consultants Ltd, Sheraton House, Castle Park, Cambridge, England CB3 0AX*

Carver, G., *Atmospheric Modelling Group, Department of Chemistry, University of Cambridge, Lensfield Road, Cambridge, England CB2 1EP*

Cooper, S.P., *Department of Mechanical and Aerospace Engineering, Rutgers University, PO Box 909, Piscataway, NJ, USA*

Crampin, S., *British Geological Survey, NERC, Murchison House, West Mains Road, Edinburgh, Scotland EH9 3LA*

Curtis, A.R., *ARC Scientific Ltd, 257 Woodstock Road, Oxford, England OX2 7AE*

Dalziel, S., *Department of Applied Mathematics and Theoretical Physics, University of Cambridge, Silver Street, Cambridge, England CB3 9EW*

Davies, A.M., *Proudman Oceanographic Laboratory, NERC, Bidston Observatory, Birkenhead, Merseyside, England L43 7RA*

Derwent, R.G., *Department of the Environment, Air Quality Division, Romney House, 43 Marsham Street, London, England SW1P 3PY*

Dolman, A.J., *Institute of Hydrology, NERC, Maclean Building, Crowmarsh Gifford, Wallingford, Oxfordshire, England OX10 8BB*

Eeles, C.W.O., *Institute of Hydrology, NERC, Maclean Building, Crowmarsh Gifford, Wallingford, Oxfordshire, England OX10 8BB*

Egan, S.S., *Department of Geology, University of Keele, Keele, Staffordshire, England ST5 5BG*

Flather, R., *Proudman Oceanographic Laboratory, NERC, Bidston Observatory, Birkenhead, Merseyside, England L43 7RA*

Fowler, D., *Institute of Terrestrial Ecology, NERC, Bush Estate, Penicuik, Midlothian, Scotland EH26 0QB*

Gadian, A., *Department of Physics, UMIST, (University of Manchester Institute of Science and Technology) PO Box 88, Manchester, England M60 1QD*

Grzonka, R.B., *British Aerospace Plc, Sowerby Research Centre, PO Box 5, Filton, Bristol, England BS12 7QW*

Hardisty, J., *Department of Geography and Earth Resources, University of Hull, Cottingham Road, Hull, England HU6 7RX*

Hibberd, S., *Department of Theoretical Mechanics, University of Nottingham, University Park, Nottingham, England NG7 2RD*

Hoggart, C.R., *Sir William Halcrow and Partners Ltd, Burderop Park, Swindon, Wiltshire, England SN4 0QD*

Hough, A.M., *AEA Environment and Energy, Harwell Laboratory, Didcot, Oxfordshire, England OX11 0RA*

Hunt, J.C.R., *Department of Applied Mathematics and Theoretical Physics, University of Cambridge*
and
Cambridge Environmental Research Consultants Ltd, Sheraton House, Castle Park, Cambridge, England CB3 0AX

Jakeman, A.J., *Centre for Resource and Environmental Studies, Australian National University, Canberra, Australia ACT 2601*

James, I.N., *Department of Meteorology, University of Reading, 2, Earley Gate, Whiteknights, PO Box 239, Reading, England RG6 2AU*

James, P.M., *Department of Meteorology, University of Reading, 2, Earley Gate, Whiteknights, PO Box 239, Reading, England RG6 2AU*

Jeffers, J.N.R., *Applied Statistics Research Unit, Mathematical Institute, The University, Canterbury, Kent, England CT2 7NF*

Johnson, C.E., *AEA Environment and Energy, Harwell Laboratory, Didcot, Oxfordshire, England OX11 0RA*

Linden, P.F., *Department of Applied Mathematics and Theoretical Physics, University of Cambridge*
and
Cambridge Environmental Research Consultants Ltd, Sheraton House, Castle Park, Cambridge, England CB3 0AX

Littlewood, I.G., *Institute of Hydrology, NERC, Maclean Building, Crowmarsh Gifford, Wallingford, Oxfordshire, England OX10 8BB*

Mason, D., *NERC Unit for Thematic Information Systems, Department of Geography, University of Reading, Whiteknights, PO Box 227, Reading, England RG6 2AB*

Matthews, G.P., *Department of Environmental Sciences, Polytechnic South West, Drake Circus, Plymouth, Devon, England PL4 8AA*

McCartney, M., *Ministry of Agriculture, Fisheries and Food, Fisheries Laboratory, Lowestoft, Norfolk, England NR33 0HT*

Norton, W.A., *Department of Applied Mathematics and Theoretical Physics, University of Cambridge, Silver Street, Cambridge, England CB3 9EW*

Osment, J., *Sir William Halcrow and Partners Ltd, Burderop Park, Swindon, Wiltshire, England SN4 0QD*

Perkins, R.J., *Department of Applied Mathematics and Theoretical Physics, University of Cambridge, Silver Street, Cambridge, England CB3 9EW*

Phillips, S.J., *Sir William Halcrow and Partners Ltd, Burderop Park, Swindon, Wiltshire, England SN4 0QD*

Proctor, R., *Proudman Oceanographic Laboratory, NERC, Bidston Observatory, Birkenhead, Merseyside, England L43 7RA*

Radford, P.J., *Plymouth Marine Laboratory, NERC, Prospect Place, West Hoe, Plymouth, Devon, England PL1 3DH*

Reeve, D.E., *Sir William Halcrow and Partners Ltd, Burderop Park, Swindon, Wiltshire, England SN4 0QD*

Riddle, A.M., *ICI Group Environmental Laboratory, Freshwater Quarry, Brixham, Devon, England TQ5 8BA*

Scott, E.M., *Department of Statistics, University of Glasgow, University Gardens, Glasgow, Scotland G12 8QW*

Smith, R.I., *Institute of Terrestrial Ecology, NERC, Bush Estate, Penicuik, Midlothian, Scotland EH26 0QB*

Spearing, M., *Department of Environmental Sciences, Polytechnic South West, Drake Circus, Plymouth, Devon, England PL4 8AA*

Taylor, J., *Centre for Resource and Environmental Studies, Australian National University, Canberra, Australia ACT 2601*

Tett, S.F.B., *Departments of Physics and Meteorology, University of Edinburgh, The King's Buildings, Edinburgh, Scotland EH9 3JZ*

Wadge, G., *NERC Unit for Thematic Information Systems, Department of Geography, University of Reading, Whiteknights, PO Box 227, Reading, England RG6 2AB*

Wakes, S.J., *Department of Theoretical Mechanics, University of Nottingham, University Park, Nottingham, England NG7 2RD*

Webb, D.J., *Institute of Oceanographic Sciences Deacon Laboratory, NERC, Brook Road, Wormley, Godalming, Surrey, England GU8 5UB*

Whitehead, P.G., *Institute of Hydrology, NERC, Maclean Building, Crowmarsh Gifford, Wallingford, Oxfordshire, England OX10 8BB*

Wolf, J., *Proudman Oceanographic Laboratory, NERC, Bidston Observatory, Birkenhead, Merseyside, England L43 7RA*

Young, P.A.V., *NERC Unit for Thematic Information Systems, Department of Geography, University of Reading, Whiteknights, PO Box 227, Reading, England RG6 2AB*

FRAM - the Fine Resolution Antarctic Model

D.J. Webb

Institute of Oceanographic Sciences Deacon Laboratory

1. Introduction

Attempts to develop realistic models of the ocean circulation have been hampered for many years by problems due to the small scale of many important ocean features. In marked contrast to the atmosphere, where the major high and low pressure regions have scales of a thousand or more kilometres, the major meso-scale eddies of the ocean have diameters of only one to two hundred kilometres. Similarly the major winds of the atmosphere, the jet streams, may be hundreds of kilometres wide, but in the ocean the major currents, like the Gulf Stream, are only thirty to fifty kilometres wide.

The fine grid size means that the storage requirements of ocean models are very large. Because of the fine grid, the model timesteps also have to be short so the amount of computer power required to run the model ocean over any significant period becomes very large.

The limited amount of computational power available to ocean modellers has meant that until recently, models of the major oceans and of the global ocean have had to use grid sizes of 100km or more. Such models have been unable to represent the important field of meso-scale eddies or to realistically represent the major ocean currents.

In recent years the introduction of powerful vector processor computers, such as the CRAY X-MP installed at the SERC Rutherford Appleton Laboratory (RAL), has meant that at last oceanographers have been able to run realistic eddy resolving models of the major oceans.

In the US there have been two main studies. The US Community Model (Bryan and Holland, 1990), covers the North Atlantic between 15°S and 65°N, with a north-south resolution of 1/3°, an east-west resolution of 2/5° and with 30 levels in the vertical. The global model of Semtner and Chervin (1988) used a 1/2° grid in both the north-south and east-west directions with 20 levels in the vertical.

In the UK it was decided that the most efficient use of our resources would be to concentrate on the Southern Ocean. The physics of the region is very different from that of the other ocean basins primarily because of the lack of eastern and western boundaries. As a result the main balance determining the flow is expected to be that between the eastward wind stress, driving the main Circumpolar Current, and the pressure forces which differ on either side of ocean topography (Munk and Palmén, 1951). In contrast, in other ocean basins, the flow is dominated by the balance between the torque applied to the ocean by the curl of the wind stress and the change in angular momentum produced by north-south flow of water in the open ocean. Mass is conserved through the narrow boundary currents, such as the Gulf Stream, which run north-south along the western boundary of each ocean.

The Circumpolar Current, which dominates the Southern Ocean, also provides an important connecting link which transports and mixes water masses between the other major oceans of the world. In the north-south direction it acts as a barrier to the transport of heat between the tropics and Antarctica.

The eddy resolving models produce large quantities of data and special arrangements have to be made to ensure that it is all properly analysed. To achieve this goal, the FRAM project has been organised as a group effort of the UK ocean modelling community. The day to day details of running the model are dealt with by a small core team based at the Institute of Oceanographic Sciences Deacon Laboratory (IOSDL). The analysis of the physics occurring in different parts of the model is the responsibility of principal investigators based at the Universities of Southampton, Oxford, Exeter, East Anglia, Cambridge, Imperial College London and at IOSDL.

This paper concentrates on the structure of the model, the problems of running it on the CRAY X-MP and on some of the early, very realistic, results.

2. Structure of the model

The code of the FRAM model, like those of Bryan and Holland (1990) and Semtner and Chervin (1988) is based on that developed by Bryan (1969), Semtner (1974) and Cox (1984). It splits the ocean into a grid along lines of latitude, longitude and depth and sets up arrays containing the temperature, salinity and horizontal velocity at each grid point. The equations representing their evolution in time are

$$\frac{dT}{dt} + \underline{u}\cdot\nabla T + w\frac{dT}{dz} = A_h\nabla^2 T + K_h\frac{d^2T}{dz^2} \qquad (2.1)$$

$$\frac{dS}{dt} + \underline{u}\cdot\nabla S + w\frac{dS}{dt} = A_h\nabla^2 S + K_h\frac{d^2S}{dz^2} \qquad (2.2)$$

$$\frac{d\underline{u}}{dt} + \underline{u}\cdot\nabla\underline{u} + w\frac{d\underline{u}}{dz} - \underline{f}\,\underline{u} = A_v\nabla^2\underline{u} + K_v\frac{d^2\underline{u}}{dz^2} + \nabla p \tag{2.3}$$

where: T and S are temperature and salinity and $\underline{u}$ the horizontal velocity; $\underline{f}$ is the vertical coriolis vector with magnitude $2\ \Omega \sin\theta$, where Ω is the Earth's rotation rate and θ the latitude; p is the pressure; A_h and K_h are the horizontal and vertical diffusion coefficients; and A_v and K_v the horizontal and vertical viscosity coefficients. The vertical velocity w is determined by assuming the ocean is incompressible

$$\underline{\nabla}\cdot\underline{u} + \frac{dw}{dz} = 0 \tag{2.4}$$

These equations are cast in finite difference form, using the values of variables at neighbouring grid points. Central differences are used for the grad operator and the five point form of the div grad operator.

Because models such as FRAM are only just resolving many of the important physical features of the circulation, it is important that the arrangement of grid points is chosen to represent the development of short wavelength features as accurately as possible. For ocean models where the grid spacing is similar to or larger than the main horizontal scale (the Rossby radius: Gill, 1982), the best choice is the Arakawa-B grid (Messinger and Arakawa, 1976). In this grid, at each horizontal level the velocities $\underline{u}$ are defined at the corners and the temperature and salinity field at the centre of each box formed by the grid lines.

To reduce the amount of computation needed for each model run, the timestep used by the model should be as long as possible. With a finite difference grid the upper limit dt is determined by the CFL condition (Messinger and Arakawa, 1976)

$$dt * c < dx \tag{2.5}$$

where dx is the grid spacing and c the speed of the fastest wave. In the ocean the fastest wave is the surface tidal wave with a speed near 200 $m.s^{-1}$. The next fastest is the first internal mode with a speed of 2 to 3 $m.s^{-1}$. Because the tidal wave has only a minor effect on the large scale ocean circulation, the model filters it out by placing an imaginary rigid lid on the ocean. With this approximation the timestep is 2400s, without it, the timestep would be 24s.

The price paid is that we now have to solve an additional (elliptic) stream function equation for the vertically averaged velocity. In the FRAM model, because of the irregular boundaries and the large matrices involved, the equation has to be solved by relaxation. However this requires many hundred of iterations of the solution each timestep. As the number of iterations needed increases roughly proportionally to the number of horizontal gridpoints, future large eddy-resolving models will use an alternative more efficient method (Killworth et al, 1989). This solves the tidal part of the equations using a short timestep and the remainder using the long timestep.

3. Storage and operation of the model

The success of the Cox code for the large ocean models is a result of three main features. The first is a good finite difference scheme, the second an efficient use of storage and the third a highly vectorisable loop structure. It was realised early on that the very large size of ocean models would mean that the main variable arrays would have to be stored on disk and the data only brought into core when required for computation.

The model therefore splits the ocean into slabs, each of which contains all the variables for a given line of latitude at a given timestep. There are three such slab datasets, each stored on a separate disk. At any instant the program is reading from the disks corresponding to the last and current timesteps and writing new results to the disk for the next timestep (Fig. 1). Buffering is used to overlap processing and input-output.

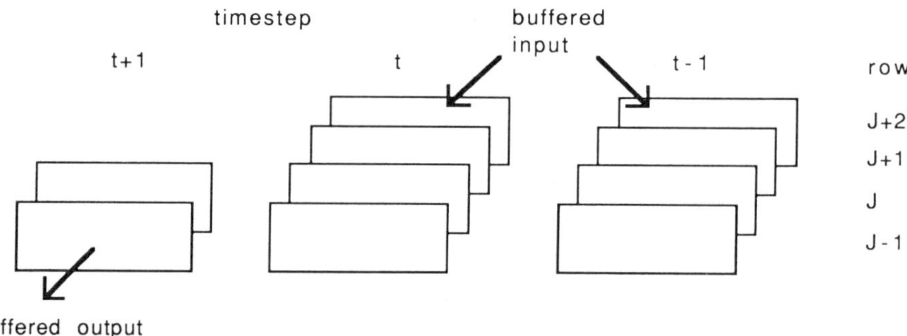

Fig. 1: The arrangement of slabs in core. At any one time slabs J-1, J and J+1 from timesteps t and t-1 are in core and being used to calculate the new variables for slab J at timestep t+1. In the background the J+2 slabs from timesteps t and t-1 are buffered in from disk and the J-1 slab for timestep t+1 is buffered out.

On the RAL CRAY X-MP this scheme works well for a single processor, the I/O wait time being negligible. However with more processors the I/O wait time becomes significant as the 10 mbyte/sec disk transfer rates cannot keep up with the CPU data requirements. We overcame this problem, when the COS operating system was in use, by placing one of the slab files on SSD and using software disk striping to split the other two over six separate disk units.

The loop structure of the model means that it runs very efficiently on the CRAY architecture. Polynomials of the type

```
DEN(I,K) = ((T(I,K)+A)*T(I,K)+B)*T(I,K)
         + ((S(I,K)+C)*S(I,K)+D)*T(I,K) + ...
```

used to calculate density, run at 200mflops on a single processor. Most of the code is of the form

```
TA(I,K) = TA(I,K) + A*TB(I+1,K) + B*TB(I-1,K)
                  + C*TB(I,K+1) + D*TC(I,K-1)
```

On the RAL CRAY X-MP, which had only 8mwords of memory and so a reduced number of memory banks, the slowest code was of the form

```
TA(I,K) = TA(I,K) + A(I,K)*TB(I,K)
```

This ran at only 60mflops, probably because of the large number of core to vector register transfers for each floating point operation. Overall the model ran at 95mflops with a single processor. Under the COS operating system microtasking was implemented for the most vectorisable parts of the code, speeding the model up to 240mflops with four processors. Further tests carried out under UNICOS on a larger CRAY Y-MP, with eight processors and a large SSD have reached speeds near 1gflop.

Because FRAM requires a very long run of the model, it has been automated to restart itself just before the allowed computer time is reached and after system crashes. Archive and restart datasets are also stored at regular intervals.

4. Problems of analysis

The main practical problem that hampers the analysis of the data is the large size of the model datasets. To keep track of how the model run is progressing we also need to rapidly plot large amounts of data at remote sites for analysis by the research groups.

A full archive dataset from a single timestep of the FRAM model has a size of 180mbytes. These datasets are stored on cassette tapes at RAL and can be used directly by analysis programs running on the CRAY. An example is a program for calculating the pressure field. This is not calculated directly by the FRAM model but is essential for understanding the momentum balance in the ocean.

For rapid plotting of results from the CRAY, we make use of a scheme which compresses data arrays using an ASCII format. Each model variable is represented by two or more ASCII characters (out of a subset of 64). The file header contains data on the base and range of the data. Two other ASCII characters are used as escape characters to define land and other special features.

The two character format, which gives a precision of one in 4096, is suitable for most plotting purposes. An ASCII file containing the stream function field (170,000 variables), contains 340kbytes and can be readily transferred to the remote sites over the Joint Academic Network (JANET) and University networks.

For displaying the results at the University sites, the most common system used is a UNIX-based SUN graphics workstation. Low level machine calls are used to do the plotting because high level calls were found to be too slow when used with the large data arrays.

Further analysis of the data can also be carried out on the UNIX workstations. For this purpose the archive datasets are first converted from 64-bit CRAY binary to the 32 bit IEEE floating point format used by many workstations. The converted files, each of which is 90mbytes long, are stored on Exabyte tapes using a dedicated tape unit connected to the CRAY. These are posted to the University sites where they can be read by similar tape units connected to each workstation.

The advantages of the Exabyte-based system are that each tape has a large capacity (2gbytes), the media cost is low (essential when many hundreds of datasets may be involved), and the tape transfer rate (10 mbytes/min) compares well with many optical disk units.

5. Early results from the model

The model was first initialised with the Levitus (1982) temperature and salinity dataset and zero velocity. After a few timesteps the model became unstable due to ocean modes trapped near Kerguelen and New Zealand. These appear to be driven by the interaction between bottom pressure and topography. To overcome the instability the model was restarted with a cold (-2°C) saline (36.69°/oo) motionless ocean, and the temperature and salinity fields were then dynamically relaxed to the Levitus values. This scheme was found to be stable.

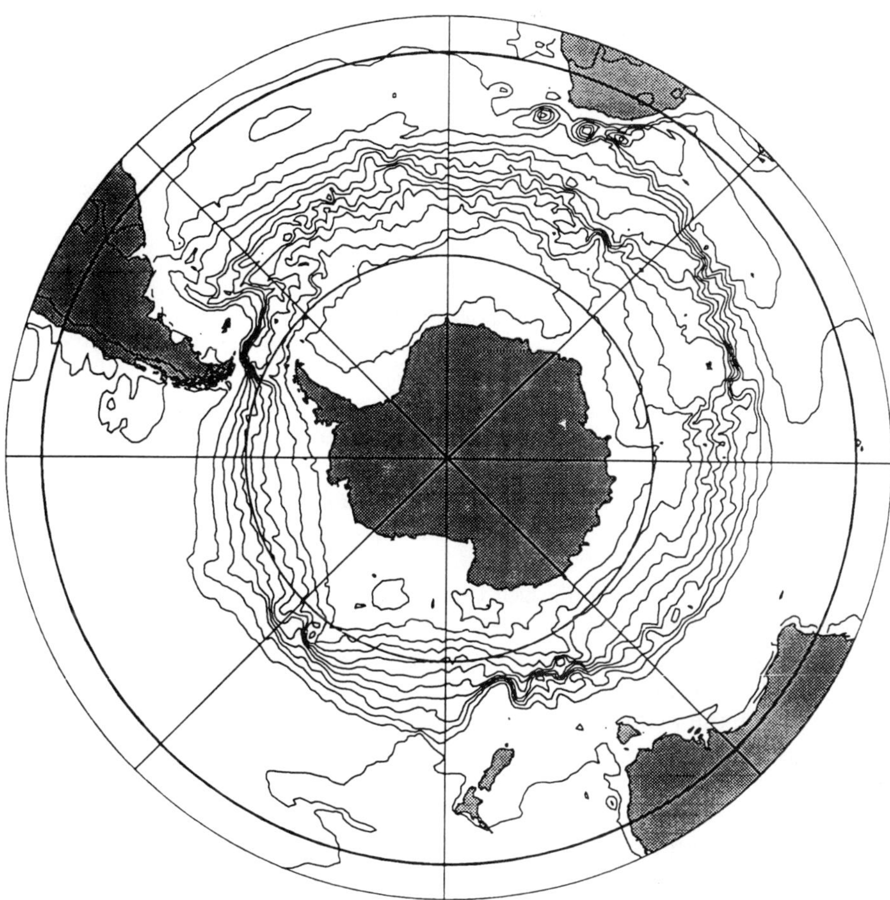

Fig. 2: The stream function, representing the vertically integrated transport, at the end of six years of the model run. The contour interval is 20Sv (1Sv equals 10^6 $m^3.s^{-1}$). The path of the main Circumpolar Current, which travels from west to east, is distorted by the bottom topography. South of Africa, the Agulhas current spawns large eddies which carry warm water from the Indian Ocean north-westwards into the South Atlantic.

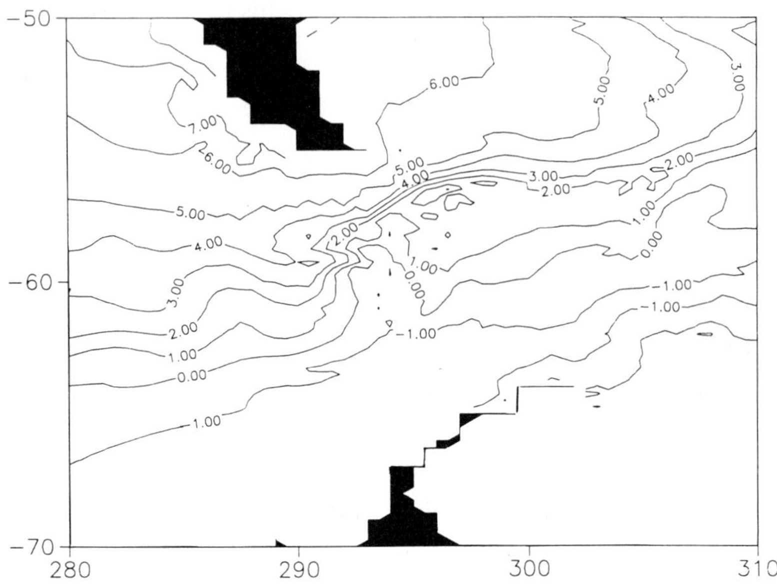

Fig. 3: The surface temperature field near Drake Passage showing an example of the sharp fronts produced by the model.

During the first two and a half years, the relaxation timescale used is six months for the near surface layers, and eighteen months for the deeper layers. From then up to six years the timescale is one year throughout. At the end of six years the ocean has a realistic temperature and salinity structure but the relaxation is so weak that it permits eddies, fronts, and other features to develop. The model is forced by annual mean winds. These are built up linearly during the third year of the run from zero to their steady value.

During the first few days of the integration, the model shows large-scale Rossby waves, but by day ten the main feature of the flow is a very weak Circumpolar Current. When the winds are added, the model responds to the changes in less than 10 days.

By the end of year six the transport through Drake Passage, the total kinetic energy of the model and other diagnostics settle down, indicating that the momentum budget of the model has reached its asymptotic state. The transport through Drake Passage is 195 x 10^6 $m^3.s^{-1}$, about 23 x 10^6 $m^3.s^{-1}$ of which is due to the wind. The observed values are near 130 x 10^6 $m^3.s^{-1}$. The difference may be due to small errors in the pressure field around abyssal topography in the model. It corresponds to an excess bottom velocity of 0.021 $m.s^{-1}$ through Drake Passage.

Although this discrepancy is disappointing, in other respects the model performs remarkably well. The Levitus data contains no sharp fronts but despite this the model produces frontal structures very similar to those observed in the ocean. This is especially true in Drake Passage (Fig. 3), where we have a good set of observations available and where the model reproduces both the observed fronts and narrow bands of high velocity currents.

Because the stratification of the Southern Ocean is so weak, most of the important circulation features show up in plots of the stream function (Fig. 2). The vertically averaged current runs along the contours representing stream function and the velocity is largest where the contours are closest together.

The model results show that topographic features dominate the path of the Circumpolar Current. One of the most interesting regions is in the South Pacific, where the Current crosses the isolated Pacific-Antarctic ridge, through the Menard, Eltanin and Udintsev fracture zones. The maximum flow is found at the southernmost Udintsev fracture zone where the current has a total transport of 100 x 10^6 $m^3.s^{-1}$, three times that of the Gulf Stream, through a channel less than 130km wide. On each side of the region, steering along lines of constant depth is evident. In addition, strong eddy activity is observed near the ridge axis, and there is only weak activity on the neighbouring abyssal plains. This is in qualitative agreement with the GEOSAT observations of surface variability.

Steering by topography is also observed to the east of Drake Passage where part of the flow forms the Falklands Current, shifting the axis of the current from 60°S to 45°S, to the north of the Kerguelen Plateau and to the south of New Zealand.

In the north, the model also produces realistic western boundary currents. The Brazil current has a realistic transport and separates from the coast at the correct latitude. The model East Australian current is slightly stronger than observed but it turns away from the coast at the correct latitude to pass north of New Zealand. It also produces the series of large eddies which are observed off Sydney and drift southwards, in good agreement with observations.

The most eddy-energetic region of the model is near the Agulhas Current off South Africa. The model shows that the flux of the current reaches 89 x 10^6 $m^3.s^{-1}$ at 35°S at the start of the recirculation region where eddies are formed. New eddies are formed roughly every 160 days and drift into the South Atlantic. Further south a second band of eddies is formed along the line of the Circumpolar Current in good agreement with observations. South of the Circumpolar Current, the model also produces realistic circulation gyres in the Weddell and Ross Seas.

The only serious numerical problem found with the model was a small plus-minus feature in parts of the stream function. Later tests with a biharmonic viscosity removed the error.

Although we tend to look on FRAM as a prognostic model, the first six years of the run can also be regarded as an assimilation of the observed (Levitus) data into a numerical model. The equations of motion have been used to sharpen up fronts and to produce a velocity field in dynamic equilibrium with the observations. The model describes features with length scales as small as 30km, matching well with present day observations. It is thus useful for generating hypotheses which can be tested at sea, and as such has been used for planning long hydrographic sections in the World Ocean Circulation Experiment (WOCE).

6. Ocean models in the 1990s

During the 1990s models like FRAM will be developed further as part of national and international programs which study the interaction of the ocean and atmosphere. The largest of the international programs, WOCE, has as its first goal the development of ocean models to predict climate change and the collection of data to test them.

The experiment includes extensive ship and satellite based surveys of the ocean circulation and accurate measurements of the transport of heat within the oceans. It naturally includes a large numerical modelling component aimed at improving our ability to predict ocean circulation and climate. Computer models of the ocean will also be used for assimilating the large amount of satellite data which will be collected during the period. The need to develop assimilation schemes arises because prediction is only possible if one has a good initial estimate of the state of the ocean. Given the large size of the ocean and the small scale of many of its important features, the only realistic way to collect the bulk of the data is through satellite surveillance of the ocean surface. By assimilating the data into numerical models of the ocean, the surface satellite measurements can also yield information on the currents at depth.

A second major international program, the Tropical Ocean Global Atmosphere (TOGA) project is aimed specifically at understanding the effect of the tropical oceans on climate. TOGA includes research on the El Niño, a coupled ocean-atmosphere phenomenon which affects the countries around the Pacific, and research on the effect of the equatorial sea surface temperature field on the global atmospheric circulation. TOGA also has a large modelling component aimed at improving the models of the tropical ocean and predicting its effect on the atmosphere.

In addition to these projects which concentrate primarily on the physical aspects of ocean circulation, attention is also turning to modelling the biological and chemical systems within the ocean. This is being done first because of their importance in understanding the productivity of the oceans and secondly because of their central role in the carbon dioxide cycle and its effect on climate. The UK is now carrying out a major Biological Ocean Fluxes Study (BOFS) which is contributing to the international Joint Global Ocean Fluxes Study (JGOFS).

These are important programs which are attracting significant amounts of scientific support and funding from the UK. The WOCE experimental program is being supported by the Natural Environment Research Council (NERC) through a Community Research Project. Ship time is already being committed and further support is to be provided through a Special Topic. Additional funding of up to £3.5m per year is coming from the Advisory Board for the Research Councils (ABRC).

The numerical modelling aspect of WOCE is being supported by NERC through the FRAM Community Research Project and Special Topic. The UK is also contributing 15% to the £600 million being spent by the European Space Agency on the ERS-1 and ERS-2 satellite programs. These satellites will have a major role during the first half of the decade, in collecting oceanographic data for the WOCE and TOGA programs. The SERC costs for developing the Along Track Scanning Radiometer (ATSR) for ERS-1 are believed to amount to an additional £5 to £10 million.

6.1 The problems for the 1990s

The research projects using computer models of the ocean that need to be carried out during the 1990s in order to achieve the aims of the WOCE, TOGA and BOFS programs include :-

(a) The use of eddy resolving models which cover the whole world to study the physics and heat transport of the different oceans and how they interact.

(b) The development of models which can be used for predicting ocean properties for periods of a few months to a few years.

(c) The development of assimilation methods and models in order to assimilate the large amount of data to be collected by the ERS-1 and

ERS-2 satellites and similar satellites to be launched by the USA and Japan.

(d) An extension of the models to include chemistry and biology in order to study the carbon cycle within the oceans. In particular researchers are interested in studying the connections between the patchiness in the distribution of plankton and other life in the ocean and the small size of many of the features of the ocean circulation.

(e) The use of such models with extended time integrations to study changes in climate over periods of hundreds of years.

6.2 The computer requirements

To give an idea of the computer power that will be needed, the requirements of the FRAM model have been extrapolated to give those required by a high resolution (10km) global model. The other projects listed above are expected to have similar or even larger requirements.

This shows that the projects will need of order 10^{16} to 10^{17} floating point instructions per model run. The core storage requirement for the run will be 72mwords, the disk or similar fast access storage requirement will be 1.2gwords and the transfer rate from disk to core will need to be over 150 mwords per second. A large region of archive storage will also be required.

In order that such model runs can be carried out in a reasonable amount of time computers will be required which can carry out 10^{11} floating point operations per second.

The storage requirement for the archived datasets could run to many thousands of gwords. These are required to analyse the results of each run. Alternatively one could require that all analysis should be carried out during each model run. However this would require more computing during each run and runs would need to be repeated as new factors came to light during the analysis.

In practice, computer speeds are unlikely to exceed 10^{10} operations per second during the first half of the decade. During this period the research will have to be of limited areas of ocean or with limited resolution. Halving the resolution reduces the core storage requirement by a factor of two, the total storage requirement by a factor of four and the total number of floating point operations by a factor of eight.

7. Conclusions

The results of the FRAM model and the other large eddy-resolving models show that given adequate resolution, present schemes for modelling the ocean can give good qualitative agreement with observations. Because they represent the ocean in such detail, the eddy resolving models are very effective for studying the physics of the ocean circulation.

Over the next decade we hope that further improvements in model physics and computer power will enable us to develop models which can be used in accurate forecasts of climate change.

8. Acknowledgements

The FRAM Community Research Project is supported by the Natural Environment Research Council (NERC). It involves researchers from the Institute of Oceanographic Sciences Deacon Laboratory (IOSDL), and from the Universities of Southampton, Oxford, London, Exeter, East Anglia, and Cambridge. Computations were performed at the Atlas Centre, Rutherford Appleton Laboratory (RAL) of the Science and Engineering Research Council (SERC).

9. References

Bryan, K. (1969). A Numerical Method for Studying the Circulation of the World Ocean. Journal of Computational Physics, 4:347-376.

Bryan, F.O. and Holland, W.R. (1990). A High Resolution Simulation of the Wind- and Thermohaline-Driven Circulation in the North Atlantic Ocean. National Centre for Atmospheric Research, PO Box 3000, Boulder, Colorado 80307, USA. (unpublished manuscript).

Cox, M. D. (1984). GFDL Ocean Group Technical Report No.1, Geophysical Fluid Dynamics Laboratory, Princeton University, Princeton, N.J. 08542, USA.

Gill, A.E. (1982). Atmosphere-Ocean Dynamics. Academic Press, New York, 662pp.

Killworth, P.D., Stainforth, D., Webb D.J. and Paterson, S.M. (1989). A Free Surface Bryan-Cox-Semtner Model. IOS Deacon Laboratory, Report No. 270.

Levitus, S. (1982). NOAA Professional Paper No.13. Geophysical Fluid Dynamics Laboratory, Princeton University, Princeton, N.J.08542, USA.

Messinger, F. and Arakawa, A. (1976). Numerical Methods used in Atmospheric Models, Vol 1. GARP Publication Series No. 17, World Meteorological Organisation, Switzerland.

Munk, W.H. and Palmén, E. (1951). Note on the Dynamics of the Antarctic Circumpolar Current. Tellus 3(1):53-55.

Semtner, A.J. (1974). An Oceanic General Circulation Model with Bottom Topography. Technical report No. 9., Department of Meteorology, University of California, Los Angeles.

Semtner, A.J. and Chervin, R.M. (1988). J. Geophys. Res., 93(c12):15502-15522

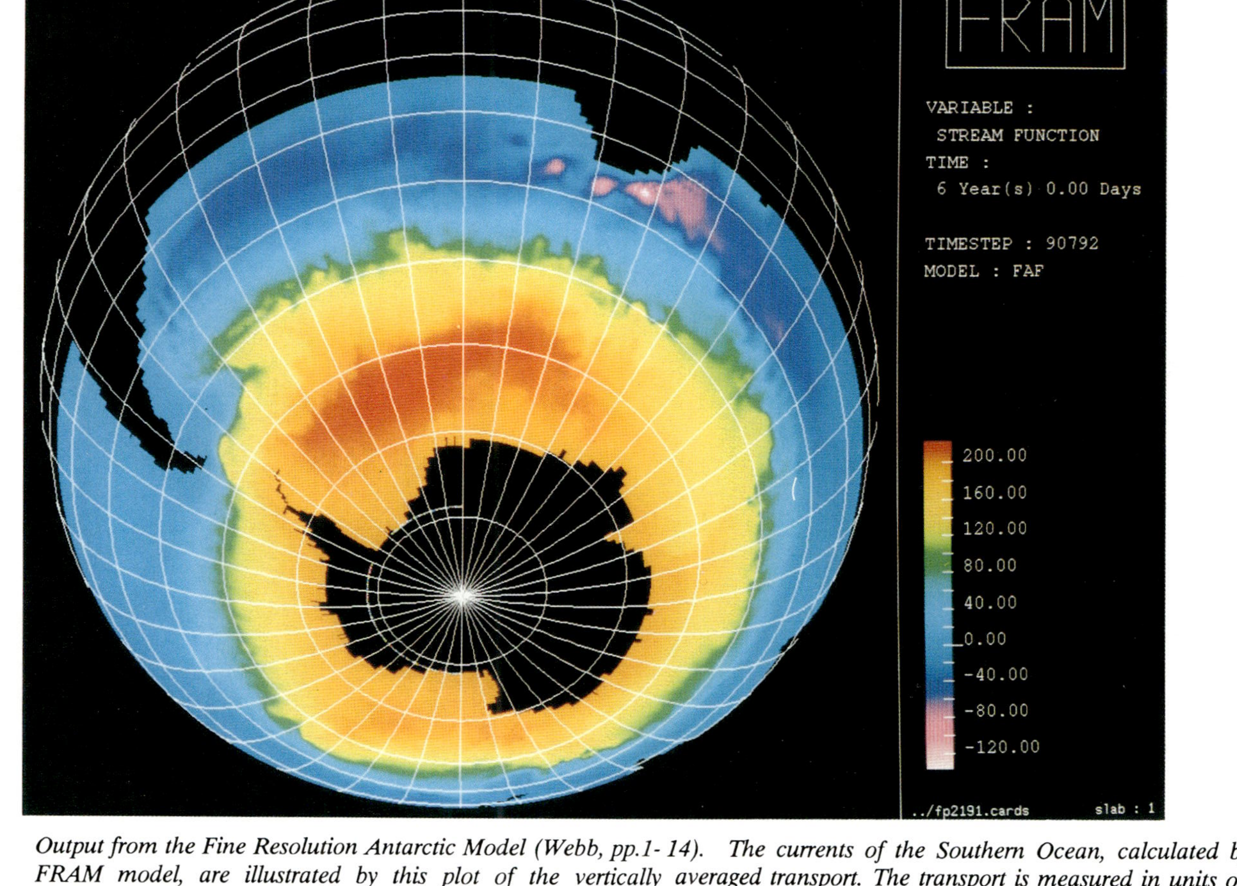

Plate 1: *Output from the Fine Resolution Antarctic Model (Webb, pp.1- 14). The currents of the Southern Ocean, calculated by the FRAM model, are illustrated by this plot of the vertically averaged transport. The transport is measured in units of* 10^{12} $m^3.s^{-1}$. *The mean currents follow the contours of constant colour and go from west to east through Drake Passage (between the southern tip of South America and the Antarctic Peninsula).*

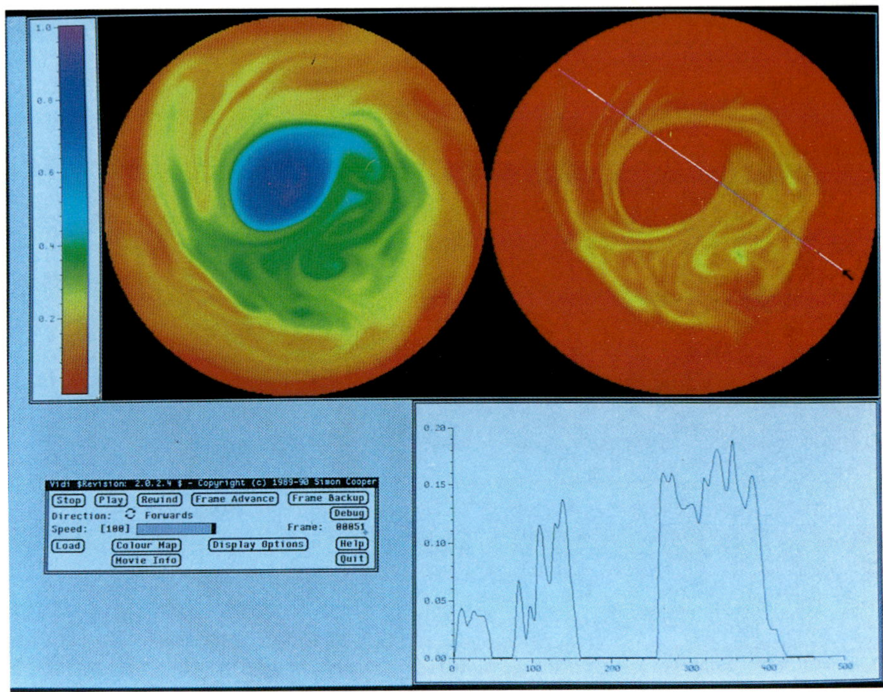

Plate 2: *Visualisation of the output of atmospheric models: an example of VIDI, the Visual Interactive Data Interface, in use (see Cooper and Norton, p.190).*

Plate 3: *The "Menger sponge", applied in computer modelling of the geometry of sandstone (see Spearing and Matthews, p.150).*

Oceanographic forecast models

R. Flather, R. Proctor, J. Wolf

Proudman Oceanographic Laboratory

1. Introduction - tides, storm surges and waves

A knowledge of and the ability to predict variations in sea-level and sea-state have long been recognized as essential for those living and working on or near the sea. Changes in sea-level are due, predominantly, to the *tides*, generated by the variations in the gravitational attraction of the Sun and Moon, and to the effects of storms. Winds and variations in atmospheric pressure associated with a storm can raise or lower sea level by several metres over a period ranging from a few hours to two or three days, producing a *storm surge*. The winds also generate *waves*, with periods up to perhaps 20s and wavelengths of several tens of metres. Such waves may be large enough to threaten offshore structures and to damage and breach coastal defences, allowing inundation of low-lying coastal land.

Storm surges are superimposed on the normal astronomical tides, and, if the range of the tide is large, much depends on whether a surge peak occurs at the same time as tidal high water or not. Coincidence of peak surge and high tide may cause flooding of coastal areas, whereas a large surge may go unnoticed if its peak occurs at low water. In addition, non-linear processes become important as the surge enters shallow water, modifying the storm surge and causing interaction between it and the tide. Interactions also occur between the surge - tide motion and surface waves.

Storm surges affect shallow seas in many parts of the world (Welander, 1961; Murty, 1984). When they occur along low-lying poorly defended coasts, the consequences can be catastrophic - for example, a surge generated by a tropical cyclone in the Bay of Bengal in November 1970 is estimated to have killed more than 200,000 people in Bangladesh (Murty, Flather and Henry, 1986). In Europe, a severe storm in the North Sea in 1953, combined with high spring tides, caused extensive coastal flooding in the Netherlands and along the east coast of England and killed about 2000 people.

Forecasts of sea levels, currents and sea states are commonly made using numerical models. Presently, in the United Kingdom, quite separate models exist for the routine prediction of storm surges (Proctor and Flather, 1983; Flather, 1984) and waves (Golding, 1983). However, the generation of waves and surges is closely related and there are several mechanisms by means of which waves and the mean flow or water level associated with tide and surge interact, each component of the total motion affecting the others.

The inclusion of the effects of waves on the surface and bottom stresses of the surge motion resulted in significant differences when compared with stresses calculated by the presently used empirical parameterizations (Wolf, Hubbert and Flather, 1988). Wave refraction is particularly important in coastal areas, for example river mouths and tidal inlets, where there may be sharp changes in current velocities and water depths over a small distance, but is not confined to these areas. The waves are refracted by changes in mean water depth and also by the temporally and spatially varying sea surface elevations and mean currents associated with tides and storm surges. Depth and current refraction have been incorporated into the WAM (Wave Modelling) Group third-generation wave model (The WAMDI Group, 1988).

Storms during the past winter season have again caused flooding and damage to property in North Wales and other parts of the country. Although they cannot be prevented, adequate warnings can at least save life and limit damage. A new surge model and a joint model for surge and wave prediction are currently being developed at the Proudman Oceanographic Laboratory (POL); these models will hopefully yield more accurate forecasts of water levels, currents and eventually waves in the next few years.

2. Model formulation

2.1 Tide-surge model

The equations governing tide and surge motion can be written in the form

$$\frac{\partial \zeta}{\partial t} + \nabla . (D\vec{q}) = 0 \tag{2.1}$$

$$\frac{\partial \vec{q}}{\partial t} + \vec{q}.\nabla\vec{q} - f\vec{k}\times\vec{q} = -g\nabla\zeta - \frac{1}{\rho}\nabla p_a + \frac{1}{\rho D}(\vec{\tau}_s - \vec{\tau}_b) + A\nabla^2\vec{q} \tag{2.2}$$

where: t denotes time; ζ elevation of the sea surface; $\vec{q}$ the depth-mean current; $\vec{\tau}_s$ the wind stress on the sea surface; $\vec{\tau}_b$ the bottom stress; p_a atmospheric pressure on the sea surface; D the total water depth ($= h+\zeta$ where h is the undisturbed depth); ρ the density of sea water, assumed uniform; g the acceleration due to gravity; f the Coriolis parameter ($= 2\varpi\sin\phi$, where ϖ is the angular speed of rotation of the Earth and ϕ the latitude); $\vec{k}$ a unit vector in the vertical; and A coefficient of horizontal diffusion.

Eq. 2.1 is the continuity equation expressing conservation of volume. In simple terms, considering a vertical column in the sea, changes in surface elevation, ζ, associated with the tide or storm surge, are related to net fluxes of water in or out of the column.

Eq. 2.2 equates the acceleration of the water (on the left side of the equation) to the forces acting on it (on the right). The meteorological forces which generate storm surges are thus wind stress and horizontal gradients of the surface atmospheric pressure. Since the wind stress in (2.1) is divided by water depth (whereas ∇p_a is not) it follows that wind stress becomes increasingly important and dominates in shallow water. Accurate wind stress estimates are, therefore, essential for good surge forecasts.

Mathematically, the problem is closed by relating bottom stress, $\vec{\tau}_b$, to the current, $\vec{q}$, using a quadratic law

$$\vec{\tau}_b = \kappa\rho\vec{q}|\vec{q}| \qquad (2.3)$$

where κ=0.0025 is a friction parameter, and relating the wind stress, $\vec{\tau}_s$, to the surface wind velocity, $\vec{W}$, also using a quadratic law

$$\vec{\tau}_s = c_D\rho_a\vec{W}|\vec{W}| \qquad (2.4)$$

where ρ_a is the density of air and c_D a drag coefficient, given by

$$c_D\times 10^3 = 0.63 + 0.066W \qquad (2.5)$$

where W is in m.s^{-1} (Smith and Banke, 1975). In fact (2.5) attempts to parameterize the varying surface roughness associated with the changing surface wave field.

Initial and boundary conditions must also be specified. These typically take the form

$$\zeta = \zeta(\vec{s},t_0), \quad q = q(\vec{s},t_0) \qquad (2.6)$$

where $\vec{s}$ is position and t=t_0 the initial data time;
on coastal boundaries,

$$q_n = 0 \qquad (2.7)$$

where q_n is the component of current along the outward-directed normal to the boundary; and either

$$\zeta = \zeta' \tag{2.8}$$

$$q_n = q_n' + \frac{c}{h}(\zeta - \zeta') \tag{2.9}$$

on an open-sea boundary, where ζ' and q_n' are specified functions of position and time including contributions due to both tide and surge and $c=\sqrt{(gh)}$.

The equations must then be solved within the region of interest with p_a and $\vec{W}$ defined as functions of time and position within the domain, and ζ' (both ζ' and q_n' if (2.9) is adopted) as a function of time and position along its open-sea boundaries. A time-stepping procedure using finite difference approximations for the equations on a regular latitude - longitude grid is employed. Details of the technique can be found in Davies and Flather (1978) and Proctor and Flather (1983). The computational mesh of the existing surge forecast model is shown in Fig. 1 with grid points of the UK Meteorological Offices' (UKMO) 15 level weather prediction model, which provides the required surface wind and pressure fields.

2.2 *Wave models*

The wave model is based on the wave energy conservation equation

$$\frac{D}{Dt}E = S = S_{in} + S_{nl} + S_{ds} + S_{bf} \tag{2.10}$$

where $E=E(\sigma,\upsilon;\vec{s},t)$ is wave energy spectral density; σ the wave frequency and υ the wave direction. The source functions, S, describe respectively the wind input, non-linear energy transfers, white-capping and bottom dissipation.

The original model solves the energy equation (2.10) for the two-dimensional wave energy spectrum, with no additional ad hoc assumptions regarding the spectral shape (The WAMDI Group, 1988). Finite difference methods are used to approximate the partial differentials. The computational effort required to compute the full nonlinear transfer term S_{nl} in an operational model is prohibitive. The parameterization of this function developed by Hasselmann and Hasselmann (1985) is used. Even so, the parameterized form still consumes about 30% of the total CPU time, for any model run.

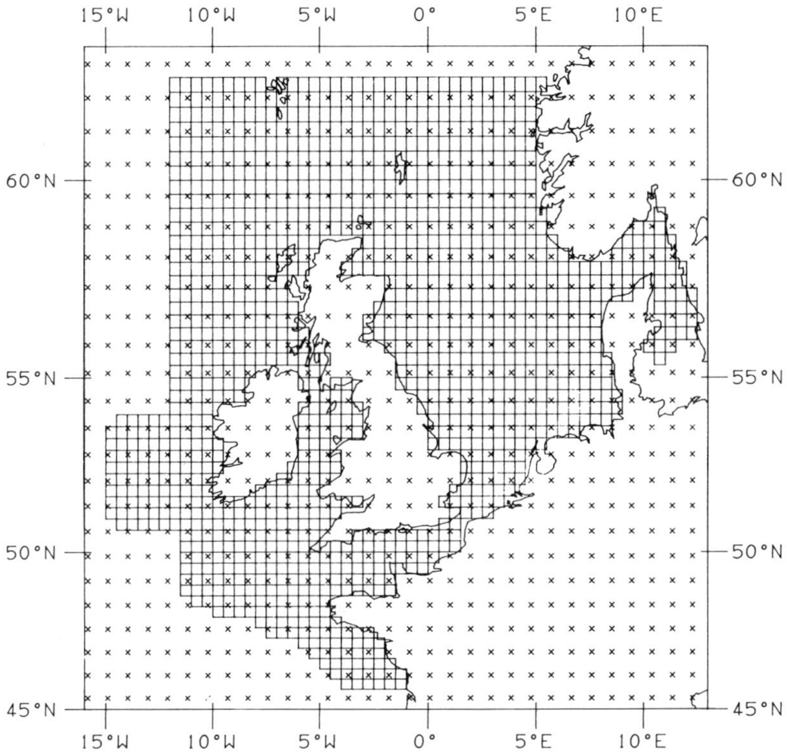

Fig. 1: Finite difference mesh of the POL surge forecast model (plotted as boxes) with wind grid points (crosses) of the UKMO 15 level weather prediction model which provides the meteorological forcing.

In the presence of currents, the fundamental conserved quantity is "wave action" (Bretherton and Garrett, 1969). In order to facilitate inclusion of refraction processes, the WAM model equations have thus been reformulated in terms of the wave action spectrum, $N=N(\sigma,\upsilon;\vec{s},t)$, where $N=E/\sigma$ (Hubbert and Wolf, 1990). The accuracy with which the model can predict refraction is dependent upon the directional resolution of the spectrum. A directional resolution of at least 15° is required; the use of a higher directional resolution may be limited by computational resource restrictions. The inclusion of depth and current refraction in a continental shelf model, similar to that shown in Fig. 1, increased the memory required by 50% and the CPU time by 43%, on a CRAY X-MP. The CPU time required without refraction is about 200s per day.

3. Surge forecasts

3.1 Development of surge forecast models

Numerical finite difference models for tides and surges were first developed in the 1960s. A simple model solving the linearised equations with meteorological forcing derived manually from synoptic charts was established by Heaps (1969). This model, with software written in Algol was designed to run on an English Electric KDF9 computer with 32K words of memory, which restricted the grid resolution to roughly 35km for a model covering the UK continental shelf. This same resolution is *still* used, see Fig. 1, though almost everything else has changed! In the early 1970s, an improved numerical scheme was introduced for the full non-linear equations and applied in tidal calculations (Flather 1976). A key step was to link the surge model with an atmospheric model, which could provide forecasts of the essential meteorological forcing for *direct* input to the surge calculation (Flather and Davies, 1976). Further development, in which tide and surge calculations were combined to represent important non-linear interactions, paved the way for real-time forecasts. The first experimental forecasts were carried out in the spring of 1978. This involved the transfer of output, produced by the UKMO weather forecast model, from their IBM 360/195 at Bracknell to a PDP 11 based "Camac" workstation at (POL) Bidston along a dial-up line. The meteorological data were then sent to the SRC (now SERC) Daresbury Laboratory across the JANET network and the surge model run on an IBM 370/165 there. The results were transmitted back to Bidston and on to Bracknell by the same route. One week of this was enough to convince all concerned that the surge forecasts would be best run at the UKMO. This was implemented during that summer and the surge model has been run there for each surge season, September to April, since. The computer requirements were quite modest; a 36 hour forecast taking about 90s CPU time on the 360/195, one of the most powerful machines of its day.

A major revision was carried out in 1982-83, when the UKMO bought a new vector processing computer, a CDC Cyber 205E, and introduced an improved atmospheric model to run on it. The calculations were reorganised and the software completely rewritten to take advantage of the capabilities of the new machine. This involved the use of explicit vector code permitted in Fortran 200. Experiments showed that best efficiency was achieved by solving the equations everywhere within a rectangular region, resulting in long vectors. Unwanted results were discarded using bit-masks. The overall size of the rectangular region, and hence the total computational effort, was reduced by "patching" - shifting parts of the sea area covered to other locations in memory which would otherwise contain only land. In this way the proportion of "useful" calculations was increased and efficiency further improved. Input and output were minimized, and the bulk of the essential data transferred using concurrent buffered I/O to and from disc. This allowed the I/O to be done whilst computations proceeded. Details can be found in Proctor and Flather (1983). The new scheme with an extended model took only 14s CPU time to run a 12 hour hindcast and 36 hour forecast, though the elapsed time including output of the results and transfer of data to the IBM 370/158 front-end for archival is about 2 minutes.

More recently, advances in the microprocessors used in PCs and workstations suggested that such machines should be capable of running surge forecasts. The use of a dedicated machine has potential advantages in allowing a more flexible approach, with forecasts run as and when required to best meet the developing circumstances and the needs of the users. This is not possible with the mainframe version which, as part of the UKMO operational suite of models, must be run to a fixed schedule. Additional more detailed local models, giving better surge estimates in areas threatened with flooding, might also be run on a dedicated system as and when needed.

Work was started during 1988 and a version of the surge forecast model, similar to that running on the Cyber, was implemented on an IBM PS/2 Model 80 (20mhz Intel 80386/80387) for the Storm Tide Warning Service in August 1989 and has been in use since then. The system uses 3270 terminal emulation to transfer the necessary meteorological data from the mainframe to the PC. The software uses standard Fortran and runs under the DOS operating system, with additional memory configured as virtual disc. DOS batch files are used to perform the required steps in the forecast procedure. These are TIDE to run the model tidal prediction; SURGE to run the tide + surge calculation and subtract the tidal solution to get the required surge component; and SAVE to update an archive of stored model results, which will eventually be used to provide input to local models and to allow supplementary calculations with, for example, adjustments to the initial conditions based on the latest observed sea levels from tide gauges. The time critical part of the procedure (the tidal prediction and archiving can be done at any convenient time) takes about 37 minutes; marginally greater than the acceptable target time of 30 minutes. The difference could be realized fairly easily with some further work, but other hardware and software developments offering an increase in performance of one to two orders of magnitude are more interesting in the longer term.

An essential part of any operational scheme is provision of routines to verify that the data on which the forecast will be based are correct. For example, if tidal input, initial model data fields and meteorological forcing are not synchronized, perhaps because the previous forecast did not run due to a hardware problem, then incorrect forecasts will be produced. The checking procedures in the PC version recognize more of the possible situations that could arise than are catered for in the mainframe version and, as a result, the system is more "robust".

3.2 *The operational surge forecast procedure*

Fig. 2 shows the organization of the surge forecast scheme. The tide-surge model, outlined above, uses output from the UKMO 15 level fine-grid weather forecast model (Gadd, 1985) to provide the necessary forcing. The surge model is run twice each day, at about 0230 and 1430 GMT. Each run comprises a 12 hour "hindcast", based on the 12 hour data assimilation calculation of the atmospheric model, and a 36 hour forecast based on the output from the following weather forecast run. The purpose of the hindcast is to provide better initial data, derived from meteorological data incorporating available observations, for the following surge forecast. Each surge model run comprises two calculations; one for tide alone and the other for tide and surge together. By this means, estimates of the surge component, including the important effects due to tide surge interaction, are derived. These can be added to tidal predictions produced by the standard harmonic method, which is more accurate than the model for sites with good tide gauges. The resulting water level predictions are used by the Storm Tide Warning Service (STWS) to issue flood warnings for ports on the east coast. STWS also forward the information by fax to the regional authorities along the south and west coasts on an advisory basis. It should be noted that the present model was not intended to provide forecasts for the south and west coasts.

The input data from the atmospheric model comprises hourly grid point values of surface atmospheric pressure and "surface" wind; components of the wind computed at the lowest level in the atmospheric model $\sigma = 0.997$, roughly 25m above sea level. In the mainframe version, tide and tide+surge calculations both extend over the full 48 hour period, with data saved every 12 hours to define initial conditions for subsequent forecasts. This allows the system to continue without the need for a "cold" start even if 3 forecasts are lost. In the PC version, a file of model-predicted tide is maintained from one forecast run to the next. In normal circumstances, the first 12 hours of data are discarded and a 12 hour model run carried out to update this file for the new hindcast - forecast period. However, the system is able to recognize and modify this automatically to deal, for example, with the case of a lost forecast leaving a 24 hour rather than a 12 hour gap. Also, the PC version can use any set of model fields, stored every hour, as initial data for a subsequent calculation.

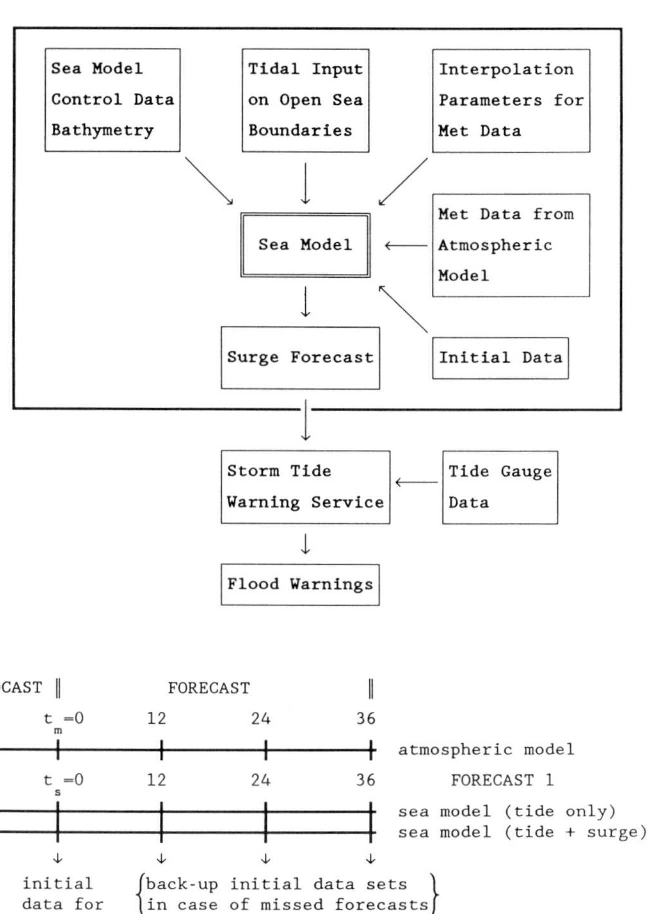

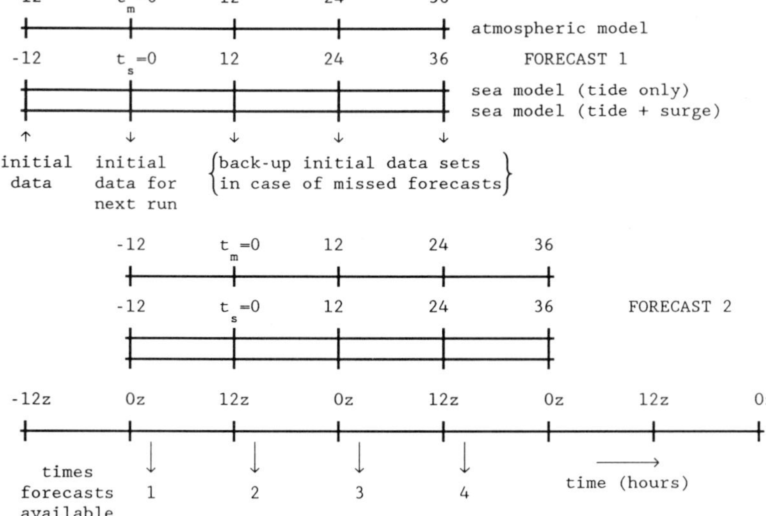

Fig. 2: Organization of the flood warning scheme and the calculations carried out using the surge model.

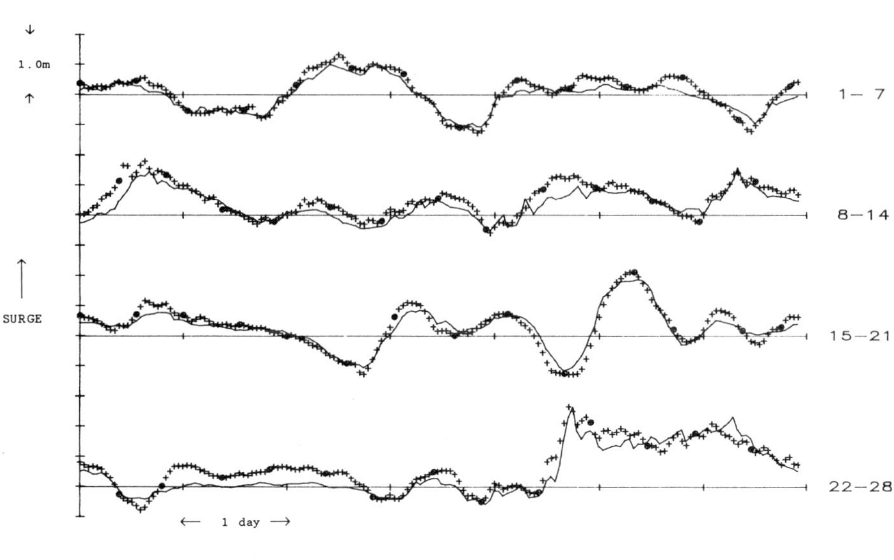

Fig. 3: Comparison between forecast (line) and observed (crosses) surges at Lowestoft in February 1990. The circles indicate the surge residual at approximately the time of tidal high water.

3.3 Surges and floods during the winter of 1989-90

To illustrate the results from the model, output for some of the storms in the winter of 1989-90 is presented. The season was particularly stormy, with significant flooding affecting the south coast in December 1989 and a major flood on the North Wales coast in February 1990 as a result of which 2000 people were evacuated from Towyn. The return periods of the maximum water levels at Holyhead and Liverpool during this event have been estimated as between 150 and 200 years (Flather and Blackman, unpublished report 1990).

Fig. 3 shows a standard monthly plot for February 1990 of the variation with time of forecast and observed surges at Lowestoft. Such plots are produced routinely at POL as part of the data verification and archiving procedure. The plotted forecast data comprise 12 hour sections from successive forecast runs, the period extracted being that most useful for warning purposes, 6 to 18 hours of the forecast. As can be seen, the forecasts for east coast ports are generally very accurate.

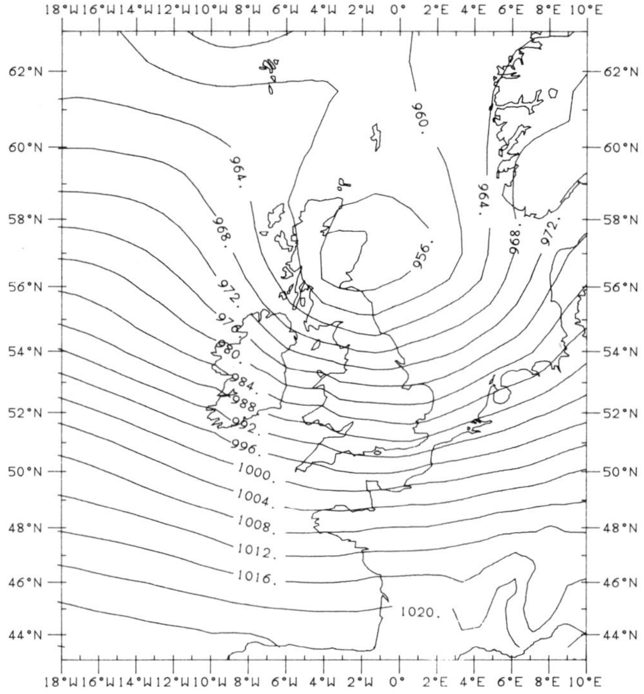

Fig. 4: Surface atmospheric pressure distribution at 0600GMT on 26 February from the UKMO atmospheric model.

Fig. 4 shows the surface atmospheric pressure distribution at 0600GMT on 26 February from the UKMO atmospheric model. A depression with central pressure about 951mb crossed Scotland during the morning, bringing severe westerly gales over the Irish Sea. Table 1 shows the surge forecast for west coast ports from the model run carried out at 0230GMT on 26 February, 8-9 hours before the flooding of Towyn, as received by STWS and sent out to west coast authorities. Fig. 5 shows the comparison of forecast and observed surges at Holyhead, Liverpool and Heysham. The forecasts in Fig. 5 are typical of those produced for the Irish Sea; the model failing to predict the large surge peaks exceeding 2.5m which occurred near low water at Liverpool and Heysham. However, the *important* predictions for the times of tidal high water are remarkably accurate. Surges at high water from model results are flagged with "H" in Table 1 and plotted as ⊖ in Fig. 5.

Model forecasts for the Bristol Channel are generally much less satisfactory. The prediction for the morning tide on 26 February at

STORM SURGE FORECAST.
RESIDUAL ELEVATIONS IN METRES.
DATA STARTS AT 0 HRS GMT 26/ 2/1990

GMT	1687 NLYN	1513 ILFR	1471 AVON	1467 MILF	1380 FISH	1293 BARM	1248 HOLY	1207 HILB	1119 HEYS	1074 WORK	1608 PMTH	1604 PLND	1645 PLYM	1610 NWHN	1524 DVER
0	0.14L	0.24L	0.15	0.25	0.28	0.29	0.30	0.32	0.38H	0.37H	0.14	0.18	0.16L	0.07H	-0.23H
100	0.14	0.47	0.17	0.32L	0.32	0.39	0.34	0.33	0.42	0.43	0.08	0.21L	0.17	0.02	-0.07
200	0.16	0.47	0.31L	0.37	0.37L	0.45	0.39	0.39	0.46	0.47	0.04	0.21	0.20	0.08	0.03
300	0.18	0.45	0.70	0.42	0.50	0.50	0.46	0.48	0.55	0.53	0.11	0.20	0.20	0.14	0.15
400	0.16	0.56	0.95	0.45	0.56	0.63L	0.53L	0.56	0.62	0.58	0.17	0.23	0.15	0.15	0.09
500	0.12	0.58	0.99	0.45	0.60	0.78	0.63	0.68	0.72	0.67	0.22L	0.28	0.13	0.17	-0.04
600	0.10H	0.50	1.03	0.41	0.59	0.85	0.73	0.85L	0.84L	0.77L	0.31	0.22	0.16H	0.19L	-0.23
700	0.09	0.37H	0.93	0.36H	0.55	0.71	0.72	1.12	1.07	0.94	0.23	0.24	0.14	0.31	-0.20L
800	0.08	0.36	0.60H	0.32	0.54H	0.67	0.71	1.31	1.38	1.09	0.28	0.26H	0.13	0.31	-0.01
900	0.07	0.32	0.30	0.28	0.49	0.73H	0.74	1.32	1.42	1.19	0.33	0.27	0.17	0.48	0.46
1000	0.06	0.24	0.29	0.25	0.58	0.77	0.72H	1.27	1.24	1.19	0.34	0.31	0.10	0.45	0.51
1100	0.04	0.37	0.40	0.23	0.52	0.76	0.70	1.22	1.18	1.08	0.37H	0.27	0.05	0.38	-0.14
1200	0.04L	0.47	0.49	0.27	0.44	0.69	0.65	1.09H	1.09H	0.89H	0.38	0.18	0.14	0.44H	-0.06H
1300	0.04	0.55L	0.60	0.27L	0.42	0.53	0.54	0.95	0.88	0.76	0.31	0.21L	0.11L	0.40	0.18
1400	-0.02	0.47	0.74	0.21	0.40	0.53	0.45	0.82	0.82	0.69	0.25	0.21	0.00	0.45	0.32
1500	-0.02	0.26	0.97L	0.10	0.34L	0.53	0.43	0.70	0.75	0.66	0.36	0.20	0.05	0.57	0.50
1600	0.00	0.11	1.08	0.05	0.24	0.40L	0.36	0.63	0.68	0.58	0.39	0.15	0.07	0.53	0.55
1700	-0.01	0.06	0.82	0.07	0.12	0.29	0.31L	0.65	0.60	0.51	0.43L	0.29	0.02	0.48	0.42
1800	-0.04H	0.15	0.54	0.07	0.15	0.19	0.21	0.62L	0.58	0.49	0.42	0.19	0.06	0.40L	0.11
1900	0.00	0.15H	0.35	0.07H	0.16	0.19	0.17	0.61	0.61L	0.47L	0.39	0.24	0.10H	0.49	0.30
2000	0.02	0.02	0.22H	0.05	0.19H	0.32	0.25	0.54	0.60	0.45	0.38	0.23H	0.08	0.52	0.73L
2100	0.03	0.07	0.06	0.09	0.23	0.32H	0.31	0.47	0.50	0.40	0.40	0.19	0.08	0.62	1.11
2200	0.02	0.18	-0.02	0.12	0.31	0.38	0.38H	0.56	0.51	0.43	0.42	0.17	0.05	0.72	1.23
2300	0.00	0.24	0.07	0.17	0.31	0.46	0.37	0.60	0.58	0.46	0.50H	0.18	0.01	0.74	0.90
0	0.02L	0.27	0.16	0.19	0.37	0.44	0.38	0.66H	0.64H	0.52	0.61	0.19	0.04	0.70H	0.65H
100	0.02	0.29L	0.23	0.20L	0.34	0.42	0.36	0.72	0.68	0.54H	0.62	0.33L	0.07L	0.57	0.60
200	0.04	0.31	0.29	0.19	0.32	0.44	0.43	0.72	0.71	0.58	0.52	0.43	0.12	0.56	0.62
300	0.07	0.25	0.38L	0.14	0.37L	0.49	0.47	0.71	0.74	0.66	0.46	0.38	0.23	0.55	0.59
400	0.11	0.12	0.51	0.13	0.37	0.54	0.51	0.79	0.79	0.69	0.45	0.29	0.22	0.54	0.60
500	0.07	0.15	0.50	0.15	0.39	0.57L	0.52L	0.88	0.84	0.73	0.46	0.29	0.13	0.56	0.64
600	0.01H	0.22	0.43	0.17	0.34	0.53	0.51	0.98	0.92	0.75	0.46L	0.27	0.07	0.54	0.52
700	-0.01	0.22H	0.40	0.19H	0.34	0.45	0.47	1.08L	1.00L	0.77L	0.47	0.22	0.05H	0.56L	0.49
800	0.01	0.21	0.38	0.18	0.39	0.44	0.47	1.08	1.05	0.78	0.46	0.19H	0.06	0.60	0.58L
900	0.06	0.18	0.30H	0.17	0.37H	0.53H	0.50	0.91	1.02	0.73	0.47	0.19	0.09	0.60	0.77
1000	0.06	0.21	0.25	0.15	0.37	0.57	0.51	0.86	0.85	0.76	0.43	0.16	0.09	0.62	0.96
1100	0.02	0.25	0.30	0.16	0.35	0.55	0.46H	0.89	0.86	0.69	0.38H	0.14	0.02	0.58	0.86
1200	0.01	0.28	0.32	0.19	0.29	0.49	0.41	0.88	0.85	0.67	0.40	0.14	0.03	0.57	0.63

Table 1. *Forecast surge elevations for the West and South coasts produced by the P. O. L. storm surge model. Initial data time of the forecast run was 0000GMT 26 February 1990 and the results were available to S. T. W. S. at about 0230GMT.*

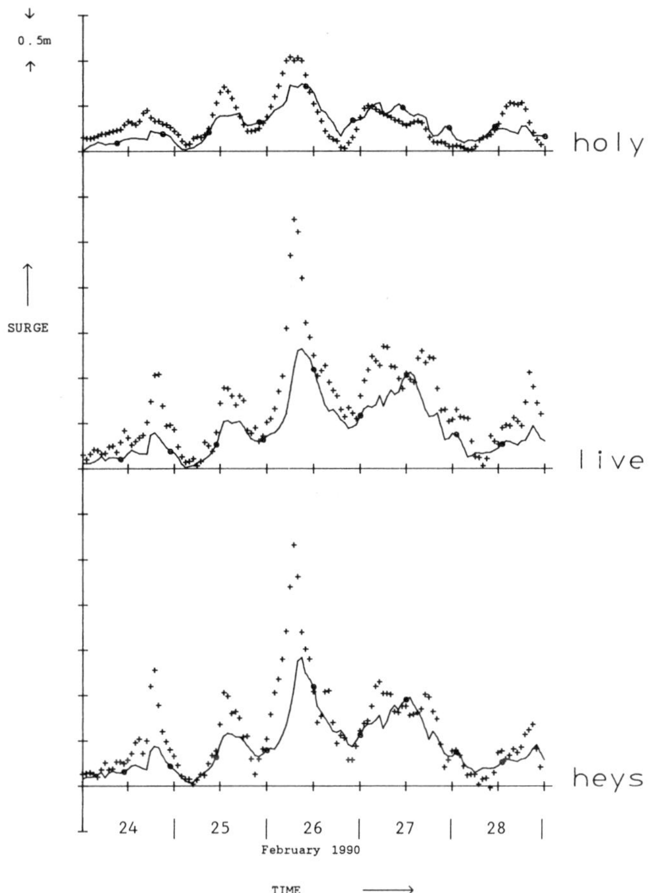

Fig. 5: Comparison between surge elevations derived from observations (crosses) and from model forecasts (lines and circles) for Holyhead, Liverpool and Heysham during the period 24-28 February 1990. Results from hours 6-18 of successive model forecasts are plotted and circles indicate the approximate time of high water.

Avonmouth was only 60cm, about half that observed. This is partly a problem of resolution; the surge model with a 35km grid simply cannot represent the essential local details of surge generation and propagation within the Bristol Channel. Likewise, the atmospheric model with a 75km grid is unable to describe the important local wind variations. With adequate resolution, 4km or less for the surge model, and accurate wind and pressure data, Proctor and Flather (1989) have shown that accurate surge forecasts can be produced even in this dynamically complex area.

4. Future prospects

A new version of the surge model to run on the CRAY Y-MP at the UKMO is under development. It is intended that this model will have a reduced grid size over the whole shelf with further grid refinements in important areas like the Bristol Channel, the Wash and Morecambe Bay. It is also intended that the new model will incorporate improved dynamics. However, it is not certain how much can be achieved within the various operational and financial constraints.

The PC version of the model will be retained and the flexibility inherent in the use of a dedicated machine exploited further. This will require some enhancement of the existing hardware to provide an increase in processor power of at least one and preferably two orders of magnitude. Possibilities include the use of RISC based workstations and additional processors, transputers and/or Intel i860 microprocessors, with the present PS/2 systems. This approach would seem most promising for the next few years and is under investigation. It seems likely that even more powerful processors and workstations will come onto the market. These machines would be capable of running wave-tide-surge models, which require the power of a CRAY, but at a small fraction of the cost.

The development of the wave-tide-surge model will continue. Experiments are in progress using the storms of the winter of 1989-90 as test cases to quantify the effects due to interactions between waves and water level and current variations associated with the tide and surge motion. The indications are that these interactions are important. Implementation of a combined wave-tide-surge model should lead to better surge forecasts and more accurate wave predictions, particularly for coastal regions.

Finally, all wave and surge forecast models depend on accurate meteorological data to drive them. It is essential, therefore, to make use of the very best available wind and pressure forecasts. The development by the UKMO of mesoscale atmospheric models, with spatial grid resolution of about 15km (Golding, 1987), is of particular interest. Should such a model become operational, it could provide meteorological forcing with the detailed resolution of local variations needed for future more accurate surge and wave models.

5. Acknowledgments

The work described in this paper including the development, maintenance and running of the operational surge forecast model was funded by the Ministry of Agriculture Fisheries and Food.

Excellent support and assistance from the UK Meteorological Office and the Storm Tide Warning Service is gratefully acknowledged.

The observed surges from Liverpool, plotted in Fig. 5, were derived from data kindly provided by the Mersey Docks and Harbour Company.

6. References

Bretherton, F.P. and Garrett, C.J.R. (1969). Wavetrains in inhomogeneous moving media. Proceedings of the Royal Society of London, A, 302:529-554.

Davies, A.M. and Flather, R.A. (1978). Application of numerical models of the north west European continental shelf and the North Sea to the computation of the storm surges of November - December 1973. Deutsche Hydrographische Zeitschrift Ergänzungsheft, A, Nr.14:51pp.

Flather, R.A. (1976). A tidal model of the north-west European continental shelf. Mémoires Société Royale des Sciences de Liège, 10:141-164.

Flather, R.A. (1984). A numerical investigation of the storm surge of 31 January and 1 February 1953 in the North Sea. Quarterly Journal of the Royal Meteorological Society, 110:591-612.

Flather, R.A. and Davies, A.M. (1976). Note on a preliminary scheme for storm surge prediction using numerical models. Quarterly Journal of the Royal Meteorological Society, 102:123-132.

Gadd, A. J. (1985). The 15-level weather prediction model. Meteorological Magazine, 114:222-226.

Golding, B. (1983). A wave prediction system for real-time sea state forecasting. Quarterly Journal of the Royal Meteorological Society, 109:393-416.

Golding, B. (1987). The UK Meteorological Office mesoscale model. Boundary Layer Meteorology, 41:97-107.

Heaps, N.S. (1969). A two-dimensional numerical sea model. Philosophical Transactions of the Royal Society of London, A, 265:93-137.

Hubbert, K.P. and Wolf, J. (1990). Numerical investigation of depth and current refraction of waves. Journal of Geophysical Research (in press).

Hasselmann, S. and Hasselmann, K. (1985). Computations and parameterizations of the nonlinear energy transfer in a gravity wave spectrum. Part 1: A new method for the efficient computation of the exact nonlinear transfer integral. Journal of Physical Oceanography, 15:1369-1377.

Murty, T.S. (1984). Storm Surges - Meteorological Ocean Tides. Canadian Bulletin of Fisheries and Aquatic Sciences, 212:897pp.

Murty, T.S., Flather, R.A. and Henry R.F., (1986). The Storm Surge Problem in the Bay of Bengal. Progress in Oceanography, 16:195-233.

Proctor, R. and Flather, R.A. (1983). Routine storm surge forecasting using numerical models: procedures and computer programs for use on the CDC Cyber 205 at the British Meteorological Office. Institute of Oceanographic Sciences, Report, No.167:171pp.

Proctor, R. and Flather, R.A. (1989). Storm surge prediction in the Bristol Channel - the floods of 13 December 1981. Continental Shelf Research, 9:889-918.

Smith S. D. and Banke E., (1975). Variation of the sea surface drag coefficient with wind speed. Quarterly Journal of the Royal Meteorological Society, 101:665-673.

The WAMDI Group (1988). The WAM Model - a third generation ocean wave prediction model. Journal of Physical Oceanography, 18:1775-1810.

Welander, P., (1961). Storm Surges. Advances in Geophysics, 8:315-379, Academic Press, New York and London.

Wolf, J., Hubbert, K.P. and Flather, R.A. (1988). A feasibility study for the development of a joint surge and wave model. Proudman Oceanographic Laboratory, Report, No.1:109pp.

The importance of algorithm design and data structure in implementing 3D hydrodynamic models on parallel-vector computers

A.M. Davies

Proudman Oceanographic Laboratory

R.B. Grzonka

NERC Computer Services

1. Introduction

The history of large scale scientific computing has been characterised by a requirement to develop numerical algorithms capable of solving large scientific problems but minimising the computational overhead required to solve the problem. Obviously if massive computational power were available and the problems to be solved were small, then any accurate numerical algorithm would be acceptable. However in most scientific problems, the application requires major computational resources on the most powerful and advanced scientific computers. In many problems long time integrations are required involving at each time step the repetition of identical tasks on new data. In these applications it is essential to make the most effective use of the computational resources available, particularly in a long time integration calculation where a small computational saving at each time step can result in a major overall saving.

Initially numerical algorithms were designed to make most effective use of a serial machine, with a single processor taking values from and returning results to a single memory. Subsequently vector processor computers were developed using pipelined functional units (e.g. CRAY-1) which were fed from a vector register. In recent years enhanced performance has been achieved by using multiprocessor versions of these computers (e.g. the CRAY X-MP/48 with four processors) also computers with multiple processors and a memory hierarchy e.g. cache, main memory, virtual memory have become available (e.g. the IBM 3090/600 with six processors).

As hardware has progressed, so new numerical algorithms to take maximum advantage of the latest computer architecture, have had to be developed. Initially on serial machines the major constraint was to minimize processor time and memory requirements. Memory use in shallow sea models was minimized by storing only the sea regions and using indirect addressing to locate adjacent grid points required in the finite difference scheme. These methods together with implicit or semi-implicit (A.D.I.) time integration schemes (Wolf, 1983) ensured that computer time and memory were minimised.

On vector processing computers (e.g. CRAY-1) and in particular the Cyber 205, maintaining a long vector length for optimal performance was particularly important. The simplest way of achieving this was to store both sea and land points, and to perform the calculation everywhere using masking to discard points on land. Although this method required more computational operations and storage (since land areas were now being stored), than earlier algorithms on serial machines, these methods appeared optimal on single-processor vector computers.

With the development of massively parallel machines (based on transputer arrays) having the order of a hundred to a thousand processors, with each processor assigned its own local memory (a distributed memory computer), new algorithms are having to be developed. At present it is not clear what will be the optimal algorithm to use on these machines, for the solution of the three-dimensional hydrodynamic equations. However, it is already clear that implicit or semi-implicit time integration methods will be more difficult to implement for parallel processing than explicit methods. Certainly on a massively parallel computer, where each grid point could be assigned to a processor, then an explicit time integration method would permit the solution to be advanced by one time step, at each point, in parallel. A combination of this together with a finite difference method which only required data from neighbouring points, reducing memory to memory access would appear optimal on such machines.

In this paper we do not consider the development of algorithms for such highly parallel computers with local memory, but rather computers with less parallelism and global (or shared) memory systems (e.g. CRAY X-MP/48, four processors; IBM 3090/600, six processors; CRAY Y-MP/832, eight processors). However bearing in mind that future developments will involve more highly parallel systems, which favour explicit type integration systems and low order difference schemes, we will develop algorithms having these characteristics.

In the next section the mathematical formulation of a three- dimensional turbulent energy model and a three-dimensional spectral model will be briefly described. In subsequent sections the computational implementation of the methods, in particular the modal method will be presented. The final sections of the paper will present some results from the models, and discuss future developments.

2. Mathematical models

2.1 Basic equations

Here to illustrate the major points in the development of the models, we will use Cartesian Coordinates and consider the linear equations. However results from "real world" models presented in the next section, are based upon models which use polar coordinates in order to take account of the curvature of the earth. (A detailed description of the solution of the non-linear equations using polar coordinates is given in Davies, 1986).

The linear hydrodynamic equations are given by

$$\frac{\partial \zeta}{\partial t} + \frac{\partial}{\partial x}\left\{h\int_o^1 U d\sigma\right\} + \frac{\partial}{\partial y}\left\{h\int_o^1 V d\sigma\right\} = 0 \tag{2.1}$$

$$\frac{\partial U}{\partial t} - \beta V = -g\frac{\partial \zeta}{\partial x} + \frac{1}{h^2}\frac{\partial}{\partial \sigma}\left(\mu\frac{\partial U}{\partial \sigma}\right) \tag{2.2}$$

$$\frac{\partial V}{\partial t} + \beta U = -g\frac{\partial \zeta}{\partial y} + \frac{1}{h^2}\frac{\partial}{\partial \sigma}\left(\mu\frac{\partial V}{\partial \sigma}\right) \tag{2.3}$$

where: t denotes time; $\sigma = z/h$ a normalized sigma coordinate; x,y,z cartesian coordinates with U,V the x,y components of velocity; g acceleration due to gravity, ß geostrophic coefficients taken as constants. Water depth h in a "real world" model varying with horizontal position. In these equations μ denotes vertical eddy viscosity, the parameterization of which we will consider in some detail later in this section.

Surface and bed boundary conditions are given by

$$\frac{-\rho}{h}\left(\mu\frac{\partial U}{\partial \sigma}\right)_o = F_s \quad , \frac{-\rho}{h}\left(\mu\frac{\partial V}{\partial \sigma}\right)_o = G_s \tag{2.4}$$

and

$$\frac{-\rho}{h}\left(\mu\frac{\partial U}{\partial \sigma}\right)_1 = F_B \quad , \frac{-\rho}{h}\left(\mu\frac{\partial V}{\partial \sigma}\right)_1 = G_B \tag{2.5}$$

In (2.4) and (2.5), ρ is density of sea water with F_S , G_S surface wind stress and F_B, G_B bed stress components.

Parameterising bed stress in terms of a linear slip conditions, gives

$$F_B = k\rho\ U_h, \quad G_B = k\rho\ V_h \tag{2.6}$$

with k a linear coefficient of bottom friction and U_h, V_h components of bottom current. Quadratic friction laws may also be used (Davies 1986).

An alternative to (2.6) is to assume a no-slip condition at the sea bed, namely

$$U = V = 0 \quad \text{at the bed} \tag{2.7}$$

2.2 *Turbulence energy closure model and numerical solution*

Calculations (Davies 1987), have shown that the formulation of vertical eddy viscosity μ is particularly important, in that the value of μ and its temporal and spatial variability influence the variation of current from sea surface to sea bed. Here we will consider the determination of μ in terms of the turbulence energy b, using a turbulence energy closure model, given by

$$\frac{\partial b}{\partial t} = \mu \left\{ \left(\frac{\partial U}{\partial z} \right)^2 + \left(\frac{\partial V}{\partial z} \right)^2 \right\} + \beta_a \frac{\partial}{\partial z} \left(\mu \frac{\partial b}{\partial z} \right) - \varepsilon \tag{2.8}$$

with β_a a laboratory determined constant.

Turbulence dissipation ϵ, is computed from

$$\varepsilon = C_1\ b^{3/2}/1 \tag{2.9}$$

with eddy viscosity μ given by,

$$\mu = C_o\ l\ b^{1/2} \tag{2.10}$$

and mixing length l expressed as

$$l = \frac{kz}{1 + kz/lo} \tag{2.11}$$

with

$$l_o = \gamma \int_{zo}^{h} b^{1/2} z \, dz \Big/ \int_{zo}^{h} b^{1/2} \, dz \tag{2.12}$$

and k=0.4 Von Karman's constant, and γ a constant in the range 0.1 to 0.4.

At the sea surface an appropriate boundary condition for turbulence energy is

$$\beta_a \left. \frac{\partial b}{\partial z} \right|_{z=h} = m_2 U_*^3 \tag{2.13}$$

with m_2 = 2.6 an empirical coefficient, and U_* = wind friction velocity.

While at the bed

$$\left. \frac{\partial b}{\partial z} \right|_{z=zo} = 0. \tag{2.14}$$

Values of constants are $\beta_a = 0.73$, $C_o = C^{1/4}$, $C_1 = C_o^3$, where C = 0.046, with z_o roughness length. A bottom boundary condition on velocity which is consistent with (2.14), is a no slip condition, given by (2.7).

The physical nature of the current profile determined with a no slip condition is such that there exists a high shear layer close to the sea bed. In order to resolve this layer, a fine finite difference grid is required close to the bed. This can be most readily accomplished using a logarithmic or log.linear transformation in the vertical and a uniform finite difference grid on this transformed coordinate, giving a fine grid close to the bed.

A suitable transformation, namely a log-linear is given by

$$\sigma = \ln(z/zo) + \frac{(z-zo)}{z_*} \Big/ \Delta \tag{2.15}$$

with

$$\Delta = \ln(h/zo) + \frac{(h-zo)}{z_*}$$

with z_* an arbitary value which can be used to determine the height above the bed over which the grid is essentially logarithmic.

The detailed mathematical transformation of the hydrodynamic and turbulence equations using equation (2.15) together with their numerical solution is given in Davies and Jones (1989) and will not be repeated here.

The principal numerical difficulty with the solution of the hydrodynamic equations is associated with the term involving the vertical eddy viscosity. If this is computed at the lower time step, then with a fine finite difference grid in the vertical, a time step of order seconds must be used, which is clearly impractical. Various numerical techniques (e.g. Dufort Frankel, Saulev, Crank Nicolson) have been used by various authors (e.g. Davies, 1989a,b; DeGoede, 1988,1989a,b) to centre the viscosity term in time giving an unconditionally stable numerical solution which may be appropriate for a vector processor (De Goede, 1989a,b).

Davies and Jones (1989) used the Crank Nicolson method in their turbulence energy model. This approach gave a tridiagonal matrix which could then be solved using the Thomas algorithm. However this technique cannot be readily vectorized (Ortega and Voigt, 1985) although a parallel algorithm can be developed by solving a system of tridiagonal matrices, where each matrix is associated with a different horizontal grid point.

Calculations (Davies and Jones, 1989) showed that accurate solutions in continental shelf sea regions (Fig. 1) required the order of forty-five grid points in the vertical on a log-linear coordinate in order to accurately resolve the bottom boundary layer. Consequently the method was computationally rather expensive, although it yielded detailed information on turbulence energy distributions and mixing intensities.

2.3 *Eddy viscosity closure and spectral solution*

A computationally inexpensive method, although an approach which contains the essential physics incorporated in the turbulence energy model, is to use a spectral/modal solution through the vertical. In this method the modes are eigenfunctions of an eddy viscosity profile $\Phi(\sigma)$, with viscosity expressed as

$$\mu = \alpha(x,y,t)\Phi(\sigma) \tag{2.16}$$

with $\alpha(x,y,t)$ determining the time variation of eddy viscosity.

The two components of velocity U and V are expanded in terms of time and horizontal space dependent coefficients $A_r(x,y,t)$, $B_r(x,y,t)$ and functions $f_r(\sigma)$ in the vertical, giving

$$U - \sum_{r-1}^{m} A_r(x,y,t)f_r(\sigma), \quad V - \sum_{r-1}^{m} B_r(x,y,t)f_r(\sigma) \tag{2.17}$$

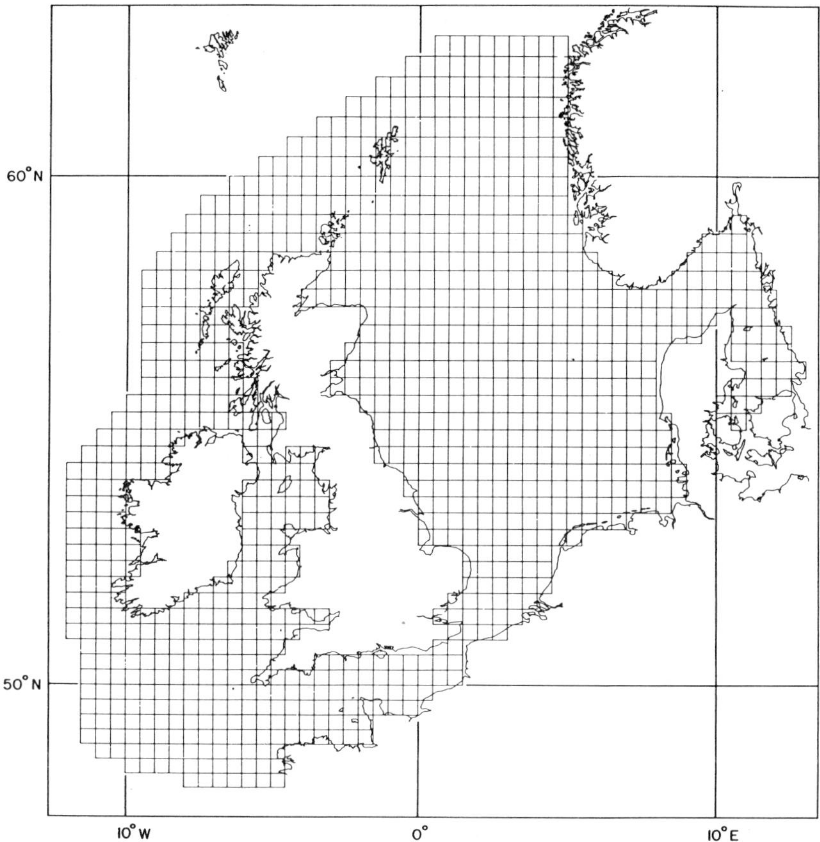

Fig. 1: Finite difference grid of the three-dimensional model.

In theory the choice of functions $f_r(\sigma)$ is arbitrary (Davies, 1983, 1987; Davies and Stephens, 1983). However, Davies (1987) showed that by using eigenfunctions (modes) of the eddy viscosity profile an uncoupled system of equations could be obtained in the linear case. Thus choosing the f_r's to be eigenfunctions of

$$\frac{d}{d\sigma}\left[\Phi\frac{df}{d\sigma}\right] = -\varepsilon f \qquad (2.18)$$

with ϵ associated eigenvalues as we will show later, does have major computational advantages.

In the case of a no slip bottom boundary condition, appropriate boundary conditions at sea surface and sea bed are

$$\left.\frac{df_r}{d\sigma}\right|_1 = 0 \quad f_r(o) = 0 \tag{2.19}$$

while for a linear slip condition

$$\left.\frac{df_r}{d\sigma}\right|_1 = 0, \quad \mu\left.\frac{df_r}{d\sigma}\right|_o = kf_r(o) \tag{2.20}$$

Also the eigenfunctions (modes) are normalized such that their surface values are unity, thus

$$f_r(1) = 1 \tag{2.21}$$

Also the eigenfunctions are orthogonal, thus

$$\int_1^o f_r\, f_k\, d\sigma = 0 \quad r \neq k \tag{2.22}$$

Applying the Galerkin method to Eq.2.1-3, they are multiplied by $f_k(\sigma)$ and integrated through the vertical. By integrating by parts the term involving the vertical eddy viscosity, surface and bed boundary conditions can be included. (A detailed derivation of the Galerkin form of the hydrodynamic equations is given in Davies (1983, 1987) and is not presented here).

By using the orthogonality property of the eigenfunctions, we obtain the Galerkin spectral form of the equations, namely.

$$\frac{\partial\zeta}{\partial t} + \sum_{r=1}^{m}\left\{\frac{\partial}{\partial x}\left[U_r h\phi_r\, a_r\right] + \frac{\partial}{\partial y}\left[V_r h\phi_r\, a_r\right]\right\} = 0 \tag{2.23}$$

$$\frac{\partial U_r}{\partial t} = \beta V_r - g\frac{\partial\zeta}{\partial x}\, a_r - U_r\frac{\alpha\epsilon_r}{h^2} + \frac{F_s}{\rho h} \tag{2.24}$$

$$\frac{\partial V_r}{\partial t} = -\beta U_r - g\frac{\partial \zeta}{\partial y} a_r - V_r \frac{\alpha \epsilon_r}{h^2} + \frac{G_s}{\rho h} \tag{2.25}$$

r=1,2,...,m

where $U_r = \frac{A_r}{\phi_r}$, $V_r = \frac{B_r}{\phi_r}$, $a_r = \int_o^1 f_r \, d\sigma$, $\phi_r = \frac{1}{\int_o^1 f_r^2 \, d\sigma}$

Eq.2.23-25 are the Galerkin form of the hydrodynamic equations, which have to be integrated forward in time subject to surface wind stress forcing, or tidal forcing along open sea boundaries. Integration of the equations yields values of surface elevation ζ and coefficients U_r, V_r, r=1,2...m from which using expansion (2.17) the current at any depth can be determined.

The Galerkin form of the hydrodynamic equations has been used very successfully on serial computers by a number of authors (e.g. Davies, 1986; Furnes, 1983; Gordon and Spaulding, 1987). In the next section we will show that the method with a basis set of modes is ideal for the new generation of multiprocessor vector computers.

3. Numerical solution of the spectral form of the hydrodynamic equations

3.1 Numerical formulation

In this section we consider the numerical solution of Eq.2.23-25 using multiprocessor vector computers, with a global memory system.

The solution of (2.23) can be readily accomplished using forward time stepping. Thus

$$\zeta_i^{(t+\tau)} = \zeta_i^{(t)} + P_i(U_r^t, V_r^t) \tag{3.1}$$

where $P_i(U_r^t, V_r^t)$ represents terms involving the summation over all modes and central differencing operations at grid point i shown in (2.23). By using only second order central differencing operations involving adjacent points the method would be ideal for a local memory machine, where on a highly parallel system each grid point could be integrated in parallel using its own local processor.

By time centring the viscosity term in (2.24), and rearranging, gives

$$U_{r,i}^{t+\tau} = \frac{\left\{ \left(U_{r,i}^{t} + \frac{\tau \alpha \varepsilon_r}{2h^2} U_{r,i} \right) + Q_i \left(\zeta_i^{t+\tau} , U_{r,i}^{t} , V_{r,i}^{t} \right) \right\}}{\left(1 - \tau \frac{\alpha \varepsilon_r}{2h^2} \right)} \tag{3.2}$$

where r=1,2,...,m

and where Q_i involves the wind stress and gradient forcing, together with the influence of the earth's rotation. By this means an unconditionally stable scheme is obtained. Again second order spatial differencing is used.

A similar equation to Eq.3.2 can be derived for the V component of velocity.

3.2 Computational solution on a multiprocessor vector computer

Multitasking: The form of Eq.2.23-25 and the solution algorithm discussed above is ideal for parallel solution using multitasking on a multiprocessor vector computer. The multitasking process can take two forms, namely macrotasking and microtasking. In macrotasking the computational overhead required to start a task and synchronize its completion with the accompanying tasks is high. Consequently in order to effectively utilize macrotasking the amount of computation within each task must be high (i.e. the task must have large granularity). Load balancing in macrotasking must be explicitly implemented by the programmer.

The solution of Eq.2.23-25 using macrotasking would be most readily accomplished at each time step by solving (2.24) and (2.25) in parallel. Since the solution of (2.24) requires the same amount of computational effort as (2.25), their parallel solution would give ideal load balancing. Solution of (2.23) using macrotasking would require domain decomposition in the horizontal, whereby the sea region is partitioned into a number of domains (i.e. two regions on a two processor computer). In the case of a complex sea region (Fig. 1) each domain should contain the same number of horizontal sea grid points at which the calculation is performed if load balance is to be maintained. Within each subdomain vectorization in the horizontal can be readily accomplished, also horizontal vectorization can be readily included in (2.24) and (2.25).

It is evident that when only two processors are available, the use of macrotasking involving the solution of (2.24) and (2.25) in parallel is probably optimal. As the number of processors increases, macrotasking involving domain decomposition in the horizontal could be applied to the solution of all the equations. For a complex sea region involving a number of land masses of variable geometry, such a horizontal domain decomposition could be very involved. Also as the number of processors increases, the granularity of each task would be reduced and the task overhead associated with macrotasking could prove to be unacceptable.

An alternative approach is to use microtasking, which involves a much smaller task overhead, and is therefore still computationally economic when the granularity of the problem is small. The implicit microtasking master-slave relationship between processors (in software only), under certain conditions, also allows load balancing to be "dynamic". Since the number of modes m used in the vertical is typically of the order of between 6 and 16 (Davies and Stephens, 1983) and each mode is independent of the others, (2.24) and (2.25) can be most readily solved by a domain decomposition in the vertical, whereby each coefficient U_r , V_r , $r = 1,2...m$ is solved in parallel on a processor. Since the computational work involved is the same for each mode the system is automatically balanced.

To utilise parallel processing in solving (2.23) would again require domain decomposition in the horizontal. However the computational effort involved in solving (2.23) is small compared with solving the 2m equations (2.24) and (2.25) particularly for large m.

Vectorisation and data organisation: Vector computers with their pipelined functional units, fed from a vector register which can access the total memory, in general yield optimum performance on long vectors of sequential elements.

On a multiprocessor vector computer an optimal way of achieving this is via a domain decomposition in the vertical, in which each mode is integrated in parallel, and vectorisation is used in the horizontal. To achieve long vectors in the horizontal, a one-dimensional rather than a two-dimensional array is used. Thus if we consider a rectangular domain of L grid points north-south and N grid points west-east, giving a total of N-GRID-POINT = N x L. Then, if the region contains only a small percentage of land, the simplest and computationally most effective means of achieving optimal performance is to perform the calculation everywhere using an integer mask of 0 or 1 to mask off the sea region. Thus on a CRAY X-MP the coding

```
CMIC$ DO     GLOBAL
      DO  1 J=1,M
      DO 2 I=1,N-GRID-POINT
            memory-access (MASK(I)*FIELD-ARRAY(I,J))
      2     CONTINUE
      1     CONTINUE
```

was found to yield optimum performance of the order of 130mflops on a single processor (Davies and Proctor, 1989) and over 500mflops (Davies et al, 1989) using four processors on a CRAY X-MP/48. In this coding, the DO GLOBAL command causes the J loop to be integrated in parallel on all available processors using microtasking. If M (the number of modes in the vertical) is chosen as an exact multiple of the number of processors, then the job can make optimal use of all processors.

For a region in which the percentage of land is high compared with the sea region, then a strip mining technique was found to be optimal (Davies et al, 1989). In this method calculations are only performed over each west-east contiguous sea region as a series of strips. Thus

```
CMIC$ DO      GLOBAL
      DO  1 J             = 1,M
      DO  2 I-STRIP       = 1,N-STRIP
      DO  3 I             = I-START(I-STRIP),I-END(I-STRIP)
          memory access (FIELD-ARRAY(I,J))
          3 CONTINUE
          2 CONTINUE
          1 CONTINUE
```

Davies et al (1989) found that when microtasking was implemented on a CRAY X-MP/48 with 32 memory banks, that memory bank conflict could be a problem when all four processor were running in parallel. The memory bank problem can be readily demonstrated on the CRAY X-MP/48 by a simple bandwidth balance argument. On that particular machine the CPU system to memory bandwidth is given by CPU memory ports (16) divided by clock cycle (1CP); a value which is significantly greater than the number of memory banks (32) divided by the bank access time (4CP). A more balanced system would be based on 64 memory banks or faster bank access. Another important parameter, particularly when memory bank access is a problem, is to maintain a high ratio of computation to memory bank access. This entails designing algorithms, data structures and coding to perform the maximum amount of computation before fetching and/or returning values to main memory. A detailed discussion of these points is given in Davies et al (1989) and will not be presented here.

It is now becoming increasingly obvious that for multiprocessor vector computers, the development of parallel algorithms, optimal data structures and some knowledge of the machine architecture is of critical importance in making the most effective use of these computers.

4. Three-dimensional hydrodynamic tidal model

In this section we briefly compare and contrast the solution of the hydrodynamic equations, using the turbulence energy model with a grid scheme in the vertical, and the eddy viscosity closure model with a modal approach in the vertical. Both models use a no-slip condition at the sea bed.

The problem considered is the calculation of M_2 tidal currents on the European Continental Shelf (Fig. 1). The turbulence energy model required 45 grid boxes in the vertical, in order to accurately determine the tidal current profile, particularly in the near bed region. The modal model however because each mode is highly sheared in the near bed region (Fig. 2, profile b) can accurately resolve the current with ten modes in the vertical.

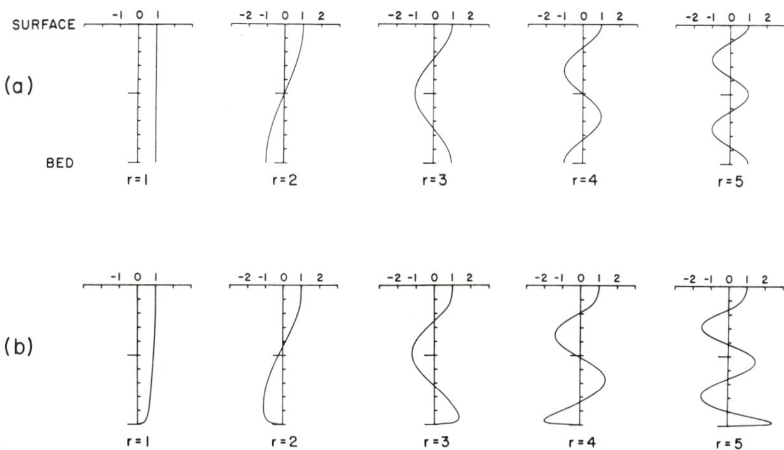

Fig. 2: Profile of the first five modes computed (a) using a slip bottom boundary condition, (b) a no-slip condition at the sea bed.

Computational performance on a CRAY X-MP/48 using a single processor was of the order of 130mflops (modal model) compared with 50mflops (turbulence energy model). Also because the modal model contained less variables, and only required ten modes compared with 45 grid boxes, then for a given horizontal resolution its memory requirement was an order of magnitude less than the turbulence energy model. It is clear from Davies et al (1989) that by using all four processors the performance of the modal model will exceed 500mflops, whereas it is not clear at present how best to develop a multi-processor version of the turbulence energy model. Davies (1989b,c) has recently shown that the turbulence energy model can be used to develop an advanced modal model which produces solutions

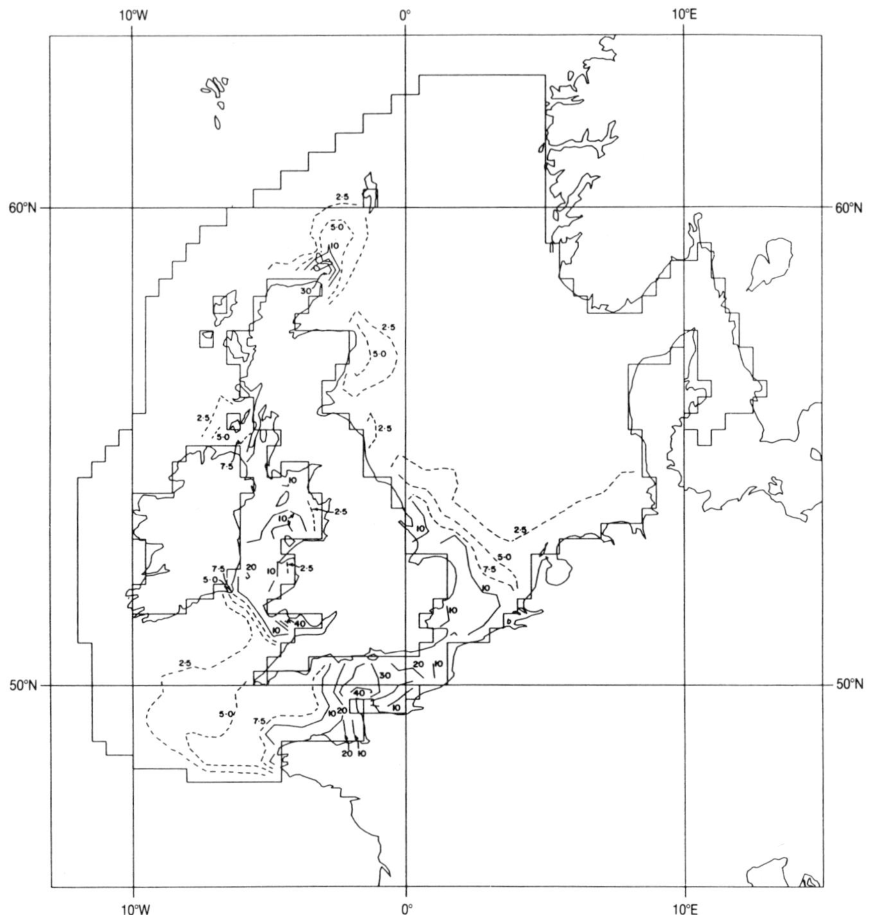

Fig. 3: Contours of surface turbulence kinetic energy ($m^2.s^{-2}$ x 10^4).

which are not significantly different from those obtained with the turbulence energy model.

Contours of turbulence energy intensity, in particular surface values (Fig. 3) show that over certain regions of the shelf, surface turbulence exceeds 0.00025 $m^2.s^{-2}$. By a simple heat balance argument based upon an annual mean heat input, regions where this surface value is exceeded will remain well mixed throughout the year, with other regions stratifying in the summer. Tidal fronts are to be expected between these two areas and in fact the position of the 0.00025 $m^2.s^{-2}$ contour is in good agreement with the positions of the tidal fronts.

A similar spatial pattern is obtained by plotting the region where the thickness of the bottom boundary layer computed with the modal model

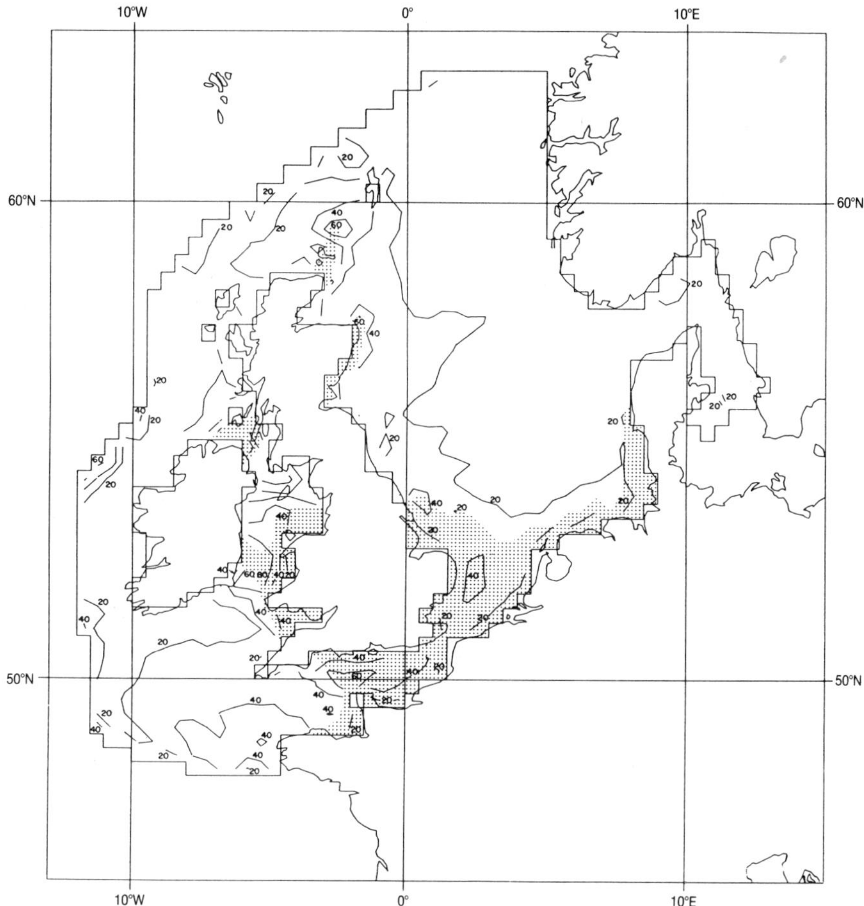

Fig. 4: Thickness of the bottom boundary layer (m). Shaded area is where boundary layer exceeds water depth.

exceeds the water depth (Fig. 4). In regions where this occurs, the tidal current is sheared throughout the water column and hence the shear production term in Eq.2.8 is non-zero and the water column can be expected to remain well mixed. This would appear to be further, though indirect, evidence that the modal model, which is computationally more efficient than the turbulence energy model is able to reproduce the major physical effects which are contained in that model.

5. Conclusions

This paper has briefly outlined the physical, mathematical and computational implementation of two types of numerical model, namely a turbulence energy and a modal model.

By a careful choice of modes (Davies 1989b), a modal model can be developed which can reproduce tidal physics to a similar level of accuracy as the turbulence model, with an associated saving in computational costs.

The mathematical form of the modal models, and the solution algorithm developed here is such that these models can be solved in a highly efficient manner on the new generation of multiprocessor vector computers. However as has been shown here, it is essential to develop suitable algorithms that not only contain load balancing, but an appropriate data structure to minimise memory access and memory bank conflicts. The development of such algorithms is crucial in making the most efficient use of the new generation of computers.

6. References

Davies, A.M. (1983). Formulation of a linear three-dimensional hydrodynamic sea model using a Galerkin-Eigenfunction method. Int. J. Num. Meth. in Fluids, 3:33-60.

Davies, A.M. (1986). A three-dimensional model of the north west European Continental Shelf, with application to the M_4 tide. Journal of Physical Oceanography, 16:797-813.

Davies, A.M. (1987). Spectral models in Continental Shelf Sea Oceanography. In: Three-Dimensional Coastal Ocean Models, N.S. Heaps, editor. A.G.U.

Davies, A.M. (1989a). On the accuracy of finite difference and spectral methods for computing tidal and wind wave current profiles. To appear in Int. J. Num. Meth. in Fluids.

Davies, A.M. (1989b). On using turbulence energy models to develop spectral viscosity models. Submitted to Continental Shelf Research.

Davies, A.M. (1989c). On the importance of time varying eddy viscosity in generating higher tidal harmonics. To appear in Journal of Geophysical Research.

Davies, A.M. and Stephens, C.V. (1983). Comparison of the finite difference and Galerkin methods as applied to the solution of the hydrodynamic equations. Appl. Math. Modelling, 1:226-240.

Davies, A.M. and Proctor, R. (1989). Developing and optimising a 3D-spectral/finite difference hydrodynamic model for the CRAY X-MP. To appear in Computers and Fluids.

Davies, A.M. and Jones, J.E. (1989). Application of a three-dimensional turbulence energy model to the determination of tidal currents on the Northwest European Continental Shelf. To appear in Journal of Geophysical Research.

Davies, A.M., Grzonka, R.B. and Stephens, C.V. (1989). Implementation of a three-dimensional hydrodynamic numerical model using parallel processing on a CRAY X-MP series computer. To appear in Advances in Parallel Computing.

De Goede, E.D. (1988). Finite difference methods for the three-dimensional hydrodynamic equations. Report NM-R8813, CWI, Amsterdam.

De Goede, E.D. (1989). A computational model for three-dimensional shallow water flows on the Alliant FX/4 Supercomputer 32:43-49.

Furnes, G.K. (1983). A three-dimensional numerical sea model with eddy viscosity varying piecewise linearly in the vertical. Continental Shelf Research, 2:231-242.

Gordon, R.B. and Spaulding, M.L. (1987). Numerical simulations of the tidal-and wind-driven circulation in Narragansett Bay. Estuarine, Coastal and Shelf Science, 24:611-636.

Ortega, J.M. and Voigt, R.G. (1985). Solution of partial differential equations on vector and parallel computers. Society for Industrial and Applied Mathematics.

Wolf, J. (1983). A comparison of semi-implicit with an explicit scheme in a three-dimensional hydrodynamic model. Continental Shelf Research, 2:215-229.

Spill model for the protection of the Forth estuary

A.M. Riddle

ICI Group Environmental Laboratory, Brixham

1. Introduction

The ICI works at Grangemouth in Scotland manufactures many compounds for sale internally to other parts of the company and to external customers. The different processes give rise to liquid effluent waste products and cooling water which are collected and discharged, under consent from the Forth River Purification Board, to the Forth estuary at the position shown on the location map of Fig. 1. Modelling of the Grangemouth discharge has been carried out at the ICI Brixham Laboratory. Initially a one-dimensional full estuary model was constructed and subsequently a localised two-dimensional plume model was developed. This plume model has been used to form the spill model described in this paper. The need for such a hazard assessment model was subsequently reinforced by the contamination of the Rhine resulting from the Sandoz warehouse fire.

The model is linked to fish toxicity data for each compound to predict the extent of toxic zones in the estuary resulting from a spill. The approach uses a Gaussian spread model for rapid computation over a short timescale (up to one tidal cycle) of the contaminated patch movement and dilution; this gives a good assessment as to the magnitude of any problem. Other work using Gaussian models is presented by Lewis (1982) and Keating (1986). Neely and Lutz (1984) have applied a similar approach to modelling chemical spills in a river. Csanady (1973) and Bowden (1983) give details of the method.

A second model using a random walk particle-tracking method is available to give a more detailed and longer term assessment. Chu and Gardner (1986) describe the random walk method as applied to an estuary but neglect the effects of diffusion, whereas Thomson (1987) describes this type of modelling as applied to atmospheric dispersion of pollutants and Webb (1982) applied the method to the spread and transport of Caesium-137 in the Irish sea.

The aim of the work was to produce a model for two uses, within the factory at Grangemouth:-

(i) As an emergency tool to predict the likely impact on the Forth estuary of a chemical spill in the factory.

(ii) As a planning tool for assessing the hazard posed by new feedstocks, intermediates and products on the site; the model assessments are being used in planning how and in what quantities the material should be stored in the works.

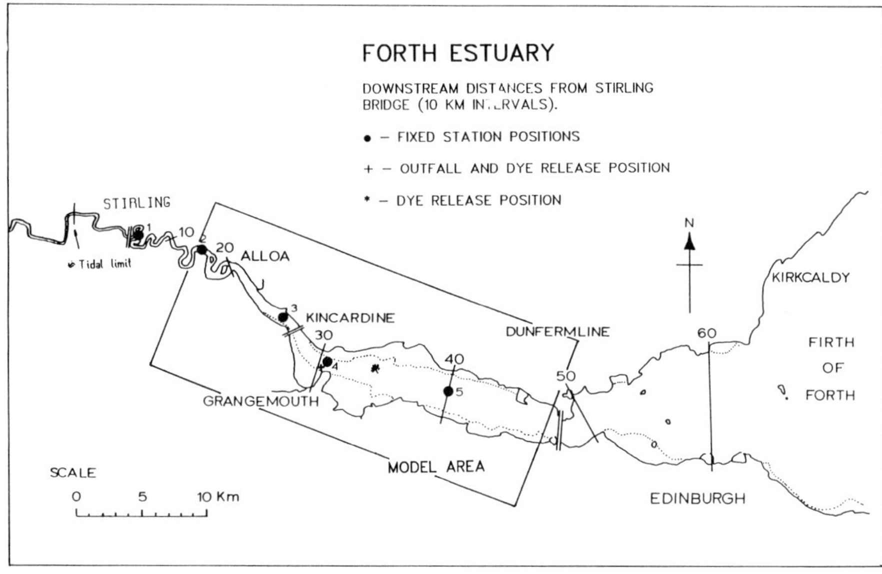

Fig. 1: Forth estuary with model area and survey station positions.

The model runs on a microVAX II computer at ICI Grangemouth. In the Gaussian form for emergency work, it will give hazard predictions in less than three minutes producing printed and graphical output.

This paper covers the theoretical basis for the computer model, the surveys carried out in the estuary, calibration and verification of the model and finally some model results.

2. Theory

The area being modelled is covered by a 200 metre grid extending from Alloa in the west to South Queensferry in the east (see Fig. 1). The grid is used to define the following data in the estuary: depth (relative to mean tide level); the peak flood tide currents and directions; and the residual currents and directions. The data have been obtained from the following sources:-

(i) Depths - Admiralty charts for the Forth estuary and survey echo soundings taken from the July 1987 survey.

(ii) Tidal currents - Admiralty charts for the Forth estuary and current meter data from the 1979 survey (positions shown on Fig. 1), also dye patch movements from surveys in 1986, 1987 and 1988 (Fig. 1).

(iii) Residual currents - Derived from average river inputs to the Forth estuary.

2.1 Transport processes

The model tracks the movement of an element of waste in the estuary as it is moved by the combined effects of the tidal current, residual current and wind driven current. The calculations are done on a timestepping basis, calculating the parameters for the next timestep from the position of the element at the end of the previous timestep.

The water depths, peak flood tidal current and residual current at the required position, are obtained from the values defined at the four corners of the appropriate grid square using two applications of linear interpolation. The instantaneous tidal current is obtained by combining the peak flood tide (C_a), with a harmonic representation of the tidal variation over time, and with a factor which relates the tidal current to the tidal range

$$C_i = C_a . f . (a_1 \cos(st) + a_2 \cos(2st) + \ldots \quad b_1 \sin(st) + b_2 \sin(2st) + \ldots) \quad (2.1)$$

where C_i is the instantaneous tidal current, f is the tidal range factor, s is the tidal frequency, t time and a_i, b_i, the coefficients from a Fourier fit to tidal current data (6 harmonic components used to simulate the current). Fig. 2a shows the tidal curve that has been used in the model. These data were measured at Grangemouth. The instantaneous water height is obtained in an analogous manner from the data for tidal height at Rosyth (see Fig. 2b). Four harmonic components have proved adequate for simulating the water surface variations.

The residual currents are very small in comparison to the mean tidal current (< 0.001 and $0.5\ \mathrm{m.s^{-1}}$ respectively at Grangemouth) and have an insignificant effect on transport over a period of 1-2 tides, so no allowance was made for changes in residual flow resulting from changes in freshwater discharges from the rivers. Increased river discharge would cause increased flushing and hence conditions better than predicted. Field survey work has shown that the wind generated currents are small compared to the tidal currents in the Forth estuary; the wind current has been represented by 0.5% of the wind speed, based on observations in the Forth and in the Dart in South Devon. The overall displacement of an element of waste is the vector sum of the tidal, residual and wind effects taken over a timestep

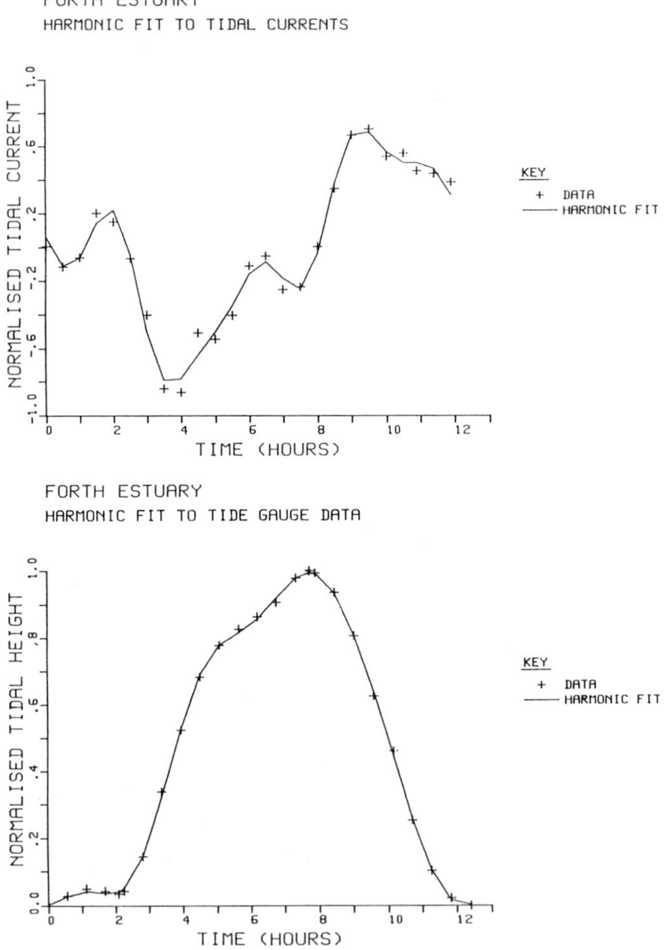

Fig. 2: Tidal currents measured at Grangemouth and water surface heights measured at Rosyth.

which is small enough to ensure the waste element cannot jump past any grid squares. The following equation is used

$$t_s = 0.5\frac{dx}{c} \tag{2.2}$$

where: t_s is the timestep; dx the grid size; and c is the maximum tidal current over the area.

2.2 *Dispersion processes*

For the Gaussian model, the spread of a patch is based on Fickian diffusion with current shear in the direction of the tidal flow. The formulae used for the patch spread are (Bowden, 1965; Talbot and Talbot, 1974; and Lewis, 1979)

$$\left(\frac{w}{4}\right)^2 = \left(\frac{w_o}{4}\right)^2 + 2k_h t \tag{2.3}$$

$$\left(\frac{d}{2}\right)^2 = \left(\frac{d_o}{2}\right)^2 + 2k_z t$$

$$\left(\frac{l}{4}\right)^2 = \left(\frac{l_o}{4}\right)^2 + 2k_h t + \frac{Sk_z t^3}{28}$$

where d,w,l are the depth, width and length of the waste patch and the subscript o designates the original patch size. k_z and k_h are the vertical and horizontal mixing coefficients, S the vertical shear in the direction of the tidal flow and t the time since the formation of the patch.

If the calculated vertical standard deviation of the patch (=d/2) exceeds 0.8 times the water depth then mixing is said to be complete over depth and the patch depth is set equal to the water depth (Pasquill, 1975, pages 328-329). The spread of the patch is easily related to dilution of the material.

For the random walk model, a number of particles related to the volume and concentration of the spilt material are released initially in the spill area. Each particle is moved by the transport mechanisms described above. The spread is determined by giving each particle an additional displacement to represent the effect of turbulent diffusion. Mixing coefficients for the random walk simulations are taken as those measured from the dye experiments thus neglecting Taylor's statistical theory of turbulence (Taylor,

1935) for the dispersion of particles in a fixed frame of reference. In practice this assumption is acceptable in the Forth estuary since large scale plume oscillations are not observed. The diffusion effect on each particle is calculated as follows (Webb, 1982):

$\sqrt{(6.k_z.t_s)}.R1$	in the vertical direction;
$\sqrt{(6.k_h.t_s)}.R2$	in the horizontal direction along the tidal current direction;
$\sqrt{(6.k_h.t_s)}.R3$	in the horizontal direction across the tidal current direction.

Where R1, R2 and R3 are random numbers in the range -1 to +1. The effect of shear is incorporated using the vertical current profile (Van Veen, 1938)

$$c = C_i \left(1 - \frac{z}{h}\right)^n \qquad (2.4)$$

where c is the tidal current at depth z from the estuary surface, h is the water depth and n = 0.192 is taken from the measurements of Van Veen (1938).

At the edges of the model grid it is assumed that if a particle is transported out of the modelled area then it is lost to the calculations. For this reason the model could not be used for discharges which are close to the edge of the model (the requirement is that the area covered by the model must be sufficiently large to ensure that particles can only be taken out of the area by the residual current, not directly by the tidal current).

At the land, seabed and sea surface boundaries particles are not allowed to be "lost". This is achieved by re-calculating the diffusion effects with different random numbers.

3. Initial patch formation

If a chemical is spilt in an unprotected area and finds its way into the factory drains, it flows to the pumping station at Dalgrain and is mixed with the normal works effluent. This takes approximately 20 minutes. It is then pumped down the outfall to the diffuser section on the bed of the estuary.

The model can simulate "instantaneous" spills or longer term releases (e.g. a leaking valve). As the material reaches the pumping station it is mixed with the effluent in the sump. It is assumed that complete mixing takes place in the sump before the effluent is pumped down the outfall pipeline. The concentration (C_t) of spilt material in the sump can be calculated from

$$C_t = C_1\ (1-\exp(-\frac{vt}{V})) \qquad 0 \le t \le t_{sp} \qquad (3.1)$$

$$C_t = C_1\ (1-\exp(-v\frac{t_{sp}}{V}))\ \exp(-v\frac{(t-t_{sp})}{V}) \qquad t > t_{sp}$$

$$C_1 = \frac{w_{sp}}{(t_{sp}v)}$$

where: w_{sp} is the weight of material spilt; C_t is the sump concentration at time t; v the pump rate; V the sump volume; and t_{sp} the duration of the spill.

The effluent is buoyant relative to the estuary waters so as it discharges at the diffuser the plume rises to the estuary surface entraining water in the process. The initial dilution of the effluent from the diffuser to the estuary surface is calculated using the stagnant water dilutions of Abraham (1963) and the flowing water correction of Agg and Wakeford (1972).

The fish toxicity of the spilt material is used to compute the initial toxic zone in the estuary. The 96 hour LC50 to rainbow trout is currently being used as the toxicity measure with a short term safe concentration being taken as one tenth of the LC50. The subsequent movement and dilution of the material is then computed using the methods described in Section 2 above.

4. Survey work

The model is based on data from the Forth estuary. The Brixham Laboratory with support from the Grangemouth Works has carried out annual field surveys in the estuary since 1970. The surveys providing data for this model are shown in Table 1; the types of data collected are also shown on this table. The studies were designed to collect the basic data required to set up the site specific model of the Forth and to provide information for verification of the model.

Table 1. *Field surveys providing data for the model setup and verification.*

Date	Survey description
25-26 June 1979	Spring tide - 5 fixed stations sampled half hourly throughout the tidal cycle and through the water column for current, salinity, temperature suspended solids and various chemical parameters.
2-3 July 1979	Neap tide - as above
* 7 July 1986	Flood and ebb tide dye experiments.
* 8 July 1986	Flood and ebb tide dye experiments.
* 21 July 1987	HW slack tide - dye experiment. Echo sounding - Kincardine to Alloa.
* 22 July 1987	Flood tide dye tracking Kincardine to Alloa.
* 23 July 1987	Flood tide dye experiment.
* 13 July 1988	Flood tide dye experiment.
* 14 July 1988	Ebb tide dye experiment.

* *denotes data used for model validation*

Table 2. *Mixing coefficients from dye experiments.*

Date	Time	Tide	K_h	K_z	s
7.7.86	0504	Ebb	-	0.0029	-
7.7.86	1127	Flood	-	0.00081	-
8.7.86	0600	Ebb	0.18	0.0041	0.0043
8.7.86	1212	Flood	0.05	0.00085	-
21.7.87	1215	HW slack	0.16	-	0.13
13.7.88	1120	Flood	0.75	0.0047	0.030
14.7.88	0500	Ebb	0.85	0.0	0.075

Table 3. *Model sensitivity to varying parameter values using a fixed simulation time of 1 hour.*

Parameter	% change in parameter		% change in solution	
Tidal range	+10	-10	-2.2	+2.4
Wind speed	+100	-100	0.0	+0.2
Vertical mixing	+10	-10	-3.9	+4.6
Horizontal mixing	+10	-10	-3.4	+4.1
Shear	+10	-10	-4.3	+4.3

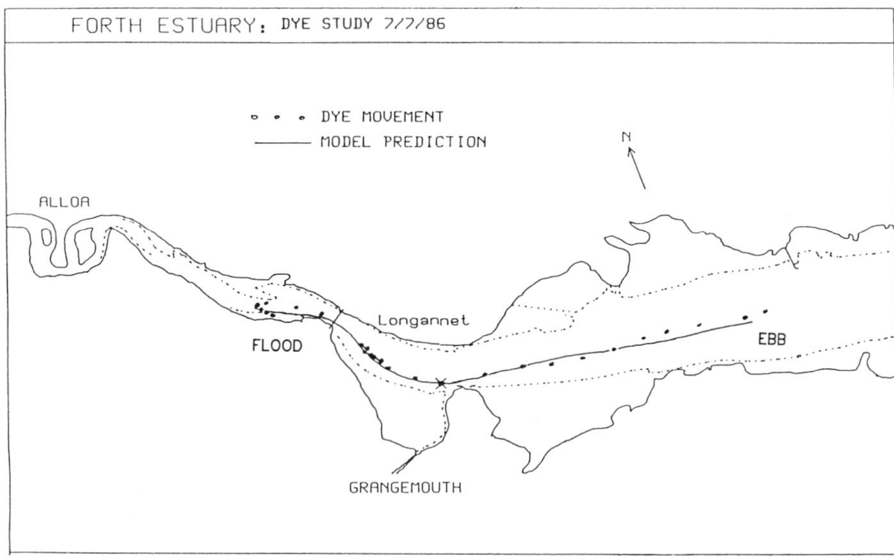

Fig. 3: Measured and predicted dye patch movements on the flood and ebb tides on 7/7/86.

5. Data analysis

The dye experiments have been analysed to obtain vertical and horizontal mixing coefficients (k_z and k_h) and the vertical current shear (s) in the direction of the tide. Equations (2.1) have been used to obtain the coefficients from the measured patch sizes. The results show that mixing in the estuary is variable (see Table 2), ranging from 0.05 to 0.85 $m^2.s^{-1}$ in the horizontal and from 0.0008 to 0.0047 $m^2.s^{-1}$ in the vertical. Vertical current shear values ranged from 0.0043 to 0.13 s^{-1}.

Low mixing and shear values have been chosen for the operational model so that simulations err on the side of poor mixing and therefore represent the worst possible situation for inhibiting the patch dilution. The values used are:

Horizontal mixing ($m^2.s^{-1}$)	0.1
Vertical mixing ($m^2.s^{-1}$)	0.0008
Vertical shear (s^{-1})	0.03

FORTH ESTUARY MODEL

MODEL SIMULATION OF LENGTHS 13/7/88

KEY
PATCH LENGTHS (PREDICTED)
x MEASURED VALUES

LENGTH OF DYE PATCH (KM)

TIME (HOURS)

FORTH ESTUARY MODEL

MODEL SIMULATION OF DEPTHS 13/7/88

KEY
PATCH DEPTH (PREDICTED)
x MEASURED VALUES

DEPTH OF DYE PATCH (M)

TIME (HOURS)

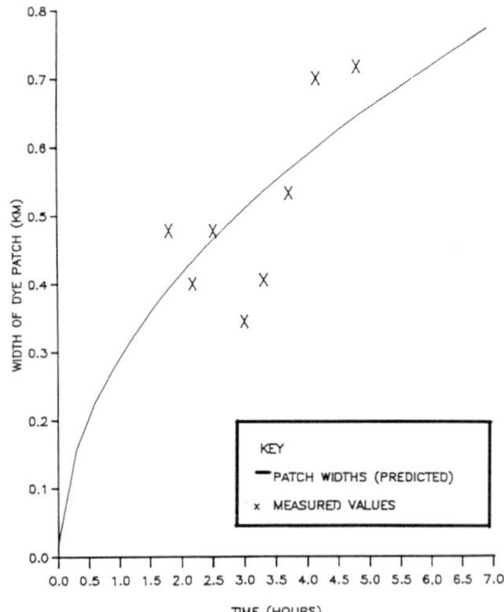

Fig. 4: Measured and predicted dye patch lengths, widths and depths on 13/7/88.

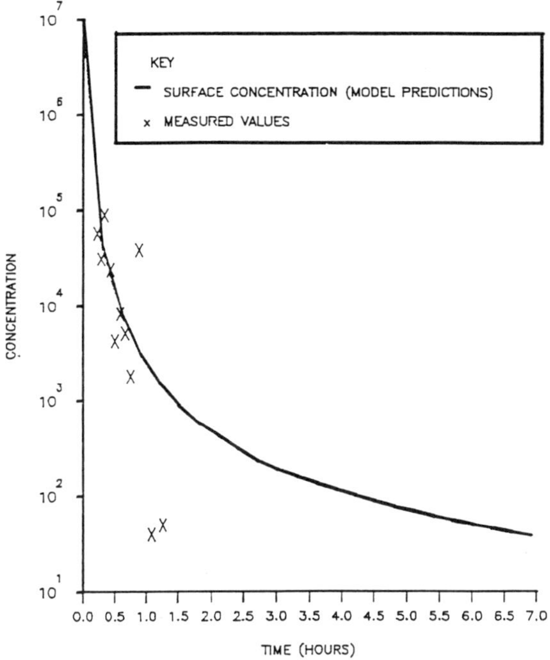

Fig. 5: Measured and predicted dye patch centre concentrations on 13/7/88.

6. Model verification

6.1 Gaussian model

Dye movement and dilution experiments have been used to verify the model predictions. The relevant studies in Table 1 have been marked with an asterisk. The model has been run to simulate each of these dye studies, and predictions of the patch movement, patch spread (horizontal and vertical) and patch dilution have been compared with the measured values. Fig. 3 shows the predicted and observed dye movements for 7.7.86 on the flood and ebb tides. Figs. 4 and 5 show the predicted and observed dye patch sizes and concentration plotted against time for 13.7.88. A qualitative assessment shows reasonably good agreement between the predictions and measurements has been obtained for all of the dye experiments.

6.2 Random walk model

Extensive tests have been made comparing the results of the random walk simulation and the Gaussian simulation for the following standard test situations :-

steady current: advection test to verify the transport section of the model for a constant current.
tidal current: tidal excursion test to verify the simulation of the tidal changes in current.
tidal range: tidal excursion variation test to verify the response of the model to changes in tidal range.
no current spread: pure mixing test to verify the plume spread sections of the model.
no current spread with shear: mixing with shear test to verify the modelling of the vertical current shear effect.

7. Model accuracy, sensitivity and limitations

The model is representing, in a simplified form, the complex processes that occur in an estuary. For example, the mixing and shear coefficients are assumed to be uniform over the area which is certainly not the case. The overall accuracy of the modelling assumptions can only be gauged against real conditions. The verification studies shown above indicate the degree of agreement obtained.

The sensitivity of the model to the tidal range, wind speed, mixing coefficients and shear has been investigated by running the model for a standard period of 1 hour and varying each parameter individually by + or - 10%. The comparisons are shown in Table 3 and give the percent change in the predicted spill concentration (after 1 hour) due to the parameter change. These results show that the tidal range is more important than the wind, but the mixing is most important. For this reason, low mixing values are used in the operational model in order to predict the worst case situation.

The model allows for degradation of a pollutant if the decay rate is known, but does not cater for chemical change within the estuary, nor is it suitable for representing the movement of oils or heavy particulate material.

The Gaussian model is limited to short time scales (of the order of one tidal cycle), because as the patch becomes large the edges of the patch are influenced by different tidal streams and reflection by the coastline. However, for the operational model this limitation is acceptable since situations where the toxic patch starts to overlap the land are clearly

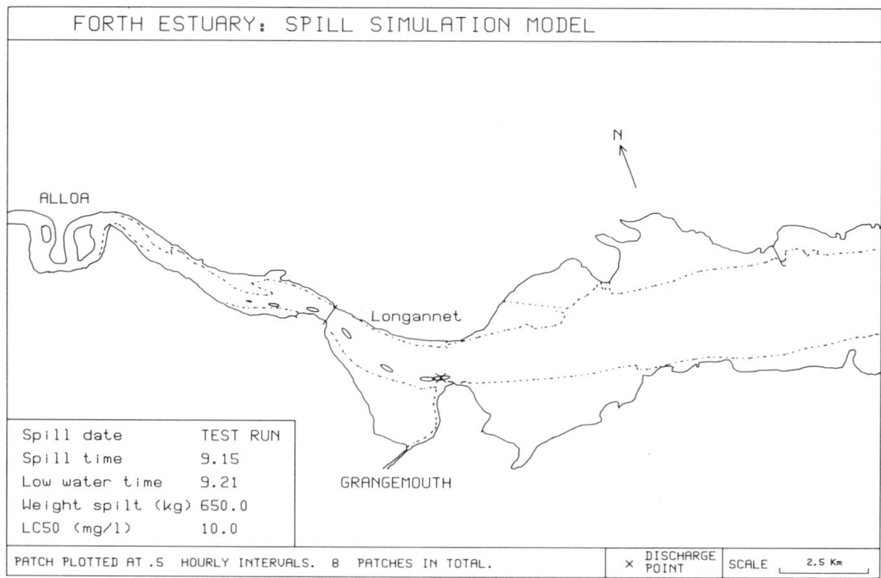

Fig. 6: Graphical output for a test run of the model.

unacceptable even if not modelled accurately. The random walk model overcomes the limitations of the Gaussian model but takes too long to run on the microVAX II computer to be useful in an emergency situation.

8. Results

A hypothetical model run is shown to indicate the use of the model in practice. The key parameters required by the model are tidal range, wind, spill time and low water time, weight of material spilt, the duration over which the spill occurs and the LC50 for the material. The results of the model run are presented on Table 4 and show that the predicted impact on the estuary lasts for 3.1 hours before the material is diluted to below the safe concentration level and shows the predicted movement and size of the toxic zone. A graphical presentation of the results is shown in Fig. 6 where the ellipses represent the safe level concentration. These ellipses are plotted at hourly intervals from the start of the spill.

Table 4. Sample output from the model.

FORTH ESTUARY EFFLUENT SPILL DISPERSION MODEL
==

ICI BRIXHAM LABORATORY - MODEL VERSION VERSION 2.01 : DATE 12-MAR-1990

Model run at 11:07 on 9-APR-90

Reading the mesh point data files

Enter tidal Range (m.) ?	5
Is the wind known ? (Y/N)	Y
Enter wind speed (miles/hour) and direction (deg from north)	5 90
Spill date ?	TEST RUN
Enter spill time (HR.MI, 24 hr. clock)	0915
Enter low water time (HR.MI, 24 hr cl.)	0921
Enter spill duration (hours)	0.0
Enter weight spilt (kg)	650
Enter maximum safe concentration (%)	0.0001

Equivalent LC50 value (mg/l) is 0.100E+02

SIGNIFICANT SPILL -- maximum concentration in the estuary
is LC50 concentration times 5.9
Toxicity in the estuary should last
less than 12 hours

INITIAL PATCH PARAMETERS

LENGTH	=	493.3 M
WIDTH	=	110.9 M
DEPTH	=	3.7 M
TIME TO FORM	=	0.8 HOURS
PEAK CONC.N	=	.61E+02 MG/L

Patch tracking calculations ...

SPILL ON : TEST RUN

PATCH MOVEMENT INFORMATION :

TIME (Hrs)	LOCATION (Km)		DEPTH (m)	DIR (deg)	SPEED (m/s)	WIDTH (Km)	LENGTH (Km)	DEPTH (m)	CONCENTRATION Surface	Bottom
2.50	12.05	2.88	3.3	82.	0.04	0.111	0.493	3.7	0.609E+02	0.000E+00
3.00	11.66	2.85	3.9	270.	0.43	0.114	0.426	3.4	0.525E+02	0.195E+02
3.50	10.52	3.16	4.4	298.	0.80	0.117	0.396	4.0	0.372E+02	0.148E+02
4.00	9.39	4.14	6.7	321.	0.73	0.113	0.377	4.6	0.270E+02	0.502E+01
4.50	8.44	4.80	7.9	281.	0.67	0.102	0.354	5.0	0.196E+02	0.283E+01
5.00	7.34	4.93	6.4	277.	0.50	0.080	0.302	5.5	0.143E+02	0.479E+01
5.50	6.66	5.03	6.5	280.	0.31	0.031	0.132	5.9	0.105E+02	0.406E+01
5.59	6.57	5.04	6.4	282.	0.29	0.000	0.000	0.0	0.100E+02	0.396E+01

PATCH MOVEMENT SUMMARY

PATCH BECOMES SAFE AFTER 3.1 HOURS IN THE ESTUARY:
THAT IS 5.6 HOURS AFTER THE SPILL

TOTAL MOVEMENT OF THE MATERIAL	=	6.16 KM
DOWNSTREAM MOVEMENT	=	0.49 KM
UPSTREAM MOVEMENT	=	5.67 KM

Plot results ? (Y/N) Y

FORTRAN STOP

9. Conclusions

A computer model has been built to assess the impact on the Forth estuary of a chemical spill in the ICI Factory at Grangemouth. The model is based on field survey data which has been collected over the period 1979-1988. Surveys in 1986, 1987 and 1988 have collected data from dye spread experiments for verification of the model. Model simulations of these dye experiments show good agreement with the observations.

The model has been installed in the Grangemouth works of ICI and is in use as an emergency tool for assessing the likely effect on the estuary of a chemical spill. The model is also being used for assessment of precautions needed for new chemicals in the works.

10. Acknowledgements

I would like to thank personnel at Grangemouth for their help on the survey and computer side, particularly Mr G. Broomfield and Dr R. Williams. I would also like to thank Dr A. Webb of the Forth River Purification Board for helpful discussions on the random walk models. Finally I would like to thank my colleagues at Brixham for all the survey work and many fruitful discussions.

11. References

Abraham, G. (1963). Jet diffusion in stagnant ambient fluid. Delft Hydraulics Lab. Pub. No.29, July 1963.

Agg, A.R. and Wakeford, A.C. (1972). Field studies of jet dilution of sewage at sea outfalls. Inst. of Public Health Engs. Journal, 71:26-153.

Bowden, K.F. (1965). Horizontal mixing in the sea due to a shearing current. J Fluid Mech., 21(2):83-85.

Bowden, K.F. (1983). Physical oceanography of coastal waters. Ellis Horwood Ltd., Chichester, England.

Chu, W.S. and Gardner, S. (1986). A two-dimensional particle tracking estuarine transport model. Water Resources Bulletin, 22(2):183-189.

Csanady, G.T. (1973). Turbulent diffusion in the environment. D Reidel Publishing Co., Dordrecht, Holland.

Keeting, T. (1987). Mathematical modelling of sea outfall performance. The Public Health Engineer, 14(6):53-58.

Lewis, R.E. (1979). Modelling the dispersion of pollutants from marine outfalls. In: Conf. on Mathematical Modelling of Turbulent Diffusion in the Environment, G.J. Harris, editor. Academic Press, pp.441-457.

Lewis, R.E. (1982). Outfalls designed to computer's order. Surveyor 159(4683):12-13.

Neely, W.B. and Lutz, R.W. (1985). Estimating exposure from a chemical spilled into a river. Journal of Hazardous Materials, 10:33-41.

Pasquill, F. and Smith, F.B. (1983). Atmospheric diffusion. Ellis Horwood Ltd., Chichester, England (3rd edition).

Talbot, J.W. and Talbot, G.A. (1974). Diffusion in shallow seas and in English coastal and estuarine waters. Rapp. P. V. Rene. Cons. Int. Explor. Mer. 167:93-110.

Taylor, G.I. (1935). Statistical theory of turbulence. Proc. Roy. Soc. A, 151(1-4):421pp.

Thomson, D.J. (1987). Random walk models for atmospheric dispersion. Meteorological Magazine, 116:142-150.

Van Veen, J. (1938). Water movements in the Straights of Dover. J. Cons. Int. Explor. Mer., 13:7-38.

Webb, A.J. (1982). A random walk model of the dispersion of caesium-137 in the Irish Sea. Thesis for MSc Univ. Wales Feb. 1982.

Particle tracking models for pollutant dispersion

C.M. Allen

University of Lancaster

1. Introduction

There has been much research and many attempts to produce models which explain and predict the dispersion of pollutants in rivers and estuaries.

Computer modelling of the dispersion of pollutants in rivers and estuaries has traditionally utilised standard advection-diffusion equations based on work of Taylor (1953, 1954) on dispersion in pipes. These equations are often used in the form

$$\frac{\partial C}{\partial t} + \underline{U}.\nabla C = K\nabla^2 C \tag{1.1}$$

where: C, $\underline{U}$ are the mean concentration and velocity; and K is an "effective longitudinal dispersion coefficient", which represents the effects of the turbulent fluctuation cross-product term $\overline{\underline{u}c}$ where c,$\underline{u}$ are the fluctuations in concentration and velocity.

These equations have usually been implemented numerically using finite-difference techniques to produce output of the mean concentration (either a spatially or time-averaged value). The analytical solution of Eq. 1.1 for the one-dimensional case where x is the longitudinal distance, with a constant value of the longitudinal dispersion coefficient K, gives a Gaussian distribution of the mean concentration such as

$$C(x,t) = \frac{M}{2(\pi Kt)^{1/2}} \exp\left[-\frac{(x-Ut)^2}{4Kt}\right] \tag{1.2}$$

where: M is the total mass of pollutant; and U is the x component of the mean velocity $\underline{U}$.

The use of (1.1) and (1.2) does not usually give successful predictions of the distributions of pollutants actually measured in rivers and estuaries. These rarely show Gaussian profiles but more usually have markedly skewed distributions. There are several reasons why this is the case - mainly because of the complications in natural environmental flows which were not present in the pipe flows studied by Taylor.

Many models of pollutant dispersion are based directly on Eq. 1.1 where C is an average concentration, either over a specific space or time interval. In a river C may be a lateral or vertical average; in an estuary it may be an average over a tidal cycle.

In practical problems of dispersion of pollutants in the environment, whether releases are controlled or accidental, it is usually more important to be able to predict the peak concentrations of pollutants which are likely to occur at a particular location and time, in order to make a proper assessment of any hazard potential. Models which calculate spatially or time averaged mean concentration values will only give realistic estimates of the actual concentration if the fluctuations in concentration are small in comparison with the mean concentration values or if the concentration is statistically steady (Chatwin and Allen, 1985).

2. Newer methods of modelling

More recently, various different types of models have been developed. Versions of the advection-diffusion equation are used by the water industry for practical problems of dispersion in rivers, with different specifications for the form of the dispersion coefficient and its relationship with flow and mixing conditions. In one model developed for estuarine mixing by Park and James (1988), a time varying formulation for the dispersion coefficient is used. However, these modifications do not overcome all the problems inherent in the use of these equations for river and estuarine flows; factors such as numerical dispersion in the model can still be problematic.

A different approach to modelling is based on a Lagrangian type formulation where the pollutant is represented by a large number of particles (up to several tens of thousands) whose paths are followed as they move through the fluid. These models have been developed solely as Lagrangian models by the author to simulate dispersion in two-dimensional rectangular channels (Allen, 1982) and also as a combined Lagrangian-Eulerian model of pollutant dispersion in estuaries (Krohn et al, 1987) where the particles are superimposed on a grid. These types of models seek to represent the physics of the turbulence processes affecting dispersion more clearly than the advection-diffusion equations.

Whereas the particle tracking approach seeks to specify the smaller scale processes in order to model dispersion, the "Aggregated Dead Zone" model (Beer and Young, 1983) is based on the assumption that the observed dispersion in natural streams is the result of exchange of pollutant with storage or "dead zones" in a reach and models the dispersion on the larger reach scale of the river which may be several kilometres long. Smith (1987) has tried to reconcile the different approaches by deriving a theoretical dispersion model of the dead zone type from the detailed hydrodynamics of the fluid flow. He has also developed a modified version of the advection-diffusion equation which incorporates a "delay-diffusion" term to take account of the contraction in a pollutant cloud which can occur at some stage of the oscillatory cycle in an estuary (Smith, 1982).

3. Particle tracking models

3.1 Model formulation

From a physical and computational point of view, the particle tracking types of models seem to offer distinct advantages for problems of pollutant dispersion in rivers and estuaries.

In the simplest type of particle tracking (or random walk) model, the paths of the particles representing a cloud of pollutant are tracked by the computer as they move through the fluid. The initial source geometry of the pollutant can be specified by a line source or point source of particles and as an instantaneous or steady release, and the particles move independently of each other in a series of jumps. At each position in the flow, the particle is given velocity components which include both the mean velocity components and turbulent fluctuations (whose sign is determined by a

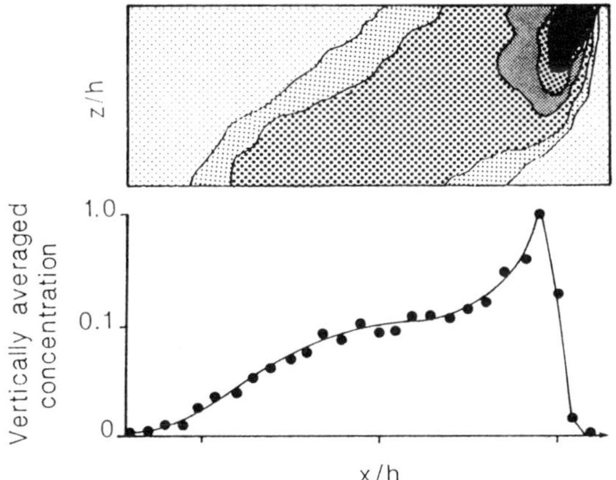

Fig. 1: Typical concentration output of a random walk model for a steady flow in a rectangular channel, a short time after the release of a line source of particles. Upper, variation in concentration over a vertical section of depth h (where z/h is the vertical axis and x/h is the longitudinal axis). The darkest tones denote the highest concentration values. Lower, vertically averaged concentration profile.

random number generator). The mean velocity can be specified either as a logarithmic profile to simulate the types of steady velocity profiles observed in rivers or as an oscillatory profile to simulate estuarine velocities. In this way, a detailed picture of the concentration of a pollutant in space and time can be built up.

The concentration distribution for a steady flow after an initial release of a line source of particles is shown in Fig. 1. The non-Gaussian distribution in the vertically averaged concentration can be clearly seen together with the variation of concentration with depth in the vertical. Different runs of the same simulation give virtually the same vertically averaged concentration distribution but show significant variability in the two-dimensional picture. Fig. 2 illustrates the variability apparent in profiles of concentration with depth for similar vertically averaged concentration values.

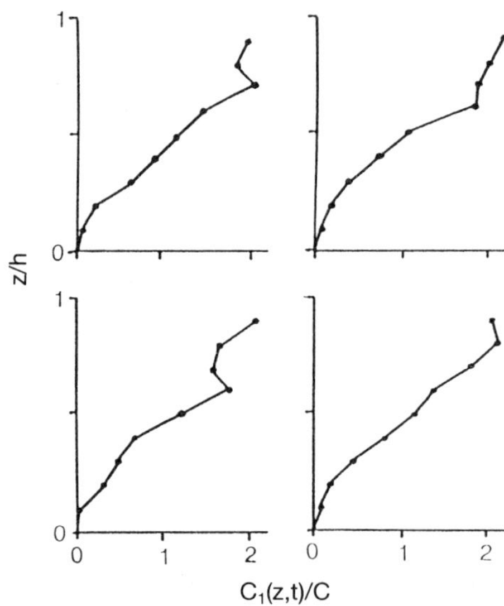

Fig. 2: Variation of concentration $C_1(z,t)/C$ with depth for several runs of the same model simulation. C is the vertically averaged concentration.

3.2 *Parameter specification*

The particle tracking models require input of the turbulent velocity components in order to predict the resulting dispersion. Existing models have used turbulent velocity components derived from laboratory experiments in rectangular channels (Sullivan, 1974). These experiments

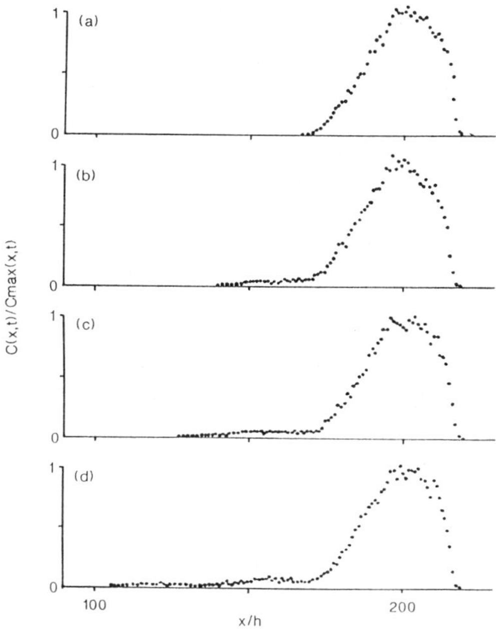

Fig. 3: Vertically averaged concentration profiles showing the effects of different proportions of "dead zones" along the channel bed. The proportion of "dead zones" is given as a percentage. (a) 0% ; (b) 10% ; (c) 20% ; (d) 30% .

showed that the vertical turbulent velocity fluctuations were approximately constant with depth.

Flows in natural rivers show a higher degree of complexity due to irregular cross-section, meandering etc which are not apparent in straight channels. For this reason, measurements of turbulent velocity components have been made in the River Severn (Heslop and Allen, 1989), in order to investigate the likely variability in and relationship between the components. Preliminary analysis of the data shows that the magnitude of the turbulent velocities tends to be greater nearer the river bed than higher in the water column and also shows cross-sectional variability with higher values on the outside of bends in the river.

There is also strong evidence from dye tracing experiments that there are important effects due to "dead zones" - areas on the bed or the side of a river where pollutants can be trapped for times which are long compared with the transport of the majority of the pollutant along that stretch of the river (Wallis et al, 1989).

One of the versatilities of the particle tracking model is that these types of processes can be simulated in a more realistic way than the normal parameterisation used in most models. Particles close to the bed can become trapped in "dead zones" for a period of time and then subsequently released back into the flow. Examples of the concentration profiles obtained for an increasing proportion of "dead zone volume" are illustrated in Fig. 3 where the presence of a higher proportion of dead zones produces an increasing tail on the concentration distribution.

Models which track large clouds of pollutants for times corresponding to several tidal cycles require large numbers of particles (up to 100,000) in order to obtain reliable statistics on the resulting concentration distribution and its associated statistical parameters, such as variance, skewness and kurtosis. For shorter run times and less stringent requirements for precision of the concentration output, then model simulations require much smaller numbers of particles for adequate concentration distributions.

4. Computing requirements

4.1 Scalar or parallel processing

Particle tracking models which are run on scalar processors contain a subroutine to calculate the trajectory of each particle. This subroutine would then be repeated for the total number of particles which are required in the simulation (Fig. 4a).

The inherent parallel structure of random walk models is such that they are eminently suitable for running on parallel processor machines. When the particle tracking models are implemented on parallel processor machines, the code is rewritten to contain master and slave programs. The master program controls the input and output for the model and the communications with the slave processors and the slave subroutines contain the code for the particle trajectory (Fig. 4b). Each slave processor then carries out the simulation for a group of particles, thus enabling several particles to be tracked simultaneously on different processors.

4.2 Mainframe or microcomputers

One of the versatile aspects of particle tracking models is that the computing requirements can be easily be tailored to the scale of the models being run without significant rewriting of the computer code. Modelling has been carried using Fortran on several mainframe and microcomputers. The run times for one particular model simulation are shown in Table 1. All the jobs are run in scalar processing format with the exception of the simulation on the Meiko transputer system which was rewritten in parallel code. On

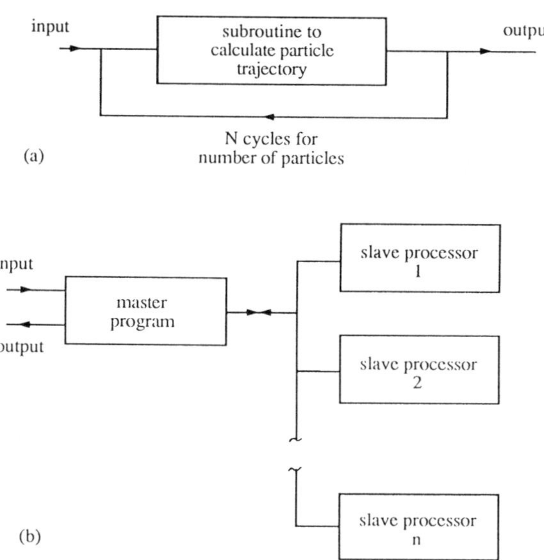

Fig. 4: Configurations of computer simulations for: (a) scalar processor; and (b) parallel processor computers.

Table 1: *Run times for a particular model simulation.*

Computer	*Run-time(secs)**
Toshiba 3200 PC with Maths coprocessor	2446
Toshiba 3200 PC with Transtech transputer board	242
Univ. Lancaster - Sequent Symmetry	273
Manchester Computer Centre - Amdahl	16
Univ. Lancaster - Meiko transputer system using domain of 16 processors	53

* In all of the above cases, with the exception of the Amdahl, the run-time is roughly equivalent to elapsed time. For the Amdahl, the time given is CPU time; the protocol for job queuing, input and output means that the elapsed time is likely to be 300-600 secs depending on usage of the machine.

the Meiko system, the Fortran code is linked to the occam software which runs the different processors, using a series of fortnet subroutines (Allen and Leck, 1989).

One point of particular note is the comparison between the Toshiba PC-compatible with a slot-in transputer board and that of the mainframe Sequent Symmetry. Significant advantages in ease of operation are obtained by the use of a dedicated, portable microcomputer system for these model simulations.

Use of the Meiko system has considerable advantages when the models are run using much larger (up to 100,000) numbers of particles. In this case the maximum use is being made of the fast processor speed without the restrictions of the slower communications between processors.

5. Future developments

There are two main areas in which particle tracking models will make progress. For further development of parallel processing versions of these models, the efficiency of the models in computational terms needs to be optimised. This can be done by choosing the number of processors for a particular model simulation in order to utilise the fast processing speed of each transputer without limiting the speed of the simulation because of the time taken for communication between processors.

The other, and from a scientific point of view more important, area is that of parameter specification in the particle tracking models. For practical problems involving dispersion of pollutants, it is vital to investigate the behaviour of the turbulent velocity fluctuations which are used as input to the models. This integration of model and experimental work is in progress, both for dispersion in rivers (Heslop and Allen, 1989) and for a large compound channel experiment (Guymer et al, 1990) where detailed turbulence and dye dispersion experiments have been carried out under more controlled conditions than are possible in natural rivers.

Preliminary comparisons of model simulation with dye dispersion experiments carried out in the River Severn show good qualitative agreement between model and experiment (Allen and Heslop, 1990) for straight sections of the river, but for meander sections it is likely that the models will need to be extended to three dimensions in order to incorporate the lateral variations in the flow structure.

6. Conclusions

Particle tracking models offer exciting opportunities for computer simulation of dispersion processes in rivers and estuaries. However, the models are still at an early stage in their development compared with the research effort which has been invested into numerical models of the advection-diffusion equations.

The advent of transputer-based computers offers the increased computer power needed to run more complex types of particle tracking models, but it is still vital to have integrated model/experimental research programmes since the fundamental input to these models is a proper understanding of the physical processes responsible for dispersion in rivers and estuaries.

7. Acknowledgements

I should like to thank Graeme Hughes for computer support on the Meiko transputer system.

8. References

Allen, C.M. (1982). Numerical simulation of contaminant dispersion in estuary flows. Proc. R. Soc. Lond. A 381:179-192.

Allen, C.M. and Heslop, S.E. (1990). Dispersion in larger rivers using random walk models. (in preparation).

Allen, R.J. and Heck, L. (1989). Fortnet: A parallel fortran harness for porting application codes to transputer arrays. Proc. Int. Conf. on Applications of Transputers, Liverpool, July 1989.

Beer, T. and Young, P.C. (1983). Longitudinal dispersion in natural streams. J. Env. Eng. ASCE. 109(5):1049-1067.

Chatwin, P.C. and Allen, C.M. (1985). Mathematical models of dispersion in rivers and estuaries. Ann. Rev. Fluid Mech. 17:119-149.

Guymer, I., Brockie, N.J.W. and Allen, C.M. (1990). Towards random walk models in a large scale laboratory facility. Proc. Int. Conf. on Physical Modelling of Transport and Dispersion, M.I.T., August 1990.

Heslop, S.E. and Allen, C.M. (1989). Turbulence and dispersion in larger UK rivers. Proc. International Association for Hydraulic Research Congress, Ottawa, August 1989.

Krohn, J., Duwe, K. and Pfeiffer K.D. (1987). A high resolution 3D model system for baroclinic estuarine dynamics and passive pollutant dispersion. In: Three-dimensional models of marine and estuarine dynamics, J.C.J. Nihoul and B.M. Jamart, editors. Elsevier, pp.555-572.

Park, J.K. and James, A. (1988). Time-varying turbulent mixing in a stratified estuary and the application to a Lagrangian 2D model. Est. Coastal Shelf Sci. 27:503-520.

Smith, R. (1982). Contaminant dispersion in oscillatory flows. J. Fluid Mech. 114:379-398.

Smith, R. (1987). A two-equation model for contaminant dispersion in natural streams. J. Fluid Mech. 178:257-278.

Sullivan, P.J. (1974). Instantaneous velocity and length scales in a turbulent shear flow. Advances in Geophysics, 18A:213-223.

Taylor, G.I. (1953). Dispersion of soluble matter in solvent flowing slowly through a tube. Proc. R. Soc. Lond. A, 219:186-203.

Taylor, G.I. (1954). The dispersion of matter in turbulent flow through a pipe. Proc. R. Soc. Lond. A, 223:446-468.

Wallis, S.G., Young, P.C. and Beven, K.J. (1989). Experimental investigation of the aggregated dead zone model for longitudinal solute transport in stream channels. Proceedings of the Institution of Civil Engineers, Part 2, 87(March):1-22.

Physically-based modelling of catchment hydrology: a likelihood approach to reducing predictive uncertainty

A.M. Binley and K.J. Beven

University of Lancaster

1. Introduction

In 1856 Henri Darcy, a French engineer, published his observations of flow rates through laboratory samples of porous media. This led to the widely known and accepted Darcy's Law which states that a linear relationship exists between a fluid volume flux density and the pressure gradient causing that flow, the coefficient of proportionality being labelled the hydraulic conductivity.

Today, many models of catchment hydrology employ Darcy's Law and similar empirical relationships within a mass continuity balance framework. Models of this type are often referred to as physically-based as the parameters are, in theory, physically understood and measurable. Distributed physically-based models account for spatial (and temporal) variability of model parameters and usually consist of a series of partial differential equations, which are often highly non-linear. The Système Hydrologique Européen (SHE) and the Institute of Hydrology Distributed Model (IHDM) are examples of such model types and employ finite difference and finite element methods to solve the boundary value problem of catchment hydrology. In order to satisfy stability and convergence criteria, the number of grid nodes or elements used in these models is often large and thus the model parameter set is large also. In fact, even on intensively monitored research catchments, the number of parameter values is often far too great for determination by experiment, even if the experimental techniques to determine parameter values at the model grid scale existed (see discussion by Beven, 1987, 1989). Often the distributed property is neglected on data requirement grounds and the models are treated as lumped models, for example, a single "effective" Darcian hydraulic conductivity value may be used over a large area.

In principle, physically-based models do not require calibration, however, such an exercise is often essential (for example, Bathurst, 1986; Calver,

1988). Due to the computational requirements of these models the calibration period must be small and the procedure adopted relatively simple. A calibration dataset normally consists of a series of rainfall events and the respective stream discharge hydrograph. Current practice assumes that, for a given model structure, a set of parameter values exists that will provide the optimum simulation of the observations. Calibration strategies, either automatic "hill-climbing" or trial and error methods, have therefore been aimed at identifying this "optimal" parameter set in terms of some objective function or goodness of fit index. Considerable research effort has been expended in terms of overcoming some of the difficulties of this approach in terms of overparameterised models, intercorrelation between the effects of parameters, insensitive parameters, and local optima in the response surface.

There is evidence to question this belief in an optimum parameter set. We know that using different calibration periods may lead to different "optimal" sets of parameter values for the same catchment, and that starting search routines from different initial sets of parameter values may also yield different "optimal" values. Our predictions, therefore, contain an inherent uncertainty. It is our view that in applying physically-based models of catchment hydrology (and indeed most models in other environmental science areas) that a quantitative assessment must be made of the uncertainty in our predictions associated with the simplifications of the model and calibration procedure and uncertainty in any observations. Furthermore, we need to assess what level of information is useful in reducing this predictive uncertainty. For example, in the case of catchment hydrology, will observations of state variables, for example groundwater levels or soil moisture profiles, constrain our uncertainty bounds significantly?

Early attempts to calibrate and assess the predictive uncertainty of physically-based models have been constrained by the computing requirements of a single run of such models. The study of Rogers et al (1985) made use of the approximate uncertainty estimation techniques of Rosenblueth (1975), which require a minimal number of model runs. In a more recent study Binley et al (1989) compared the Rosenblueth method with the computationally more expensive, but more accurate, Monte Carlo simulation procedure. Their results suggest that the Rosenblueth method may be useful as an initial estimator of model uncertainties.

The aim of this paper is to propose a general framework for evaluating predictive uncertainty and reducing the uncertainty bounds. The procedure - Generalised Likelihood Uncertainty Estimation (GLUE) - is based on a different set of concepts of model calibration from those used in the past.

2. The Generalised Likelihood Uncertainty Estimation (GLUE) calibration procedure

The GLUE procedure is based on the premise that *all* model structure/parameter value sets are equally likely as simulators of the system of interest, until that likelihood is conditioned on a comparison with the observed behaviour or on some other a priori knowledge. The traditional approach is a special case of this view in which the set of parameter values giving the best objective function value is given a likelihood of 1.0 and all other parameter sets are given a likelihood of 0.0. However, the likelihood is much more realistically viewed as a fuzzy measure of model performance, that can also be used as a likelihood distribution in a Bayes equation approach to estimating uncertainties.

The GLUE procedure is based upon multiple simulations using randomly selected parameter values; it is, therefore, a variation on Monte Carlo simulation. It is necessary to define distributions from which to select the parameter values, together with any expected intercorrelation between the parameter values. The parameters here might also include values specifying the initial conditions for a simulation. This requires the a priori estimation of the likelihood of different parameter values. This may be a purely subjective choice on the part of the modeller, or may be derived from field measurements or empirical estimation procedures based on soil type or catchment characteristics. Any appropriate distribution may be used: if only a range of parameters can be suggested then values might be selected from a uniform distribution over that range; if a mean and standard deviation can be estimated then a normal distribution may be appropriate. Multiple simulations are then made. Without any additional information these represent equally likely simulations of the catchment of interest and can be used to estimate the uncertainty in the model predictions conditioned only on the initial distributions of the parameters. It has been found (Binley et al, 1989) that the distributions of predicted discharges determined in this way may be highly non-normal so that we have chosen to use the 5 and 95% quantiles as an indicator of the 90% confidence interval for the predictions rather than the mean +/- 1.667 std. dev.

However, there will often be some observations with which to determine how well each of those simulations is reproducing the behaviour of the catchment (In the case of ungauged catchments it is suggested that wherever possible a modelling study should be associated with a program of such measurements). The goodness of fit can be expressed as a further likelihood function. The choice of this likelihood function is purely subjective; the only constraint is that it should be scaled so that each likelihood is ≥ 0 and that the sum of the likelihoods over all simulations should be 1. It is clearly seen that the traditional approach is therefore an extreme case and implies that having identified the "optimal" simulation, all other simulations are given zero likelihood and there is no further uncertainty (even though that simulation may be greatly in error). Perturbation procedures in which predictive uncertainty is determined from the shape of the response surface around the optimum may be treated in this approach in terms of a different particular form of likelihood function.

It is worth noting that such likelihood measures may be based on behaviour criteria quite different from those normally employed such as the sum of the squared errors in predicted discharges. For example, it may be known that in a particular catchment overland flow has never been observed, so that all simulations producing significant amounts of overland flow will be given a likelihood of zero, *even though* those simulations might produce excellent simulations of observed stream discharges.

Likelihood measures may be combined and updated using Bayes theorem. The new likelihoods may then be used as weighting functions to estimate the uncertainty in model predictions by ordering the predicted discharge at each timestep and searching for the 5th and 95th percentiles on the resulting cumulative likelihood function. As more observed data become available, further updating of the likelihoods may be carried out.

3. Application of the GLUE procedure to a Darcian headwater

In order to evaluate the GLUE procedure we consider here the task of applying a semi-distributed Darcian flow model to a fully distributed system. The procedure adopted was two part. Firstly, a fully three-dimensional analysis of a small synthetically-generated catchment was carried out. Secondly, using a limited data set from this "observed" Darcian behaviour, a distributed model treated in a lumped manner was applied to the catchment using the GLUE methodology to assess the lumped model's predictive uncertainty and determine how the availability of additional water table elevation data might be effective in reducing the predictive uncertainty.

3.1 Headwater Description

For the purpose of this study, a large hillslope hollow of 0.25 km^2 plan area was established. The topography of the headwater is shown in Fig. 1, together with the location of a small stream at the base of the slope. The stream has a length of 60m. The soil within the headwater has variable thickness and is underlain by an impermeable bedrock. The average soil depth is 1m. The saturated hydraulic conductivity of the soil is also considered to vary spatially, albeit with a much smaller correlation length

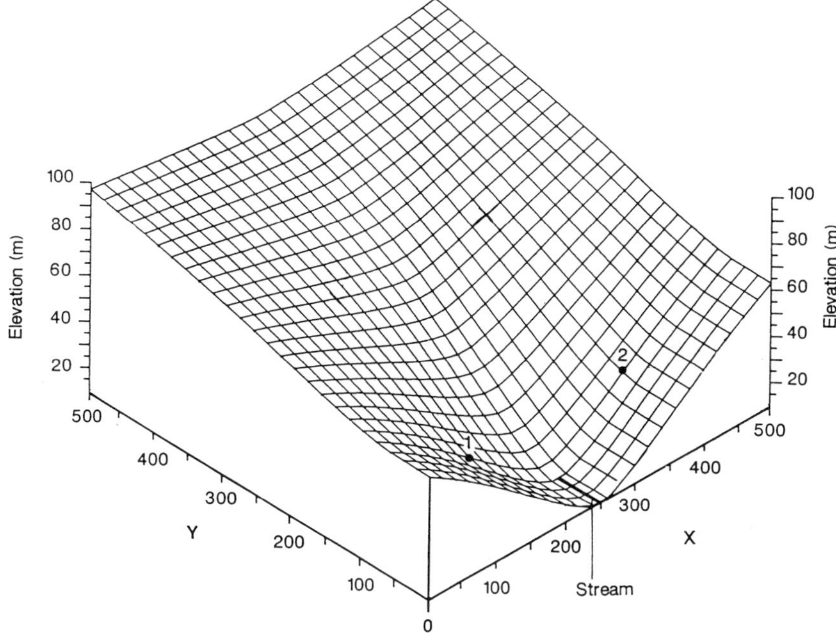

Fig. 1: Headwater topography showing location of water table fluctuation measurement points.

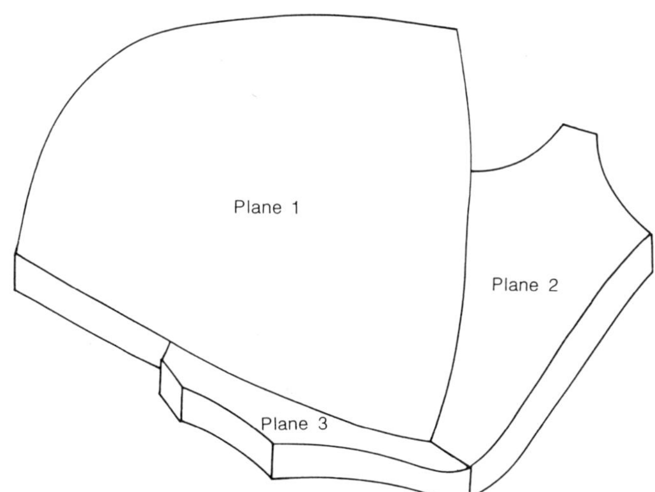

Fig. 2: IHDM discretisation of the headwater.

than that of the soil depth field. Log normally distributed conductivity fields were established at ten depths, the values were obtained from a distribution with a log variance of 1.0. The mean saturated hydraulic conductivity (K_s) of each of the ten layers varies with depth according to the exponential decline often observed in real soils (Beven, 1984). Patterns of soil depth and saturated hydraulic conductivity were obtained using a three-dimensional Turning Bands program supplied by J.R.M. Hosking, derived from that of Mantoglou and Wilson (1981).

In order to derive the Darcian response of the headwater to prescribed boundary conditions a fully three-dimensional finite element solution of the Richards equation (the general mass continuity equation for Darcian flow in saturated-unsaturated porous media) was formulated. Due to the great computational burden of the resulting highly non-linear equations, the solution was implemented on the CDC Cyber 205 at Manchester Computer Centre using the vectorised Jacobi Conjugate Gradient method of Kincaid et al (1984).

The headwater was discretised into 11856 node points (10175 elements). Each element has a maximum plan dimension of 20m and a maximum vertical thickness of 0.12m. To allow fine resolution in areas of high hydraulic gradients, that is at the soil surface and at the channel seepage face, the minimum element dimensions are 0.05m in plan and 0.02m in the vertical.

3.2 Headwater response

To remove any effects of initial conditions a "setting up" period was established, which consists of a series of rainfall events, totalling 119.0mm, over a period of 62.5 days. Prior to this period, the soil water conditions were described by a constant pressure head throughout the entire hillslope. The setting up period was discretised into 1281 various size timesteps and required approximately 13 CPU hours on the Cyber 205.

The soil water conditions at the end of this period were then adopted as initial conditions for the series of events under investigation. This series comprised of a total precipitation input of 383.8mm (equivalent to 9.6×10^4 m^3) over a period of 90 days. This simulation period was discretised into 4643 timesteps and a total of approximately 49 CPU hours were required on the Cyber 205. Further details of the three-dimensional solution may be found in Binley and Beven (1989).

3.3 *Application of the Institute of Hydrology Distributed Model (IHDM)*

The IHDM employs the same flow equations as the three-dimensional model described above. However, to reduce the computational burden, the catchment is treated as a series of independent two-dimensional hillslope planes of variable width aligned in such a way that the flows across slope are minimal. For a homogeneous catchment, both the three-dimensional model and the IHDM would yield virtually identical results. However, for the heterogeneous system considered here, treating the IHDM in a typical lumped parameter manner will lead to some discrepancy depending on the effectiveness of the lumped parameters. For the purpose of this study, the headwater was discretised into three IHDM hillslope planes, as shown in Fig. 2.

It was assumed that sufficient knowledge of the soil parameters was known to define a distribution function, equivalent to that used to generate the original spatial patterns used in the three-dimensional model. Each IHDM plane was assigned a homogeneous soil of constant depth of 1m but the soil hydraulic conductivity parameters for each plane were selected independently. It was also assumed that evapotranspiration losses, precipitation inputs, unsaturated soil water characteristics and the exponential decline of hydraulic conductivity with depth were known exactly.

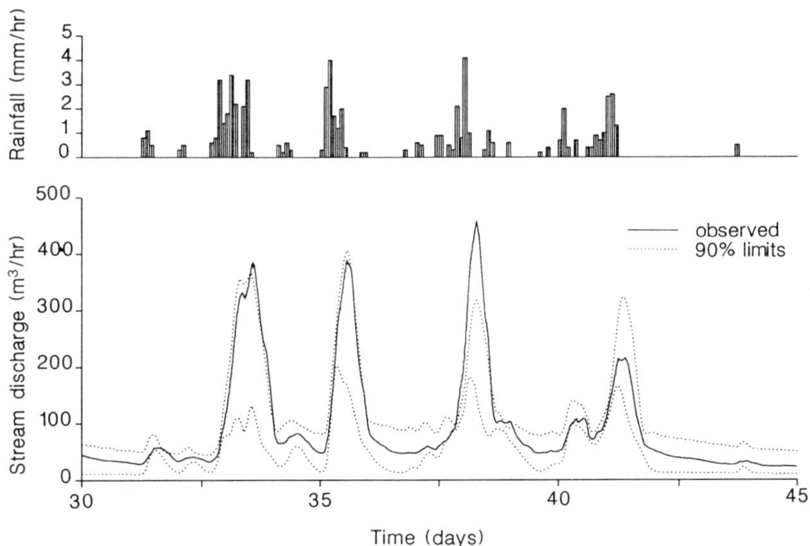

Fig. 3: Observed discharge and 90% prediction limits, between time t=30 and t=45 days, based on a priori estimates of parameters.

This is somewhat unrealistic as in practice there will be significant uncertainty in these model parameters. It should therefore be noted that some initial conditioning of the model has been applied.

Based on this a priori conditioning, the GLUE procedure was initiated by performing 200 IHDM simulations of the 90 day period, preceded by the 62.5 day setting up period. Fig. 3 shows the 90% uncertainty limits, calculated on the basis of the a priori estimates of the parameter distributions, for a 15 day period. For one event in Fig. 3 the "observed" response lies significantly outside the uncertainty bounds suggesting that the lumped parameter model is not an adequate simulator of the heterogeneous hollow. Fig. 4 shows the groundwater table fluctuation uncertainty limits at two locations in the headwater (indicated in Fig. 1) based on the same a priori model parameter estimates. The model uncertainty limits for these responses deviate little from those obtained by physical reasoning.

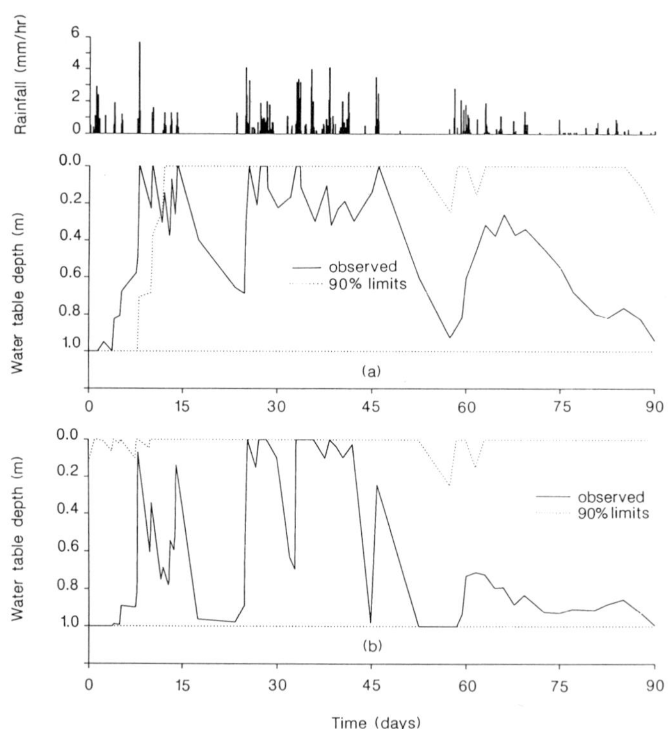

Fig. 4: Observed water table fluctuation and 90% prediction limits based on a priori estimates of parameters (a) Well 1, (b) Well 2. For location of wells see Fig. 1.

In order to test the suitability of the proposed procedure in reducing predictive uncertainty a calibration sequence lasting 30 days was abstracted from the beginning of the 90 day period and a likelihood function evaluated for the 200 realisations based on the following expression

$$L_i = \alpha_i \left(\sum_{j=1}^{m} \frac{W_j}{\sigma_{j,i}}\right)^N \tag{3.1}$$

where: m is the number of observed responses, that is, the number of discharge and water table observation sites,
L_i is the likelihood of the i'th realisation,
W_j is the weight for observation j such that $\sum W_j = 1$,
$\sigma_{j,i}$ is the error variance of the i'th realisation compared with the j'th observation,
α_i is the weight of the i'th realisation such that $\sum L_i = 1$,
N is the likelihood shape factor.

The likelihood shape factor in (3.1) controls the shape of the likelihood distribution. For example, N=0 is equivalent to a uniform likelihood distribution, $N \rightarrow \infty$ is equivalent to the traditional "optimum" solution. Note that for all realisations considered non-behavioural $L_i = 0$. For the study presented here the "worst" 50% of the realisations were considered non-behavioural. As stated earlier, this cut-off may be determined on more physical reasoning in practice.

Using only stream discharge data from the calibration period (that is, m=1 in Eq. 3.1), Fig. 5 shows the conditioned stream discharge uncertainty limits for the same period in Fig. 3. From the nature of the operation the uncertainty bounds have been reduced, however, the poorly simulated response at approximately t=38 days in Fig. 5 still lies significantly above the 95% limit. For comparison with the traditional calibration approach, Fig. 5 also shows the response of the "optimum" single IHDM realisation. It is worth noting that the likelihood for this realisation is not significantly different from a number of other simulations using completely different parameter sets.

A minimal reduction in water table elevation uncertainty bounds was also observed, see Fig. 6. Also shown in Fig. 6 is a comparison of the water table fluctuations resulting from the "optimum" realisation based on discharge conditioning and the "mean" response from the 200 realisations weighted according to the likelihood of each realisation. Both responses are poor in comparison with the observed behaviour, however, the prediction based on a likelihood weighted mean is superior to that of the single highest likelihood.

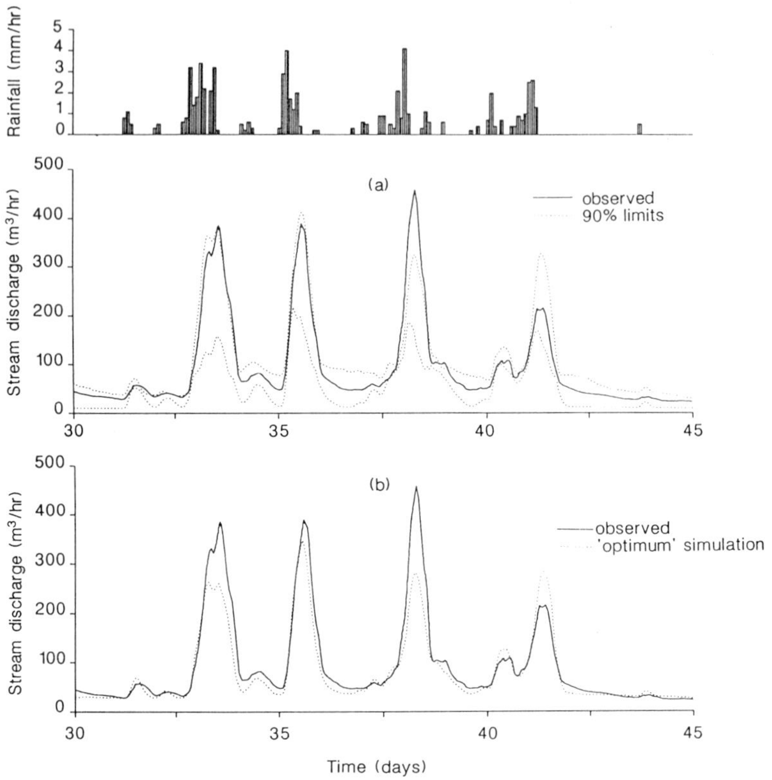

Fig. 5: Observed discharge, 90% prediction limits and "optimum" single prediction, between time $t=30$ and $t=45$ days, after likelihood conditioning.

Further tests have been made on the usefulness of combining the groundwater table and discharge data in evaluating the likelihood distributions, however, it appears that, for the case considered, the lumped parameter model is such a poor simulator of the fully distributed system that such efforts are fruitless. This has direct implications for current modelling practice.

If one is to apply calibration techniques based on likelihood estimation then the shape of the likelihood distribution, with respect to the model parameters, is clearly of interest. Of particular importance is the difference, if any, between the two parameter subsets classified as behaviour and non-behaviour type. In Fig. 7 the two cumulative frequency diagrams for each of the three hillslope planes are presented. Using the Kolmogorov-

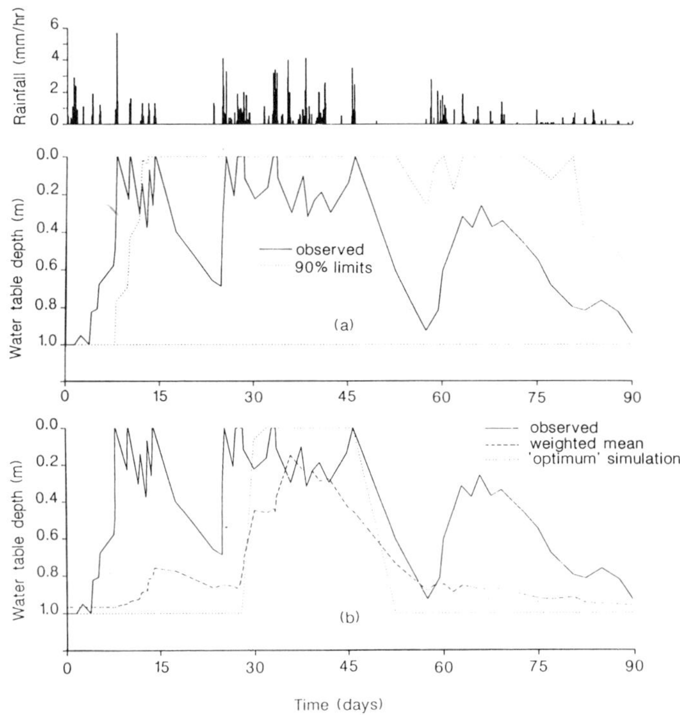

Fig. 6: Observed water table fluctuation, 90% prediction limits, "optimum" prediction and weighted mean prediction, after likelihood conditioning, for well 1.

Smirnov D statistic, the non-behaviour and behaviour parameter distributions were found to be significantly different at the 5% level for planes 1 and 3 (Fig. 2 identifies the plane numbering). This suggests reasonable grounds for evaluating posterior parameter distributions from the likelihood functions using the Bayes equation. It is worth noting that Hornberger and Spear (1981) and Hornberger et al (1985) have used this type of approach based on a likelihood distribution to evaluate the sensitivity of different parameters in complex models.

3.4 Implementation of the GLUE procedures on a parallel computer

The GLUE procedure described is ideally suited to parallel processing, in particular local memory MIMD computers such as those based on the Inmos transputer. As a relatively significant amount of calculation is required for

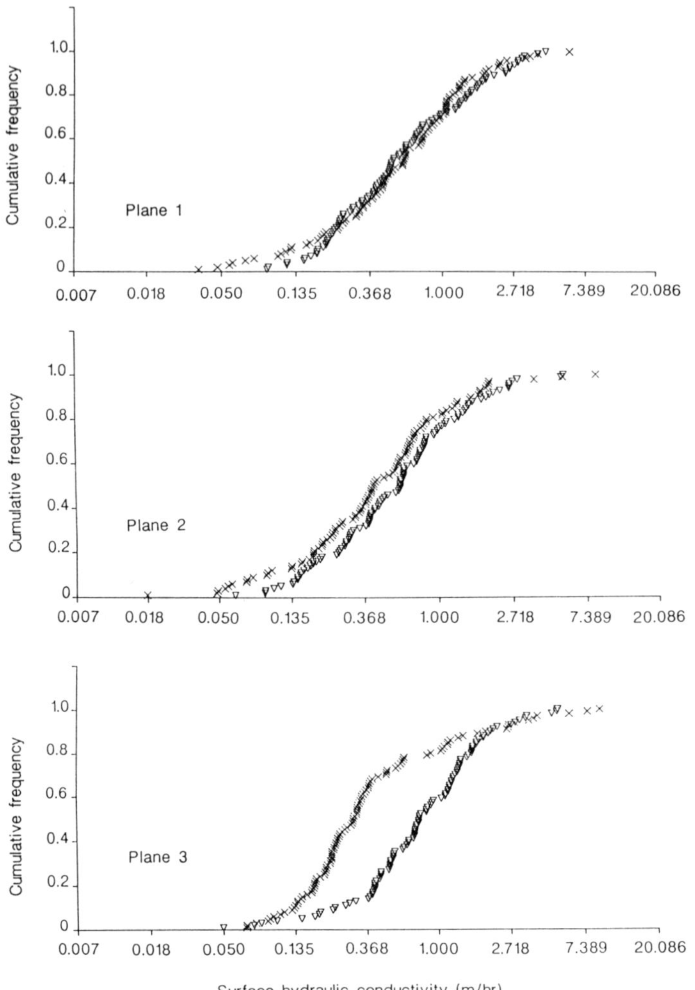

Fig. 7: A comparison of cumulative frequency diagrams for IHDM planes 1, 2 and 3 shown in Fig. 2. Cross indicates behavioural, triangle non-behavioural simulations.

each Monte Carlo realisation, communication overheads are minimal and load balancing is relatively straightforward. In the runs summarised above for example, although each realisation required 30 hours of computing time on a 10 mips computer, 20 such realisations could be made concurrently on the "Environmental Science Domain" of the Meiko Computing Surface at

Lancaster. It is also worth noting that once the base runs have been made, all the consequent operations of calculating and combining different likelihood functions are linear and require relatively little computer time.

4. In conclusion

A generalised procedure has been proposed for the estimation and reduction of model predictive uncertainty. It is clear that, for the example considered here, little benefit is gained by knowledge of discharge and water tables, given accurate information about soil parameter distributions and the errors inherent in trying to reproduce the behaviour of the three-dimensional model by a three plane IHDM with homogeneous "effective" parameter values. The wide confidence limits, particularly for the water table levels suggests that in this case the results are dominated by model structure errors, even where a Darcian model is atttempting to simulate a Darcian "reality". This work will have interesting implications for the application of physically-based models to field data sets, where it is not known if Darcian flow equations are appropriate and where similar problems of model discretisation and specification of "effective" parameters may be encountered. We are currently involved in applying the IHDM in a GLUE framework to a field data set from a small catchment in Svartberget, Sweden, which has been made available by K. Bishop of the University of Cambridge.

5. Acknowledgement

This work was supported by The Natural Environment Research Council (Grant GR3/6264).

6. References

Bathurst, J.C. (1986). Physically-based distributed modelling of an upland catchment using the Système Hydrologique Européen, J.Hydrol., 87:79-102.

Beven, K.J. (1984). Infiltration into a class of vertically non-uniform soils, Hydrol. Sci. J, 29:425-434.

Beven, K.J. (1987). Towards a new paradigm in hydrology, in Water for the Future, IAHS Pub. 165.

Beven, K.J. (1989). Changing ideas in Hydrology. The Case of Physically Based Models, J. Hydrology, 105:157-172.

Binley, A.M. and Beven, K.J. (1989). Modelling heterogeneous Darcian headwaters. Proc. 2nd National Hydrology Symposium, British Hydrological Society, Sheffield.

Binley, A.M., Beven, K.J., Calver, A. and Watts, L.G. (1989). Changing responses in hydrology: Evaluating physically-based predictions in the light of uncertainty, submitted to Water Resour. Res.

Calver, A. (1988). Calibration, sensitivity and validation of a physically-based rainfall-runoff model, J. Hydrol., 103:102-115.

Hornberger, G.M. and Spear, R.C. (1981). An approach to the preliminary analysis of environmental systems. J. Environ. Management, 12:7-18.

Hornberger, G.M., Beven, K.J., Cosby, B.J. and Sappington, D.E. (1985). Shenandoah Watershed Study: calibration of a topography-based variable contributing area hydrological model to a small forested catchment. Water Resour. Res., 21:1841-1850.

Kincaid, D.R., Oppe, T.C. and Young, D.M. (1984). ITPACKV 2C User's Guide, Rept. CNA-191, Centre for Numerical Analysis, Univ. Texas at Austin, U.S.A.

Mantoglou, A. and Wilson, J.L. (1981). Simulation of random fields with the turning bands method, Tech. Rept. 264, Dept. Civil Eng., M.I.T., Mass., USA.

Rogers, C.C.M., Beven, K.J., Morris, E.M. and Anderson, M.G. (1985). Sensitivity analysis, calibration and predictive uncertainty of the Institute of Hydrology Distributed Model, J. Hydrology, 81:179-191.

Rosenblueth, E. (1975). Point Estimates for probability moments, Proc. Nat. Acad. Sci. USA, 72(10):3812-3814.

Catchment-scale rainfall-runoff event modelling and dynamic hydrograph separation using times series analysis techniques

A.J. Jakeman

Australian National University

I.G. Littlewood, P.G. Whitehead

Institute of Hydrology

1. Introduction

The aim of this paper is to present a framework for constructing a catchment-scale rainfall-runoff model which possibly has widespread applications in hydrology. An example is given for a small (0.72 km^2) upland catchment in Wales; an excellent model-fit and hydrograph separation into quick and slow flow components are presented on the basis of hourly time interval data. We omit the detail of the modelling technique (in the spirit of the Conference) and concentrate, rather, on placing it in the context of problems often encountered when modelling complex environmental systems. The plan of the paper is first to outline the essential steps involved in the construction of a credible computational model for solving an environmental problem. Subsequently, the approach to our particular hydrological problem is considered step-by-step within this scheme, whilst commenting on the general issues and pitfalls which can be encountered in the model construction procedure. The validity of the model is demonstrated by applying it to a period of record not employed for model calibration.

A particular feature of the model is that it allows (in many cases) the separation of a hydrograph into quick and slow flow components. Because the general form of the model we employ can be conceptualised as a number of linear reservoirs acting in series and/or parallel (each with a characteristic response time), and because the dynamics of these linear reservoirs can be expressed as components of the instantaneous unit hydrograph, it is possible to resolve the hydrograph into component flows. For the catchment we investigate the "best" configuration of linear reservoirs is two acting in parallel, one with a short characteristic response time (quick flow) and the other with a long response time (slow flow).

Preliminary results of using this hydrograph separation technique indicate that hydrographs for streams draining nearby catchments in the same single land-use headwater area of a river basin can comprise quite different proportions of quick and slow flow. The differences in flow component proportions between catchments are not always as might be expected from a comparison of one or two obvious parameters like catchment area and slope; clearly, factors such as the type, depth and spatial distribution of soils within a catchment are important additional controls of the magnitude and dynamics of quick and slow flow components of total streamflow.

A motivation for the development of the modelling technique we present was the desire to learn more about the provenance, flow pathways and residence times of stream water for runoff event stream chemistry modelling. Whilst our technique does not attempt to answer all these questions it does provide a method of classifying hydrograph behaviour in terms of characteristic peak magnitudes and response times. The exact meaning of such response times for stream chemistry modelling has yet to be established.

The paper concludes with a short discussion of the opportunities and advantages of the modelling approach presented.

2. Steps in construction of computer models

The framework for the paper follows the basic steps involved in the building of a computational mathematical model. These steps can be organised in the following way:

(a) definition of modelling objectives,
(b) specification of
 -system of interest and its influencing environment
 -data
 -other prior knowledge including model structure and parameters
(c) selection of model family by combining (a) and (b),
(d) identification of model structure,
(e) estimation of model parameters,
(f) choice of algorithmic implementation and computing requirements,
(g) verification including diagnostic checking,
(h) quantification of uncertainty,
(i) validation on other data sets.

These steps have been discussed by Jakeman (1989), for example, in general terms. In the next section the hydrological problem of rainfall-runoff event modelling is used to illustrate issues which arise at each of these stages.

3. Modelling of catchment-scale rainfall-runoff events and hydrograph separation

3.1 Definition of objectives

When minimisation of predictive uncertainty is important in a modelling exercise, a temptation to avoid in general is explicit or implicit over-specification of the objectives. For our hydrological problem the objective is

> to determine the catchment-scale dynamic relationship between climatic variables (especially rainfall) and streamflow and, if possible, its component flows, at close time intervals (of the order of one hour) using data that can be made available from minimally instrumented catchments.

Notice that we are not interested directly in simulating streamflow response to land use changes or in inferring hydrological response in different parts of a catchment. This simplifies the problem considerably because it obviates the need to incorporate directly any assumptions about the spatial distribution of catchment properties. By avoiding a physiographic description, a lumped parameter dynamic relationship can be sought which requires considerably less assumed physical knowledge and input data to drive the model.

We are also not just interested in the simple problem of predicting or interpolating streamflow records under existing conditions. If this were the case, then we could try a purely non-causal stochastic or statistical approach such as building an autoregressive moving-average model after enforcing some sort of stationarity in the streamflow data. One of our aims is to simulate streamflow for any hypothetical set of climatic inputs.

3.2 Prior knowledge

In the specific example, the system of interest is the catchment including the stream down to its flow measurement point. Its influencing environment or connections are also relatively easy to specify - they are the climatic inputs to the catchment.

The prior knowledge that can be available includes, firstly, some set of observational data: (possibly short) time series of rainfall, of some other climatic variables (usually temperature), of streamflow and perhaps some groundwater level measurements. Fig. 1 displays typical time series of rainfall and streamflow incorporating about 10 events over 400 hours from a small, humid region, catchment in Wales. Secondly, there are for

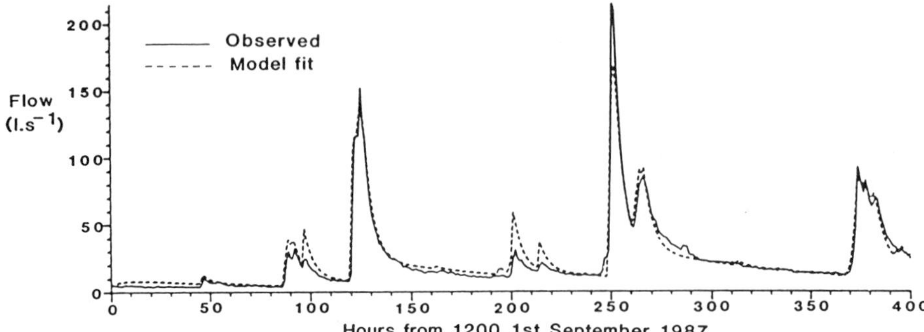

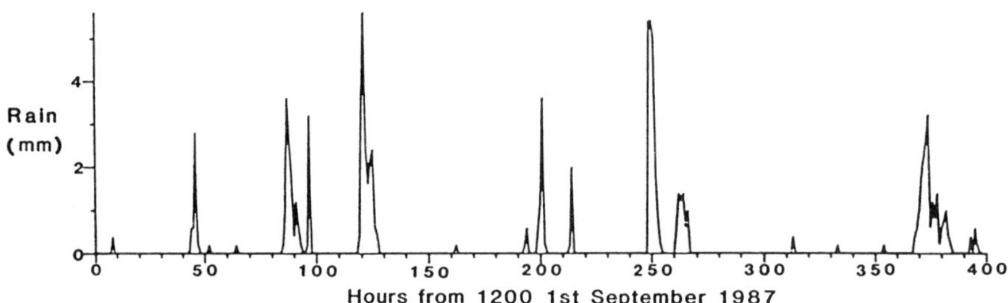

Fig. 1: Hourly rainfall and streamflow, and calibration model fit.

evaluation many simple catchment-scale models of the generation of rainfall excess, i.e. that part of the rainfall which contributes to streamflow and is not lost to evapotranspiration. Thirdly, there is a relatively strong and successful literature on the unit hydrograph concept which assumes that there is a linear integral convolution relationship between rainfall excess and streamflow. Finally, there are physical laws. Conservation of mass yields a water balance equation such that, if the volume of water in the catchment is the same at the end of an accounting period as it was at the beginning, there are equivalences between the volumes of rainfall minus evapotranspiration, rainfall excess and streamflow. It is also a physical fact, as demonstrated in Fig. 1, that in most catchments, the response of streamflow to rainfall recedes from a peak in an exponential fashion. This excludes systems with extremely non-linear, long-term storage behaviour such as karst systems (Jakeman et al, 1984). The exponentially decaying response feature suggests ways in which the solution can be approximated but this aspect is deferred until Section 3.4.

3.3 Selection of model family

The basic issue here is how to strike a balance between the level of determinism and stochasticity in the model family. The first can be addressed by deciding which are the key variables (and their associations) to which model output is sensitive, and which are the appropriate spatial dimensions and discretisations. For the hydrological problem, rainfall is obviously necessary and, if a seasonally independent model is required, then temperature (as a surrogate for evapotranspiration) will also be important.

It is advantageous to appeal to a stochastic representation, even if this means only incorporating an error term with the deterministic relationship. One advantage of stochastic model families is that they can be assessed according to formal criteria for their statistical applicability to the information available. They can be used to assess the hypothesised level of determinism and to develop an efficient parameterisation by concentrating on describing the observational behaviour. This is achieved by assuming that the attribute of modelling interest at time step k is a sum of two components: a function of the observation history and causal process parameters until time step k-1 and a purely random component. Then a probabilistic description of the parameters can be postulated, allowing assessment of the tradeoff to be determined between model complexity (especially causality) and model-fit to observations.

We abbreviate our discussion by dismissing the relationship between rainfall and rainfall excess, although this is by no means a simple problem to model. Thus, using our knowledge in Sections 3.1 and 3.2, we could be led to speculating a model family for the relationship between rainfall excess and streamflow of the form

$$x_k = h_k u_0 + h_{k-1} u_1 + h_{k-2} u_2 + \dots + h_0 u_k \qquad (3.1)$$

$$y_k = x_k + \xi_k \qquad (3.2)$$

The first equation is a linear function at time k of the rainfall excess history (u_k, u_{k-1}, ..., u_0) and the unit hydrograph parameters (h_0, ..., h_k) which define the response of streamflow to an instantaneous input of one unit of rainfall excess (i.e. the impulse response function or instantaneous unit hydrograph, IUH). The second equation is a simple stochastic relationship to adjust the model predicted streamflow x_k by a random component ξ_k to yield observed streamflow. Thus, analysis of the properties of ξ_k helps to reveal the appropriateness of the deterministic description (3.1). However, the theoretical properties of the model family must be considered first and this is one of the functions of the next stage in constructing a computational mathematical model.

3.4 Identification of model structure

The major consideration in going from a model family to a specific model is the degree of ill-posedness (Hadamard, 1932). Formally, this includes whether a solution for the parameters $h_0, h_1, ..., h_k$ exists and is unique, but in practice it mainly means the degree to which the parameters are stable to perturbations in the input data $u_0, u_1, ..., u_k$. It is well known (e.g. Jakeman and Young, 1984) that small errors in $u_0, u_1, ..., u_k$ in (3.1) lead to wild oscillations in $h_0, h_1, ...,h_k$. Indeed, ill-posedness is a widespread characteristic of many traditional formulations of flow and transport processes in the environment (see Dietrich et al, 1988). The phenomenon is also known as the problem of identifiability. The main way of dealing with it is to replace the formulation with a well-posed one by restricting the solution set. For example, smoothness constraints may be imposed on the data, the solution or both. However, the simplest way is to approximate the solution $h_0, h_1, ..., h_k$ with a parameterisation of low order.

The prior physical knowledge we have for the hydrological problem suggests that approximation of the unit hydrograph could be made by a sum of exponential-like functions. Fortunately, rational transfer functions (e.g. Box and Jenkins, 1976; Young, 1984) in the backward shift operator yield solutions which are a discrete approximation to exponentials (Jakeman and Young, 1984). Thus (3.1) can be approximated by

$$x_k = - a_1 x_{k-1} - \dots - a_n x_{k-n} + b_0 u_k + \dots + b_m u_{k-m} \qquad (3.3)$$

For such a parameterisation, the next stage in the procedure is to identify the model structure, i.e. the order of the lags m and n on the rainfall excess input (u) and modelled streamflow (x) respectively.

This part of the identification procedure can be reduced to a parameter estimation exercise for each of the possible options for the values of m and n. In each exercise we compute statistics of model uncertainty and model-fit. We select the average relative parameter error (ARPE) and coefficient of determination (R^2) to specify a tradeoff between model complexity/uncertainty and fit. For linear models of the form (3.3) with (3.2), the tradeoff can easily be judged as that value of the set (m,n) at which any increase in either m, n or both yields no substantial improvement in model fit but quite substantial deterioration in average relative parameter error. Table 1 demonstrates this tradeoff for fitting the model of rainfall excess and streamflow associated with the datasets in Fig. 1. Rainfall excess was calculated according to a model in Whitehead et al (1979) which involves only one parameter estimated by trial and error. The optimal model order is m = 1 and n = 2, and this structure is insensitive to the parameter value in the model for deriving rainfall excess. This model structure, which we believe is the most appropriate (identifiably) for a range of catchments, permits automatic separation of the hydrograph into quick and slow response flow components but this aspect of our model is deferred until Section 3.7.

Table 1: *Order identification results for a model - (3.2),(3.3) - of a catchment in Wales.*

Parameterisation (n,m)	%ARPE	R^2
(1,0)	.0319	.8191
(2,0)	21.2976	.8206
*(2,1)	.0214	.9455
(3,0)	.3637	.8449
(3,1)	.8216	.9014
(3,2)	38.1116	.9078

3.5 *Estimation of model parameters*

Formal estimation involves some choice of objective function, such as minimisation of a loss function or maximisation of probability, to yield parameters with desirable properties. The choice of estimation method depends on the properties required and the type of model and data involved.

For the hydrological problem, we prefer a method that is consistent, unbiased and statistically efficient so that accurate and relatively low variance estimates can be obtained even though the time series of rainfall and streamflow available may be short. The method should also yield an indication of the variance of the parameters in order to allow objective comparison among competing model orders m and n. The final but most important property required of the method is that it is stable to data and model errors ξ_k when the order of any exponential decay in the hydrograph is close to zero. This occurs when the base flow component of a stream hydrograph recedes extremely slowly.

3.6 *Algorithmic implementation and computing requirements*

Care must be taken at this stage because of the limitations of most computers which do not operate on real numbers but on a finite subset, usually a floating point number system. Anderssen and De Hoog (1983) provide a lucid exposition of some of the problems in ensuring acceptable computational performance.

For computational estimation of the parameters in (3.3), we have selected a simple refined instrumental variable (SRIV) algorithm to satisfy all the stated requirements. Jakeman et al (1990) argue the case for SRIV. Only a (fully) refined instrumental variable (RIV) algorithm competes seriously with this choice. The RIV is asymptotically more efficient than SRIV which in any case is adequate. However, the RIV is computationally more complex, which slows its convergence on short data sets. It also requires that the model error ξ_k be of constant variance, which may not be appropriate when attempting to model different streamflow peaks of widely varying magnitude. Traditional algorithms, such as basic instrumental variables and least squares, have difficulty in identifying an exponential component of the parameterisation with slow decay.

The major computational effort involved in any of the IV algorithms is the effective inversion at each iteration of a matrix of the order of the number of parameters. The number of parameters and iterations required are both invariably less than 10 so that implementation on microcomputers is entirely feasible.

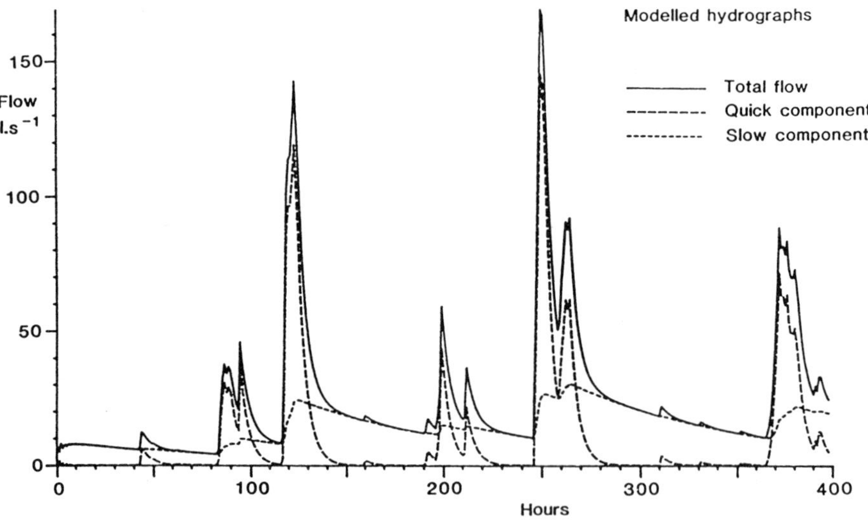

Fig. 2: Hydrograph separation for the calibration period.

3.7 *Verification*

It was indicated at the identification stage that there are two necessary conditions with respect to model credibility. These are that the model output yields an adequate reproduction of measured behaviour and that the model parameters are estimated with sufficient confidence to yield behaviour for future uses with acceptable accuracy. Fig. 1 also shows our fit of the estimated model to the streamflow data. As displayed in Table 1, the fit explains almost 95% of the variance of streamflow. Table 1 also shows that the average relative parameter error, where error can be interpreted as variance, is a healthy .02%.

Primary consideration at this stage should also be given to judgement of whether or not any physical properties which can be inferred from the model are plausible. The major physical catchment-scale characteristics inferred from the estimated hydrological model can be summarised as follows:

Number (n) of identifiable/separable flow components	2
Time constant of quick flow pathway (hours)	4.4
Time constant of slow flow pathway (hours)	99.5
Quick flow contribution to stream flow volume (%)	39
Slow flow contribution to stream flow volume (%)	61
Quick flow contribution to peak of the IUH (%)	92
Slow flow contribution to peak of the IUH (%)	8

The hydrograph separation into quick and slow flow components shown in Fig. 2 accords well with rule-of-thumb baseflow separation but both methodologies remain to be assessed on the basis of satisfactory field tracer studies.

Other diagnostic checks which should be performed at this stage relate to the model residuals. These include checking that they are essentially zero-mean and not cross-correlated with causal variables of potential influence on observational (streamflow) behaviour. Some models and estimation techniques may require additional properties for the residuals, such as independence (not autocorrelated) and constant variance (stationarity). Selection of the simple refined instrumental variable algorithm to implement the method of parameter estimation does not require these added restrictions.

3.8 *Quantification of uncertainty*

To assess confidence in any use of a model, knowledge is required of the confidence in model parameters. For some linear models, closed relationships between parameter and predictive confidence are available, while for others including (3.2) - (3.3) and for non-linear models, some form of sensitivity analysis, using Monte Carlo simulation for instance, may be required. The instrumental variable algorithms naturally yield knowledge of the variability of parameter estimates in linear dynamic models of the form (3.2) - (3.3). Whitehead and Young (1979) illustrate how to use Monte Carlo simulation to generate streamflow variability associated with this knowledge.

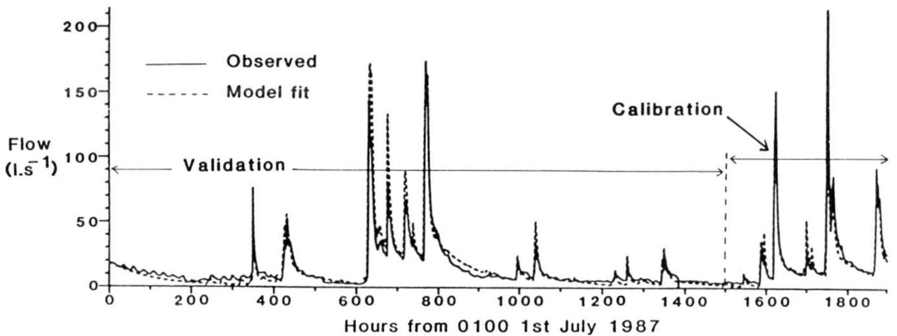

Fig. 3: Model-fit for the validation and calibration periods.

3.9 Validation

As with most of the steps in a systematic approach to computational modelling, there may be considerable overlap and iteration among this and other stages. Thus, the verification step is part of the necessary procedure in establishing the validity of a model. However, the most convincing assessment of the validity of a model is its performance in reproducing observational behaviour not used previously in the model construction. In seeking to explore the validity of the catchment-scale rainfall-streamflow model, we applied the model estimated on 400 hours of time series data to the preceding 1500 hours of rainfall. This period encompasses a wide range of high, low and intermediate rainfall events as well as a relatively long period of baseflow uninterrupted by rainfall events .

The results of the validation exercise, shown in Fig. 3, are encouraging and augur well that the model will perform well on other data sets from the catchment where: rainfall is measured in a similar fashion; the same range of rainfall events is encountered; and other climatic conditions, especially temperature, are similar to those of the summer period investigated.

4. Concluding remarks

Any well-identified model relating catchment rainfall and other climatic variables to streamflow can have several uses. These concern inference of any one of the following given the other two: system inputs (rainfall), system outputs (streamflow) and system model characteristics at catchment scale (see those listed in Section 3.7). Jakeman et al (1990) argue these uses in more detail but they include the following: interpolation of catchment streamflow records; prediction of catchment streamflow under climatic changes; inference of rainfall records; derivation of unit hydrographs for engineering hydrology and flood warning; and design of hydrometric networks. In addition, if the modelling can identify separate exponential decays associated with conceptual storages, then the following uses of such hydrograph separation can be added: systematic characterisation of catchment-scale rainfall-streamflow processes for other conceptual modelling such as water quality; hydrograph synthesis by relating hydrograph properties derived from instrumented catchments to physical catchment characteristics; knowledge of baseflow dynamics to infer catchment storage/wetness, which may in turn be capable of spatial distribution by applying other physical models.

There are some noteworthy advantages of using the computational approach described in this paper for modelling rainfall-runoff events.

(a) There is no penalty in terms of requiring other information apart from rainfall and streamflow (and perhaps temperature), e.g. topographic information is not required. Indeed, the low parameterisation of the transfer function representation combined with the SRIV estimation allows for reduction in the length of rainfall-streamflow time series required for analysis.

(b) In contrast to non-parametric approaches (e.g. Mays and Taur, 1982) which solve the discretised convolution equation (3.1) and apply constraints to stabilise the solution, the SRIV algorithm yields a guide to the sensitivity of model parameters since an estimate of the covariance matrix is a natural by-product of the estimation procedure.

(c) There is no need to select "clean", single-peaked events for analysis, as required by some unit hydrograph methods.

(d) Hydrograph separation is an integral part of model identification and not, as in some methods which derive a unit hydrograph for quickflow only, a necessary prerequisite.

(e) Hydrograph separation in our method is dynamically sensitive to rainfall and its recent history, unlike many of the popular methods (reviewed and evaluated, for example, by Nathan and McMahon, in press) which perform baseflow separation using only the streamflow record.

5. References

Anderssen, R.S. and De Hoog, F.R. (1983). The nature of numerical processes. Mathematical Scientist, 8:115-141.

Box, G.E.P. and Jenkins, G.M. (1976). Time series analysis: forecasting and control. (Revised edition). Holden Day, San Francisco.

Dietrich, C.R., Jakeman, A.J. and Ghassemi, F. (1989). Modelling for surface and subsurface hydrological management. In: Proceedings International Association for Mathematics and Computers in Simulation, 12th World Congress on Scientific Computation, July 18-22 1988, Paris, France. IMACS, Volume 2:396-398.

Hadamard, J. (1932). Le probleme de Cauchy et les equations aux derivees partielles lineaires hyperboliques. Hermann, Paris.

Jakeman, A.J. (1989). Features and principles of a systems approach to environmental modelling. In: Proceedings International Association for Mathematics and Computers in Simulation, 12th World Congress on Scientific Computation, Paris, July 18-22 1988, Paris, France. IMACS Volume 2:229-234

Jakeman, A.J. and Young, P.C. (1984). Recursive filtering and smoothing procedures for the inversion of ill-posed causal problems. Utilitas Mathematica, 25:351-376.

Jakeman, A.J., Greenway, M.A. and Jennings, J.N. (1984). Time series methods for the prediction of streamflow in a karst drainage system. Journal of Hydrology (NZ), 23:21-33.

Jakeman, A.J., Littlewood, I.G. and Whitehead, P.G. (1990). Computation of the instantaneous unit hydrograph and identifiable component flows with application to two small upland catchments. Journal of Hydrology (in press).

Mays, L.W. and Taur, C.K. (1982). Unit hydrographs via non-linear programming. Water Resources Research, 18(4):744-752.

Nathan, R.J. and McMahon (1990). Evaluation of automated techniques for baseflow and recession analyses. Water Resources Research (in press).

Whitehead, P.G., Young, P.C. and Hornberger, G.M. (1979). A systems model of stream flow and water quality in the Bedford-Ouse River - 1. Stream flow modelling. Water Research, 13:1159-1169.

Whitehead, P.G. and Young, P.C. (1979). Water quality in river systems - Monte Carlo analysis. Water Resources Research, 15:451-459.

Young, P.C. (1984). Recursive Estimation and Time Series Analysis - An Introduction. Springer-Verlag, New York.

Simulation of catchment runoff by conceptual model

C. Eeles

Institute of Hydrology

1. Introduction

Research into the lumped conceptual model and its applications has been carried out at the Institute of Hydrology since 1969. The objective of this type of model is to simulate the rainfall/runoff regime for a geographical area such as a catchment, or river basin, using physical concepts which apply to the whole of the area. These concepts are expressed as mathematical equations arranged in a modular structure making up the model algorithm. During its development the model has passed through a number of different concepts and modular structures. The present basic model has five stores representing the catchment vegetation cover, the soil surface layer, soil profile, groundwater, and the surface channel system for water in transit. Variations on this theme have been widely used in different models from the 1960s onwards. This has coincided with the mainframe computer becoming available as a research tool enabling the storage and dynamic vector processing of large amounts of data.

The development and application of the lumped conceptual model for the simulation of the rainfall/runoff regime of a catchment has reached a level where it has become a useful tool in hydrological research and water resources evaluation. This has led to the development and marketing of a software package (called HYRROM) by the Institute for general application. All environmental models have an element of "lumping", or integration over an area, in which their concepts, even if they nominally apply to a point on a grid, are applied to an area. With the "lumped" type of model the concepts are mathematical abstractions from reality, which are not designed to apply to a point in the catchment but represent the physical processes over an area. This may cover the whole of a catchment or be an area of vegetation, etc., within that catchment which is then used to weight the output from a particular module of the conceptual model.

This flexibility of concept of the environment within a catchment allows the development of a basic modular structure which can be easily varied according to the complexity of the system to be simulated and the objective of the simulation. The flexibility of the system can be judged from the recent applications of the model which have contained parametric vectors with elements ranging from 9 to 35 in the case of a complicated land use simulation. The models can be successfully optimised through their parameters with the minimal use of computing resources, and require simple time series of data for rainfall input and evapotranspiration losses (Eeles et al, 1990).

1.1 The concepts of a catchment area

This land area is one from which it is physically possible for a proportion of the incoming rainfall to form the total flow from the basin; the resultant system through which the rainfall moves becomes holistic for modelling purposes. The proportion of the incident rainfall which is not lost to the atmosphere as evaporation and transpiration follows a complex network of flow paths between different "stores" to the outfall point of the basin. The time over which the actual transport of the water takes place can vary considerably: from minutes to decades depending on the size and physical characteristics of the basin. The upper limit of the output is set by the volume of the precipitation, but this is reduced to an "effective" amount by the losses within the catchment through direct evaporation, and transpiration from the vegetation.

The incident radiant and advected energies determine the potential volume loss from the system, but this in reality varies greatly according to a number of factors affecting the actual loss arising from a given potential energy; varying combinations of these factors can produce the same loss.

The factors involved can be described as a set which modifies the effect of potential evaporative demand. The most obvious factor is the infiltration rate of the surface soil; rain falling on a deep porous sand will quickly pass through it to depth as compared to the water ponding upon an impervious clay. In the latter case water can only move into the soil through cracks and root paths. The vegetation contributes to the loss by rainfall interception on its leaves and branches leading to direct evaporation, and also by drawing water from the soil through its root system which is then transpired through the leaf surfaces at a rate dependent on the energy available and the species. The transpiration process continues as long as the roots can extract water from the surrounding soil, and it will usually contribute more to the water loss from the system than direct evaporation. The loss to the system is therefore dependent on the following factors:-

(a) The energy available: radiation and wind advected energy.
(b) The volume and intensity of rainfall; its spatial and temporal distribution.
(c) The vegetation species; their density and distribution.
(d) The soil types; their distribution, infiltration rates and storage capacities.

The remaining water in the system, or "effective rainfall", is potentially available to contribute to the streamflow at the catchment outfall. At one extreme the bulk of the rain during a high intensity storm will contribute to the overland flow because the infiltration rate into the soil is exceeded; this is the rapid response flow output from the system. This volume of water moves directly over the surface to the dendritic system of stream channels leading to the basin outfall in a very short time. At the other extreme is a gently sloping and very permeable catchment in which the water infiltrates down through the soils and rock strata to groundwater aquifers. This movement is slow, and the water eventually emerges as "base flow" into the stream.

2. HYRROM

This is an acronym for HYdrological Rainfall/RunOff Model which is basically the model outlined below, but with only nine parameters available for change and the rest fixed. The model is available as a software package for mounting on an IBM PC XT/AT or PS/2 and compatible machine and is beginning to be widely used. It was originally developed as a daily data package for use in the Windward Isles on St Lucia, but was found to have a much wider application, and has successfully simulated such catchments as the Kenwyn at Truro in the UK (area 19 km^2) over the extreme drought and subsequent recharge of 1976. It can also be used as a model using monthly data inputs with the usual restrictions applicable to any monthly model: it simulates observed data well when the inputs are changing slowly, but in rapidly changing periods the total volume is a good estimate but not for individual months within the period.

The basic model has five stores representing the vegetation cover, the soil surface layer, soil profile, groundwater store, and the surface channel store; outputs from the channel store and the groundwater store are each delayed by a period which allows them to be combined as flow from the basin. Similar types of model storage and routing have been widely used in various models. The model concepts are discussed in detail by Blackie and Eeles (1985) and are only briefly described here.

Evapotranspiration takes place from the vegetation store and the two soil stores by means of a simple multiplier function of Penman Open Water evaporation, but any standard index of evaporation potential can be used - or even actual pan evaporation data. A cosine reduction function decreases the apparent transpiration by the vegetation under moisture stress from a soil moisture deficit (SMD), which could be considered as the Penman root constant, to a greater deficit which is the "wilting point" and no further moisture loss takes place.

A partitioning function splits the "overflow" from the soil surface store by estimating the rapid response runoff, and the remaining effective rainfall is allowed to percolate to the main soil store. The partitioning function has terms which represent the decreasing infiltration with SMD and the increased runoff caused by rainfall intensity. Surface runoff is then routed through a non-linear store representing the channel flow, and delayed before it combines with the groundwater flow from the basin.

Percolation from the soil store to groundwater is estimated by another cosine reduction function which allows some transfer of water at SMDs below field capacity for the soil type, and sets a limit to the percolation rate. In HYRROM a slightly different concept is used which allows free drainage above field capacity. The observed baseflow recession curve can be used to establish the parameters in the non-linear groundwater release function, or they can be estimated by the optimisation algorithm. In the latter case errors stemming from the modular structure of the model may affect these values considerably.

The nine parameters which are available for interactive adjustment in HYRROM or automatic optimisation are:-

(a) Size of the surface vegetation and detention store.
(b) Surface runoff partitioning factor.
(c) Channel routing store output delay.
(d) Routing store index.
(e) Routing store factor.
(f) Penman open water evaporation factor.
(g) Groundwater store output delay.
(h) Groundwater store index.
(i) Groundwater store inverse factor.

These nine parameters are considered to be the minimum number to achieve a satisfactory fit to an observed hydrograph, and should at least produce an 80% correlation with daily data. For automatic optimisation a modification of the original algorithm proposed by Rosenbrock (1960) is included in the package. This finds a minimum value of the objective function by dynamic processing of the parameter vector and model functions. The function to be minimised is the root mean sum of squares of the residuals between the model simulation and the observed values to which the model is being fitted.

2.1 Data requirements

A data file is required for the above fitting process and should contain good quality daily data for rainfall and evaporation as inputs, as well as observed flows for comparison with the output simulated flow data. The minimum length of data required is a "water year", but the optimisation can be run for up to a maximum of ten years of daily data.

In prediction mode only the two sets of input data (rain and potential evaporation) are required to produce a run of flow data which can be as long as the available data series up to a maximum of fifty years. Evaporation data can be daily or monthly; in the latter case the monthly totals are distributed sinusoidally to produce daily data by HYLINK. This is a preprocessing program of the package which produces a modelling file from data files available, or direct from the Institute's hydrological database, HYDATA.

Quite often the only data available are monthly totals and these can be input to another preprocessing program, HYMONTH, which distributes the data equally over the month in order to produce a daily file for modelling. Obviously, this cannot give a realistic representation of the temporal distribution of rainfall and evaporation inputs during a month. However, this has been found to provide a satisfactory simulation of monthly totals when the rainfall/runoff is not rapidly changing, and to give a realistic total over a number of months when this is happening.

2.2 *A simulation of catchment response*

The data for the catchment area used in this simulation were defined by the outfall of the Kenwyn river at Truro in Cornwall. This catchment has an area of 19.1 km^2, and the daily rainfall and evaporation data available ran from May 1975 to December 1976 with an extreme drought in the last summer period. The simulation is shown as the dotted line in Fig. 1 compared with the solid line which shows the observed hydrograph. Both sets of data are in cumec days.

The simulation of the baseflow recessions in both summers is good; in particular the recovery from the extreme drought of 1976. The peak flow on the 21st March 1976 is underestimated by nearly a cumec. However, there was heavy snow during this period and prior to this peak the flows are underestimated which shows that snow was accumulating and not forming part of the runoff - then the melt contributed to the very high peak. This is a typical response to snow as there is no snowmelt simulation component in the model.

2.3 *Applications of the lumped model*

The model can be used for infilling missing flow data, and extending historical flow records. It can identify problems with flow data such as the

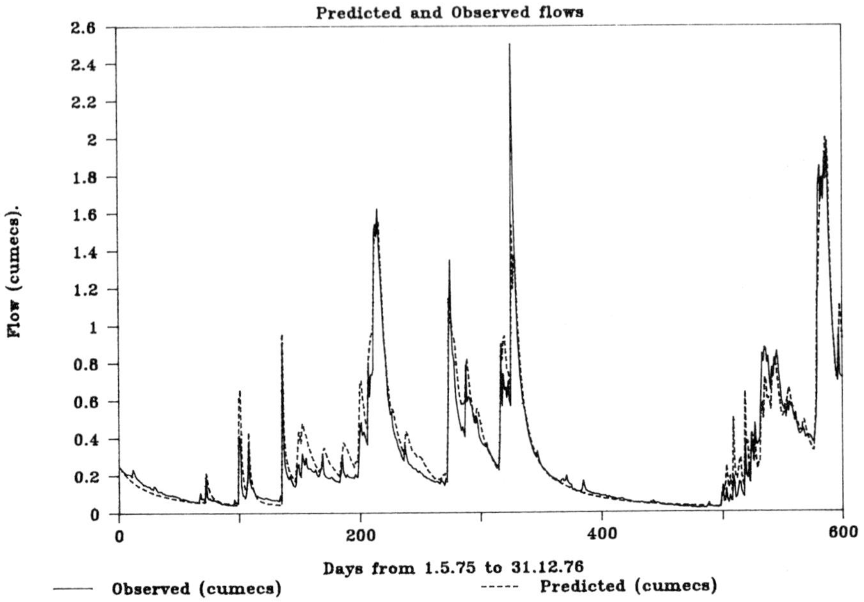

Fig. 1: Simulation example.

observed record shown in Fig. 1 where the model simulated response is greater than peak flows and their recessions. This is either caused by the inadequacy of the model or a problem with the observed data. The model has been well tested in the quality control of data and can define periods over which data may be suspect and require further investigation.

The generation of synthetic series of flow from real or stochastically generated rainfall data provides flow series which vary as the input data.

A more complex variation of HYRROM with 23 parameters has been used to isolate the effects on the hourly hydrograph of the artificial drainage changes in the surface layer of a clay catchment, Robinson et al (1990). The paper questions the apparent stationarity of catchment processes. This simulation was developed on a period with little agricultural drainage work, and then compared with the observed flow data from a later period to show the effects of the increasing network of soil drains on the magnitude of peak flows.

The most complex of these models with 35 parameters is a land use version developed to simulate the changes in yield from a system of reservoirs caused by afforestation of the catchment area. These simulations generated data over a period of 53 years with different areas of forest.

Seven sets of data were generated and then used to assess the operational and financial implications for water resource management. The basic model structure of HYRROM was used for each area of vegetation cover, and the interception and transpiration losses are interfaced with models developed by Calder and Newson (1979). All of these models have been mounted and run on an IBM PC, and are particularly efficient on the latest PS/2 series.

3. Conclusion

The general utility of these conceptual modelling packages is finding an ever increasing series of applications from simulating the effects of land use changes to the management of water resources. The models make extremely flexible use of areal concepts of the processes within a catchment, and allow the simulations to generate long series of data from relatively short calibration periods of observed data with minimal demands on computing resources.

4. References

Blackie, J.R. and Eeles, C.W.O. (1985). Lumped catchment models. In: Hydrological Forecasting, M.G.Anderson and T.P.Burt, editors. Chapter 11:311-345.

Calder, I.R. and Newson, M.D. (1979). Land use and upland water resources in Britain - a strategic look. Water Resources Bulletin, 16:1628-1639.

Eeles, C.W.O., Robinson, M. and Ward, R.C. (1990). Experimental basins and environmental models. In: Hydrological Research Basins and the Environment, Proceedings of the TNO Committee on Hydrological Research, No.44, Wageningen, Netherlands.

Robinson, M., Eeles, C. and Ward, R.C. (1990). The research basin and stationarity. In: Hydrological Research Basins and the Environment, Proceedings of the TNO Committee on Hydrological Research, No.44, Wageningen, Netherlands.

Rosenbrock, H.H. (1960). An automatic method of finding the greatest or least value of a function. Computer J., 3:175-184.

Using a single column of a GCM to explore the sensitivity to different land surface descriptions

A.J. Dolman

Institute of Hydrology

1. Introduction

General Circulation Models (GCMs) attempt to model the three-dimensional structure of atmospheric motions. On a global basis, 50% of the radiation at the top of the atmosphere reaches the surface, where about two thirds of that amount is being converted into sensible and latent heat (water vapour flux), the remainder reflected and used to warm up the surface (GARP, 1975). The land surface plays an important role in this process, partly through the positioning of the continents relative to the oceans, and partly through the fact that the surface characteristics of the land surface determine the way in which the exchange processes take place. Early experiments with GCMs have shown a great sensitivity of the global circulation patterns to changes in the land surface parameterization. Shukla and Mintz (1982) have shown that altering soil moisture initialization in a GCM has a pronounced effect on the rainfall pattern over continental land masses. Sud and Smith (1985) and Sud et al (1988) showed that changing the surface roughness in a GCM has a large influence on the horizontal convergence of moisture in the atmospheric boundary layer and on the distribution of convective rainfall. Charney et al (1977) showed how an increased albedo in the Sahel region could lead to further desertification through a feedback process whereby a reduced evaporation rate would result in lower cloudiness and lower precipitation rates. Reviews of the influence of the land surface on global circulation patterns can be found in Mintz (1984) and Rowntree (1986).

Most of these experiments are sensitivity tests of the model climate to the alteration of a single parameter such as albedo or surface roughness. Being physically not very realistic, these experiments served mainly to express the need for more accurate descriptions of the interaction of the land surface with the lowest layers of the atmosphere. Recently a number of new land surface parameterization schemes have been developed by Dickinson (1984), Sellers et al (1986) and Noilhan and Planton (1989). They

incorporate a considerable degree of complexity in their description of vegetation-atmosphere interaction. In developing these models, they are usually calibrated against micrometeorological data (e.g. Sellers et al, 1989; Pinty et al, 1989). Very few experiments exist where complete global climate simulations are made with these schemes (Sato et al, 1989).
It is not immediately obvious whether a point micrometeorological model, calibrated with single point data, performs in a satisfactory way in a large scale model. On a larger scale basically two complications arise. The first is that a larger scale model is not capable of resolving subgrid phenomena. An example of this is convective rainfall in a GCM grid. Convection in GCMs is calculated on a grid-average basis. Although a convection criterion may have been reached, the resulting precipitation is expressed as a grid-average and is consequently of very low intensity. Single point descriptions of rainfall interception loss from wetted canopies simply cannot cope with this and the result is that the GCMs currently predict interception losses which are much too high . The second problem is that sensitivity of energy partitioning to changes in certain key parameters in an energy balance model may effectively be moderated by large scale feedback processes. An example of such a phenomenon is the reduced sensitivity of evaporation to surface conductance in boundary layer models (De Bruin, 1983; McNaughton and Spriggs, 1986). Calibration and developments of new land surface parameterizations thus require the use of a large scale interactive model which can cope with the two above mentioned problems.

The use of GCMs in these studies is unlikely to be feasible as a consequence of their high computational demand. Warrilow et al (1986) developed a Single Column Model (SCM) from the UK Meteorological Office (UKMO) 11-layer GCM and used it to develop a new land surface parameterization. This paper describes experiments with that model in an attempt to develop a new, calibrated land surface description of tropical rainforest.

2. Description of single column model

The SCM was developed by Warrilow et al (1986) from the UKMO 11-layer GCM. As in the GCM, the vertical resolution is prescribed by σ, the value of the atmospheric pressure divided by the surface pressure. At present it is assumed that the grid area comprises 10000 km^2. The main feature of the SCM is its treatment of the large scale dynamics of heat and moisture advection. It is assumed that the area of interest is a circle with radius n and that there exist a constant gradient of an entity X across the area. It is then possible to write the local rate of change of X due to the large scale dynamics as (Warrilow et al, 1986)

$$\frac{\partial X}{\partial t_D} = \left(\frac{|V_n| \, (X_r - X)}{\partial n}\right) - W\frac{\partial X}{\partial p} \tag{2.1}$$

where: $|V_n|$ is the absolute value of the component of horizontal velocity; w, the vertical velocity; p the pressure; n the radius of the area; and X_r a reference value of X outside the grid. Eq. 2.1 shows that for a treatment of large scale advection it is necessary to specify V_n, W and X_r. These can be treated as random variables since they are not calculated in the column. The right hand side of (2.1) represents the vertical component of the local rate of change in entity X. This is calculated by assuming that X is a linear function of the logarithm of the atmospheric pressures between two levels. Since W is known for each level, (2.1) can be solved. Global climate data collected by Oort (1983) are used to describe the probability distribution functions of the random variables. The model is tuned to produce the correct annual rainfall through a relation of the standard deviation of the dewpoint temperature and air temperature. Gaussian distributions are assumed for all the random variables and a random number generator is used to derive actual values once a day. In order to avoid large step changes in the atmospheric variables from day to day, the actual value used on a single day is interpolated from the values of two consecutive days.

Four subroutines deal with the physical processes such as radiation, dry and moist convection and large scale (frontal) precipitation and surface and boundary layer turbulent exchange. The model is written in Fortran. Timesteps in the model are typically 300 to 900 seconds and integrations are usually performed over a full annual cycle. For the experiments described in this paper, the model was run on an IBM PS/2 machine with 25MHz processor. A complete annual cycle takes several hours CPU time.

3. Land surface scheme

The land surface scheme presently used in the UKMO GCM is shown in Fig. 1. Surface temperature is a prognostic variable and is calculated using a four layer surface-soil temperature scheme (Warrilow et al, 1986). It is designed to respond accurately to radiative forcing in the frequencies of half a day to a year. The surface temperature is an average single value representative of a mixture of bare soil and vegetation and is equal to the temperature of the first thin soil layer. Fluxes of latent and sensible heat are calculated from the basic transfer equations, relating the flux to a gradient and a transfer coefficient. The equations are solved through the use of a bulk transfer coefficient which is dependent on the stability of the

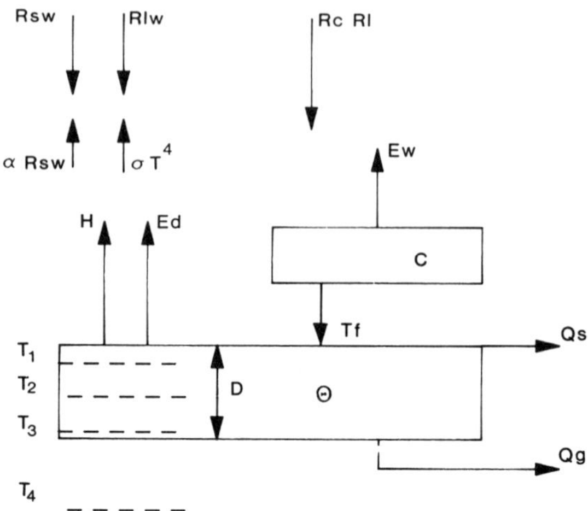

Fig. 1: Schematic diagram of the land surface scheme as used in UKMO GCM and in the original version of the SCM (Redrawn from Warrilow and Buckley, 1989). Rsw, Rlw short and longwave radiation; α albedo; Rc and Rl convective and large scale precipitation; Ew and Ed wet and dry canopy evaporation; H sensible heat flux; T_i temperature of the i'th soil layer; D rooting depth; Θ soil moisture content; Qs and Qg surface runoff and gravitational drainage; C canopy water content; T_f throughfall.

atmosphere and the roughness of the surface. In the case of latent heat flux (evaporation) the flux is related to the product of the gradient in atmospheric humidity over the lowest model layer and the reciprocal of the sum of an aerodynamic and surface resistance. The surface resistance is a parameter related to the water flow through the stomata. Depending on the actual moisture content of the root zone the surface resistance is increased with increasing soil moisture deficit. The canopy is represented in a rudimentary way by the interception store. The actual wet canopy evaporation rate depends on the amount of water stored on the canopy. If the grid canopy capacity is not exceeded, the remaining part of the grid is assumed to transpire freely.

4. Previous GCM experiments

Deforestation experiments with GCMs have generally shown that a reduction in the intensity of the hydrological cycle is likely to occur as a

Table 1. *Results of GCM deforestation experiments. LW refers to the results of Lean and Warrilow (1989), DHS to Dickinson and Henderson-Sellers (1988) and SNS to Shukla et al (1989). Values refer to the difference between deforested and control climate.*

	LW	DHS	SNS
evaporation (mm.day^{-1})	-0.61	-0.52	-1.36
rainfall (mm.day^{-1})	-1.34	0	-1.76
surface temperature (°C)	1.98	2 - 5	2.5

result of the cutting down of large parts of the tropical rainforest. Table 1 lists the results of three of the most recent experiments.

All three experiments show an increase in surface temperature and a corresponding decrease in evaporation. In the Dickinson and Henderson-Sellers study, no decrease in precipitation is observed. The actual values of the decrease in evaporation and rainfall and increase in surface temperature vary substantially. An obvious first requirement for a deforestation experiment is to predict the control climate correctly, and so it is interesting to compare the control results with observations made by Shuttleworth (1988a). Average observed rainfall is 7.1 mm.day^{-1} and evaporation 3.8 mm.day^{-1}. LW predict 6.6 and 3.1 mm.day^{-1} respectively, SNS 6.8 and 4.5 mm.day^{-1}, and DHS predict 9.2 mm. and 6.2 mm.day^{-1}. This variation in the control runs will affect the predicted changes as result of deforestation.

It is also interesting to compare the partitioning of total evaporation in dry and wet canopy evaporation. In LW, 71% of total evaporation is made up by the interception loss, the remainder by transpiration. DHS results suggest that 30 to 50% of total evaporation is made up by interception loss and SNS results indicate about 30%. Observations indicate that the interception loss accounts only for about 25% of the total evaporation, and that dry canopy evaporation of 75% dominates total evaporation. The most likely cause for the overestimation of wet canopy evaporation by DHS and LW is the neglect of spatial variability in the convective rainfall over the Amazon basin.

5. Single column model experiments

5.1 Wet canopy evaporation

Overestimation of interception loss is likely to be caused by the very low average rainfall intensities in a GCM grid. Typical average rainfall intensities are 0.25 $mm.hr^{-1}$ in the SCM. Average rainfall intensities as measured in the Reserva Ducke are 5.15 $mm.hr^{-1}$ (Lloyd et al, 1988). This difference in rainfall intensities causes problems in the modelling of interception loss. Typical canopy storage capacities for forest are about 1mm, for the tropical rainforest, Lloyd et al (1988) estimate a value of 0.7mm. In GCMs which take account of interception loss, the rainfall is essentially assumed to cover the complete grid. As the grid rainfall is a composite of several storms of variable intensity, duration and surface coverage, this is probably not very realistic.

Shuttleworth (1988b) devised a scheme whereby the grid average rainfall is assumed to cover only a small part of the grid if the rainfall is of convective origin. The local rainfall rate in the rain covered area is assumed to follow a negative exponential probability distribution

$$f(P_l) = \frac{\mu}{P} \exp[\frac{-\mu P_l}{P}] \qquad (5.1)$$

where: P and P_l are the grid average and local rainfall rate respectively; and μ the proportion of the grid which is wet. The average grid throughfall rate follows as

$$T = P \exp[\frac{-\mu C_m}{P}] \qquad (5.2)$$

where C_m is defined as the maximum canopy infiltration rate per timestep.

In Fig. 2 the sensitivity of grid average throughfall to changes in the fraction of the grid wetted with rainfall is shown. A factor $\mu=1$ represents a completely wetted grid, as in the case of large scale precipitation due to supersaturation. Shuttleworth (1988b) suggest a factor $\mu=0.3$ for convective precipitation in analogy with Warrilow et al (1986) who used a similar approach to estimate grid average runoff. Clearly the throughfall parameterization is sensitive to the exact choice of the value of μ. The interception loss modelled by this scheme, assuming $\mu=0.3$ for convective rainfall, is 40% of the rainfall and still compares unfavourably with the observations which suggest values around 10-12% (Shuttleworth, 1988a). The surface variables of the SCM with this scheme are given in table 2 (WJS1), together with the observations and the original scheme as developed by Warrilow et al (1986). The latter scheme assumes throughfall to occur before full canopy saturation is reached. Although the scheme appears to perform better than WJS1 it does so because it effectively allows rainfall to

Table 2. Surface variables in the SCM experiments.

	Observed	Observed	WJS1	WJS2
precipitation (mm)	2593	2425	2434	2308
interception loss (mm)	373	796	790	233
transpiration (mm)	1020	460	480	879
net radiation (w.m^{-2})	120	107	104	102
surface temperature (°C)		25.5	26.4	26.7

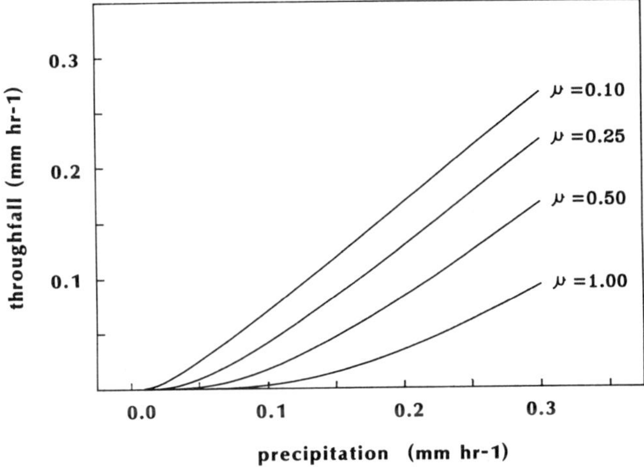

Fig. 2: Dependence of modelled grid average throughfall on μ, C_m =0.35.

run off from unsaturated canopies. Shuttleworth (1988a) reports a substantial sensitivity of modelled interception loss to the canopy storage capacity. The main reason for this is that, contrary to temperate forest interception where evaporation during the storm is important, most of the interception loss in rainforests is made up by evaporation after the rainfall has ceased. Changing the value of S, the canopy capacity from 2.5 to 0.7 mm hardly affected the modelled interception loss in the original scheme.

Fig. 3 shows the predicted frequency distribution of rainfall with time of day at Reserva Ducke, Manaus, compared with observations reported by Lloyd (1990). The dominant convective nature of Amazonian rainfall, with showers occurring predominantly in the early afternoon, is reproduced quite well by the SCM. This suggests that the grid average rainfall is correct. Given the sensitivity of modelled interception loss to the exact value of μ

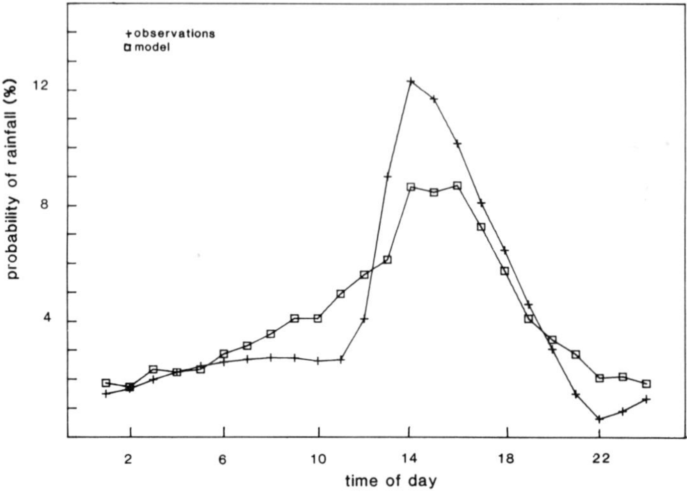

Fig. 3: Comparison of observed and predicted frequency distribution of rainfall.

(Fig. 2), and the variation in rainfall amount with time of day, it is quite likely that assuming a constant value of $\mu=0.3$ is not realistic. The Shuttleworth scheme can be extended by incorporating an explicit dependence of μ on the grid average rainfall rate according to

$$\mu = \frac{P}{P_r} \tag{5.3}$$

where P_r is an observed average rainfall rate for convective storms. From observations in the Reserva Ducke (Lloyd et al, 1988), P_r is found to be 5.15 mm hr-1. The results of this scheme (WJS2) are also given in Table 2. The result compare well with the observations. The calculated variation in μ is from 1% in the early morning to 15% in the afternoon, suggesting indeed that the original value of 0.3 was too high. The considerable variation furthermore suggests that its explicit dependence on rainfall rate adds a degree of realism to the land surface description. This is corroborated by the increased sensitivity of the modelled interception loss to a change in canopy capacity. A change in 50% in S results in a change of 27% in the modelled interception loss. Shuttleworth (1988a) reports a 25% change in interception loss due to a change of 50% in S.

5.2 *Dry canopy evaporation*

Dry canopy evaporation is modelled in most GCMs with a surface resistance which depends on the specific type of vegetation. Stand alone micrometeorological sensitivity tests usually show a strong dependence on the value of the surface resistance (Dolman et al, 1990). In Fig. 4 the sensitivity of the SCM dry canopy evaporation to the value of surface resistance is shown. Surprisingly the sensitivity is very small and the importance of the surface resistance in regulating water loss is decreased in favour of the overlying meteorological conditions.

The reduced sensitivity results from the interaction of the evaporation rate with the atmospheric boundary layer through

$$\frac{\partial q}{\partial t} = \frac{E}{\rho h} \tag{5.4}$$

which shows that the rate of change in specific humidity, q, depends on the evaporation rate, E, and the height of the boundary layer. Any increase in surface resistance and hence potential decrease in evaporation will result in a decrease in humidity and increase in surface temperature (and corresponding non-linear increase in saturated specific humidity at the surface). The resulting gradient will then be a balance between the decrease in lowest layer humidity and increase in surface saturated humidity through an increase in surface temperature.

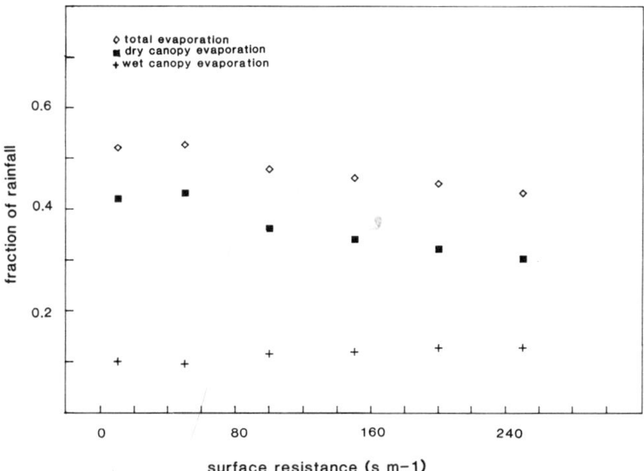

Fig. 4: Variation of predicted transpiration in August with r_s.

The present boundary layer model has a fixed boundary layer height and thus tends to present a strong feedback of the boundary layer to surface evaporation. Neglecting the sink of humidity caused by moist convection, it is essentially a closed box. Such a model tends to produce evaporation rates mainly dependent on the available energy. The relative insensitivity to changes in surface resistance suggest a similar phenomenon may be operating in the SCM.

6. Discussion

It has been shown that the incorporation of a subgrid parameterization to allow for spatially varying surface coverage of convective rainfall improves the partitioning of evaporation between wet and dry canopy evaporation. Nevertheless, the original models produced the correct total evaporation. The primary cause for this that on a larger scale the evaporation will always be constrained by the available energy. Average net radiation in the SCM varies around 105 $w.m^{-2}$, Lean and Warrilow report a figure of 147 $w.m^{-2}$. Observed net radiation is about 120 $w.m^{-2}$. As observations show that 90% of the available energy in the Amazon Basin is used to recycle rainfall through wet and dry canopy evaporation (Shuttleworth, 1988a), a correct prediction of net radiation is an obvious first requisite to be able to predict the total evaporation.

The SCM showed a remarkable low sensitivity to changes in the surface resistance. Boundary layer modelling experiments with more complex models show such a moderation to be likely, although not on the scale as observed in the SCM. It remains to be explored in further experiments whether an increased resolution and a varying boundary layer height will show a more pronounced response to changes in surface resistance. Furthermore, the neglect of entrainment at the top of the boundary layer may give a much lower sensitivity to the specification of surface resistance.

The SCM has however proved to be a very useful model in the development of new land surface parameterizations for use in GCMs. It reproduces clearly the large scale phenomena very well and the fact that it interacts with the surface in a way that micrometeorological models cannot, makes it very suitable for the purpose of developing new land surface parameterizations in GCMs.

7. Acknowledgements

The author holds a CEGB Senior Research Fellowship. The help and assistance of colleagues of the Dynamical Climatology Branch of the

Meteorological Office in transferring and implementing the Single Column Model is gratefully acknowledged.

8. References

Charney, J., Quirk, W.J., Chow, S., and Kornfield, J. (1977). A comparative study of the effects on albedo change on drought in semi-arid regions. Journal of the Atmospheric Sciences, 34:1366-1385.

De Bruin, H.A.R. (1983). A model for the Priestly-Taylor parameter **α**. Journal of Climate and Applied Meteorology, 22:572-578.

Dickinson, R.E. (1984). Modelling evapotranspiration for three dimensional climate models. In: Climate Processes and Climate Sensitivity, Geophysical Monogr., N. 29, J.E. Hansen and T. Takahashi, editors., pp.58-72. American Geophysical Union, Washington.

Dickinson, R.E. and Henderson-Sellers, A. (1988). Modelling tropical deforestation: a study of GCM land-surface parametrizations. Quarterly Journal of the Royal Meteorological Society, 114:439-462.

Dolman, A.J., Gash, J.H.C., Roberts, J.M., and Shuttleworth, W.J. (1990). Stomatal and surface conductance of tropical rainforest. Agricultural and Forest Meteorology, in press.

GARP. (1975). The physical basis of climate and climate modelling. GARP Publication Series No. 16, WMO/ICSU, Geneva.

Lean, J. and Warrilow, D.A. (1989). Simulation of the regional impact of Amazon deforestation. Nature, 342:411-413.

Lloyd, C.R., (1990). The temporal distribution of Amazonian rainfall and its implications for forest interception. Quarterly Journal of the Royal Meteorological Society, in press.

McNaughton, K.G. and Spriggs, T.W. (1986). A mixed-layer model for regional evaporation. Boundary-layer Meteorology, 34:243-262.

Mintz, Y. (1984). The sensitivity of numerically simulated climates to land surface boundary conditions. In: The global climate, J.T. Houghton, editor., pp.79-106., C.U.P. Cambridge.

Noilhan, J. and Planton, S. (1989). A simple parametrization of land surface processes for meteorological models. Monthly Weather Review, 117:536-549.

Oort, A.H. (1983). Global atmospheric circulation statistics, 1958-1973. NOAA Professional paper, 14. NOAA, Washington.

Pinty, J.-P., Mascart, P., Richard, E., and Rosset, R. (1989). An investigation of mesoscale flows induced by vegetation inhomogeneities using an evapotranspiration model calibrated against HAPEX-MOBILHY data. Journal of Applied Meteorology, 28:976-992.

Rowntree, P.R. (1986). Review of General Circulation Models as a basis for predicting the effects of vegetation change on climate. In Forests, climate and hydrology: regional impacts, E.R.C. Reynolds and F.B. Thompson, editors, pp.162-193. Kefford Press, Singapore.

Sato, N., Sellers, P.J., Randall, D.A., Schneider, E.K., Shukla, J., Kinter, J. L., Hou, Y.-T., and Albertazzi, E. (1989). Effects of implementing the Simple Biosphere model in a General Circulation Model. Journal of the Atmospheric Sciences, 46:2757-2782.

Sellers, P.J., Mintz, Y., Sud, Y.C., and Dalcher, A. (1986). A simple biosphere model (SiB) for use within general circulation models. Journal of the Atmospheric Sciences, 43:505-531.

Sellers, P.J., Shuttleworth, W.J., Dorman, J.L., Dalcher, A., and Roberts, J. M. (1989). Calibrating the Simple Biosphere Model for Amazonian tropical forest using field and remote sensing data. Part I: Average calibration with field data. Journal of Applied Meteorology, 28:727-759.

Shukla, J. and Mintz, Y. (1982). Influence of land surface evaporation on the Earth's climate. Science, 215:1498-1500.

Shukla, J., Nobre, C. and Sellers, P. (1990). Amazon deforestation and climate change. Science, 247:1322-1325.

Shuttleworth, W.J. (1988a). Evaporation from Amazonian rainforest. Proceedings of the Royal Society of London series B, 233:321-346.

Shuttleworth, W.J. (1988b). Macrohydrology - The new challenge for process hydrology. Journal of Hydrology, 100:31-56.

Sud, Y.C. and Smith, W.E. (1985). The influence of surface roughness of deserts on the July circulation. Boundary Layer Meteorology, 33:15-49.

Sud, Y.C., Shukla, J., and Mintz, Y. (1988). Influence of land surface roughness on atmospheric circulation and precipitation: a sensitivity study with a general circulation model. Journal of Applied Meteorology, 27:1036-1054.

Warrilow, D.A., Sangster, A.B., and Slingo, A. (1986). Modelling of land surface processes and their influence in European climate. DCTN 38, Bracknell, 92pp.

Warrilow, D.A. and Buckley, E. (1989). The impact of land surface processes on the moisture budget of a climate model. Annales Geophysicae, 7:439-450.

Modelling seismic wave propagation in a cracked crust

S. Crampin

British Geological Survey, Edinburgh

1. Introduction

There are two types of seismic body-wave that can propagate through elastic solids: *P*-waves with longitudinal particle vibrations, and shear waves with vibrations transverse to the direction of travel. Until very recently, almost all seismic analysis was based on *P*-waves, and the analysis of *P*-wave travel times recorded on vertical seismometers has been enormously successful in determining the overall structure of the Earth's crust (in finding oil, for example) by assuming that the blocks and layers making up the crust are isotropic. Shear waves in such isotropic rocks would have a constant velocity independent of direction of travel and polarization of the wave. In the past, the complicated wave motion of shear waves in the crust was assumed to be caused by scattering at (unspecified) inhomogeneities. As a consequence of this apparent complexity shear waves were usually neglected.

Recent technological advances, allowing seismic waves to be recorded on three-component instruments at high digital sampling rates and adequate facilities, show that instead of complexity shear waves display remarkable alignment with the local stress field. Almost all shear waves (with a few well-understood restrictions) show shear-wave splitting, where the leading split shear-wave is polarized parallel to the direction of the local compressive stress within the Earth. Such splitting is seen at the surface above small earthquakes (Crampin, 1987a), in oil company reflection surveys (Alford, 1986) where reflections from a surface source are recorded by surface recorders, and in vertical seismic profiles (VSPs) (Crampin et al, 1986) where a surface source is recorded by downwell recorders.

The major restriction on such observations is that the shear waves need to be recorded either subsurface, or at angles of incidence within the critical angle of about 35° at the surface (Booth and Crampin, 1985). Shear waves recorded at the surface outside this *shear-wave window* are severely distorted by interactions with the free surface.

Shear-wave splitting is diagnostic of some form of effective seismic anisotropy (Crampin, 1981), and the splitting in the crust is the result of

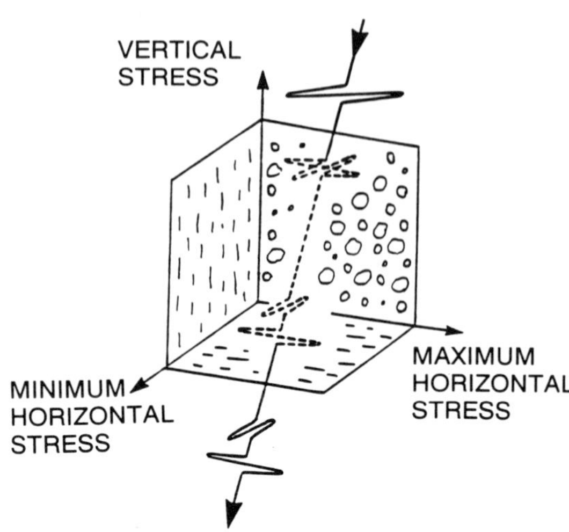

Fig. 1: Schematic illustration of shear-wave splitting in the parallel vertical stress-aligned EDA-cracks pervading the uppermost 10 to 20 km of the crust.

propagation through the distributions of fluid-filled inclusions that are known to exist in most rocks. These inclusions are the most compliant elements of the rockmass and become aligned by the regional stress-field into, typically, parallel vertical orientations, perpendicular to the direction of minimum horizontal stress. These are effectively anisotropic to seismic wave propagation and cause the observed shear-wave splitting. The distributions of aligned inclusions are known as *extensive-dilatancy anisotropy* or *EDA* (Crampin et al, 1984). The inclusions themselves are known as *EDA-cracks* because, although the inclusions may have a wide variety of physical shapes, many of the seismic effects can be modelled by distributions of thin flat parallel cracks.

Fig. 1 shows a schematic illustration of shear-wave splitting through EDA-cracks. On entering a cracked region, a shear wave, with transverse particle motion, splits into two phases with displacements parallel and perpendicular to the parallel cracks. The component polarized parallel to the cracks meets less impedance and travels faster and is less attenuated than the component polarized perpendicular to the crack faces. Such splitting inserts characteristic signatures into the wavetrain, which are now recognized in shear waves propagating through almost all rocks in the uppermost 10 to 20km of the Earth's crust. Fig. 1 also illustrates that for shear-waves propagating within about 30° or 40° of the vertical, the leading split shear-wave is polarized parallel to the strike of the cracks and parallel to the direction of maximum compressional stress.

2. General principles

Wave propagation in anisotropic solids is fundamentally different from propagation in isotropic solids and, except in isolated symmetry directions, the mathematics is complex and algebraically unaccommodating. This means that numerical modelling with computers is essential for even quite trivial investigations of wave motion in anisotropic and cracked rocks. Consequently, these advances to our understanding of shear waves have come about almost entirely because of the ability to model propagation in cracked rock numerically with computers and, in particular, to match the waveforms of observed shear waves with synthetic seismograms.

Much of the theory and many of the computer programs for calculating wave propagation in layered anisotropic models had already been developed over the previous 20 years (Crampin, 1981). Thus, when the first observations of shear-wave splitting in the crust were made above small earthquakes in Turkey (Crampin et al, 1985), suitable analysis and experience of numerical modelling of anisotropy already existed. This has meant that progress in this new branch of seismology has been particularly rapid. A major application has been recognized in investigating the internal structure of hydrocarbon reservoirs, and much of the work on seismic anisotropy is currently supported by oil and service companies.

This development has several rather unusual features. Wave propagation in anisotropic solids is quite as regular and orderly as wave propagation in isotropic solids but is fundamentally different. The behaviour is controlled by different phenomena, and requires totally different techniques to understand and evaluate the behaviour. The human mind, accustomed as it is to viewing pictures in two dimensions, has difficulties in visualizing the three-dimensional behaviour characteristic of anisotropy. Experiments in anisotropic solids are difficult to set up in the laboratory, so that almost all our understanding of the phenomena has come from numerical investigations. The computer programs have been used as laboratory tools. Particular phenomena are investigated by repeated numerical experiments until the behaviour can be understood. Although the principles of the phenomena can be usually identified analytically, understanding the detailed behaviour is the result of wide experience with such numerical experimentation.

Modelling has revealed a seemingly complicated space spanned by up to 21 elastic constants, but possessing such regular internal features, that many

(a) TRICLINIC (21)

$$\begin{array}{cccccc}
c_{1111} & c_{1122} & c_{1133} & c_{1123} & c_{1131} & c_{1112} \\
c_{1122} & c_{2222} & c_{2233} & c_{2223} & c_{2231} & c_{2212} \\
c_{1133} & c_{2233} & c_{3333} & c_{3323} & c_{3331} & c_{3312} \\
c_{1123} & c_{2223} & c_{3323} & c_{2323} & c_{2331} & c_{2312} \\
c_{1131} & c_{2231} & c_{3331} & c_{2331} & c_{3131} & c_{3112} \\
c_{1112} & c_{2212} & c_{3312} & c_{2312} & c_{3112} & c_{1212}
\end{array}$$

(b) MONOCLINIC (13)

$$\begin{array}{cccccc}
a & b & c & . & . & d \\
b & e & f & . & . & g \\
c & f & h & . & . & i \\
. & . & . & j & k & . \\
. & . & . & k & m & . \\
d & g & i & . & . & m
\end{array}$$

(c) ORTHORHOMBIC (9)

$$\begin{array}{cccccc}
a & b & c & . & . & . \\
b & d & e & . & . & . \\
c & e & f & . & . & . \\
. & . & . & g & . & . \\
. & . & . & . & h & . \\
. & . & . & . & . & i
\end{array}$$

(d) TRIGONAL (6)

$$\begin{array}{cccccc}
a & b & c & d & . & . \\
b & a & c & -d & . & . \\
c & c & e & . & . & . \\
d & -d & . & f & . & . \\
. & . & . & . & f & d \\
. & . & . & . & d & x
\end{array}$$

where $x = (a-b)/2$

(e) TETRAGONAL (6)

$$\begin{array}{cccccc}
a & b & c & . & . & . \\
b & a & c & . & . & . \\
c & c & d & . & . & . \\
. & . & . & e & . & . \\
. & . & . & . & e & . \\
. & . & . & . & . & f
\end{array}$$

(f) CUBIC (3)

$$\begin{array}{cccccc}
a & b & b & . & . & . \\
b & a & b & . & . & . \\
b & b & a & . & . & . \\
. & . & . & c & . & . \\
. & . & . & . & c & . \\
. & . & . & . & . & c
\end{array}$$

(g) HEXAGONAL (5)

$$\begin{array}{cccccc}
a & b & c & . & . & . \\
b & a & c & . & . & . \\
c & c & d & . & . & . \\
. & . & . & e & . & . \\
. & . & . & . & e & . \\
. & . & . & . & . & x
\end{array}$$

where $x = (a-b)/2$

(h) ISOTROPIC (2)

$$\begin{array}{cccccc}
a & b & b & . & . & . \\
b & a & b & . & . & . \\
b & b & a & . & . & . \\
. & . & . & x & . & . \\
. & . & . & . & x & . \\
. & . & . & . & . & x
\end{array}$$

where $x = (a-b)/2$

Fig. 2: (a) Usual representation of the symmetric fourth-order tensor of elastic constants. If x_1, x_2 and x_3 are the principal axes and the c_{ijkl} are symmetric about the main diagonal, and has triclinic symmetry. (b) to (h) are the other seven classes of anisotropic symmetry with x_1, x_2 and x_3 as principal axes. The number of independent elastic constants is in brackets following the name.

of the problems have great generality, and often resolve themselves into comparatively straightforward concepts. This paper will try to outline some of the underlying principles that have guided our exploration of this anisotropic space.

3. Computing waves in uniform anisotropic solids

The basic mathematics of elastic anisotropy has been understood since Love (1892). The elastic constants can be written as a fourth rank tensor, $\{c_{ijkl}\}$, where i,j,k,l = 1,2,3. The plane $x_m=0$, say, in such a tensor possesses mirror symmetry if $c_{ijkl} = 0$ whenever one or three of i, j, k, l are equal to m. Repeated applications of this symmetry condition demonstrates that there are just eight classes of anisotropic symmetry (with a few minor variations) involving up to 21 independent elastic constants. The elastic constants of these eight classes are given in Fig. 2, and the corresponding arrangements of symmetry planes in Fig. 3. Each anisotropic solid may have any orientation in space, and as soon as one of the simple structures in the symmetry classes in Fig. 2 is rotated to a new coordinate frame, the simple arrangement in Fig. 2 fills up and there will be 21 (not independent) elastic constants. The classic technique for deriving the velocities and polarizations of body-waves in particular directions in such anisotropic solids was by inserting appropriate direction cosines into the Kelvin-Christoffel equations (Musgrave, 1970). These yield unwieldy expressions, even for planes of mirror symmetry, and are not easy to evaluate.

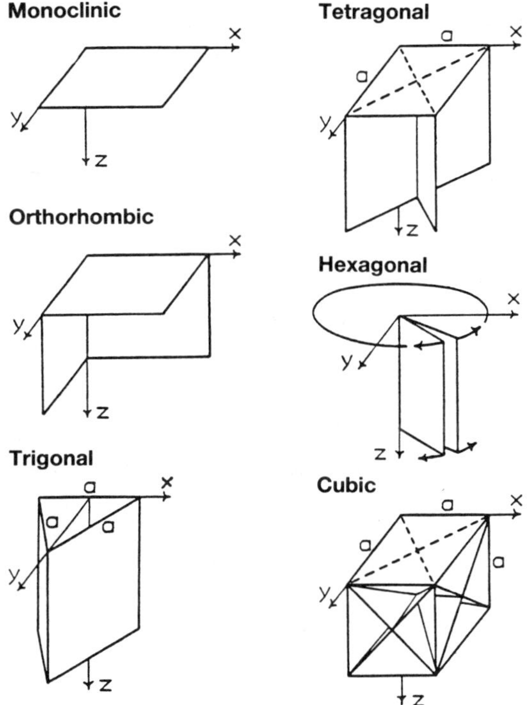

Fig. 3: Schematic illustration of the arrangement of symmetry planes of six of the eight main classes of anisotropic symmetry. The other two classes are triclinic symmetry with no symmetry planes, and isotropy with every plane a symmetry plane.

There are four procedures which have helped to resolve many of the complications of numerical computations in anisotropic models:

(a) Rotate elastic constants of each layer so that propagation is always in a fixed coordinate frame. The tensors are rotated into the coordinates of the model, so that propagation is always in the x_1 direction for investigating the properties of a homogeneous anisotropic solid, and in the $x_2=0$ plane (the sagittal plane), with the x_3-axis vertical, for multi-layered models. This use of a fixed coordinate system immediately generalizes any problem at the negligible cost (for computers) of an initial rotation of each anisotropic tensor into the fixed coordinates of the model.

First used by Crampin (1970) this simple concept of rotating tensors is the key to opening anisotropic models to computer calculations. Each program is independent of the orientation and symmetry of the particular anisotropic solid and independent of the direction of propagation through a plane layered model. With previous algebraic formulations, the complexity of any problem was dominated by the number of elastic constants in the particular class of anisotropic symmetry. Once the fixed coordinate system is adopted, the problems of the particular class of symmetry are eliminated and the overall similarities of the behaviour can be recognized.

(b) Use propagator matrices to transform boundary values between successive interfaces n, and n+1:

$$(u_1, u_2, u_3\ \sigma_{13}, \sigma_{23}, \sigma_{33})^{n+1} = D(\omega)\ (u_1, u_2, u_3, \sigma_{13}, \sigma_{23}, \sigma_{33})^{n}$$

where u_1, u_2, u_3, are the displacements; σ_{13}, σ_{23}, σ_{33} are the tangential and normal stresses at the (n+1)th and nth interfaces; and $D(\omega)$ is a 6x6 matrix dependent on the angular frequency ω, which is complex for anisotropic layers (and usually requires to be double length, or equivalent for mathematical precision). The use of propagator matrices for transforming the continuity of boundary values between plane interfaces was first used for isotropic layers by Haskell (1953). In isotropy, the motion in the sagittal plane (the plane containing *P*- and *SV*-waves) yields a real 4x4 propagator matrix which decouples from the motion in the horizontal transverse direction (*SH*-waves), which yields a real 2x2 matrix. Crampin (1970) first extended these propagator matrices to anisotropic elastic solids, yielding a complex 6x6 matrix. This was later extended to multi-layered piezoelectric media yielding a complex 8x8 matrix (Armstrong and Crampin, 1972).

(c) Use various modifications of the anisotropic reflectivity technique (Booth and Crampin, 1985; Taylor, 1987) to calculate fullwave synthetic seismograms in plane layered models, where the source is modelled by appropriate plane-wave decompositions using Bessel functions. Such fullwave solutions have been particularly important for investigating new phenomena. When we do not know exactly what to expect, *fullwave* solutions give confidence in interpreting the results. Other techniques have been and are being developed for investigating three-dimensional models, but there are as yet unresolved difficulties.

One of the problems in anisotropic models is that the energy deviates from the direction of phase propagation. In weak anisotropy, this deviation may be negligible, but in strong anisotropy, the deviation may be significant. Such deviations may be evaluated by integrating over both angle of incidence and angle of azimuth. This can be done but is computationally expensive. A method, which is computationally cheaper, is to integrate over a narrow cone of directions about the ray direction with one of the raytracing techniques. However, these are not fullwave solutions, and have not yet been adapted to cope with shear-wave *singularities* (directions in which the two split shear-waves have identical phase velocities, Crampin, 1990b). Such singularities commonly occur in directions near vertical propagation in sedimentary basins (Bush and Crampin, 1987a, 1990). Thus, raytracing techniques could give misleading results in one of the most important applications: interpreting shear-wave reflection surveys and vertical seismic profiles in sedimentary basins to monitor the internal structure of hydrocarbon reservoirs.

(d) Use complex elastic constants to model anisotropic attenuation: $c_{ijkl} = c^R_{ijkl} + c^I_{ijkl}$; where c^R_{ijkl} are real elastic constants modelling the velocity anisotropy (and shear-wave splitting); and c^I_{ijkl} are imaginary elastic constants modelling the attenuation anisotropy. This step allows modelling of both elastic and attenuation anisotropy (Crampin, 1981). Such complex elastic constants have played principally a cosmetic role in existing calculations, but are expected to be increasingly important in future investigations.

Note that these procedures are effective for weak anisotropy as is found in the Earth (velocities varying with direction by less than 10%, say). The reality of many of these concepts of wave propagation in strong anisotropy is questionable.

4. Propagation in a cracked solid

The behaviour of seismic waves in a cracked rock can be simulated by propagation through a purely elastic solid that has the same variation with direction of the velocities and attenuation as the cracked rock (Crampin, 1978). Such solids can be specified by a set of effective (complex) elastic constants (Crampin, 1984) based on the theoretical formulations of Hudson (1980, 1981). Fig. 4 shows the velocity variations through a solid containing

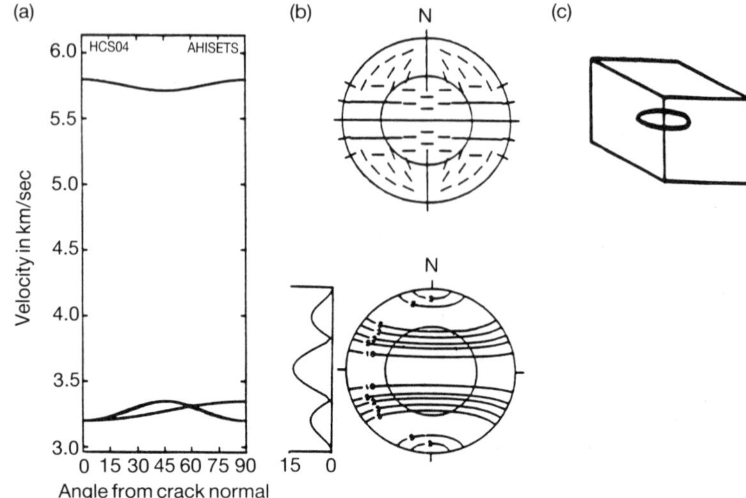

Fig. 4: The behaviour of the three body-waves, a quasi P-wave, qP, and two quasi shear-waves, qSP and qSR, in a distribution of parallel vertical liquid-filled EDA-cracks with crack density $CD = N\,a^3/V = 0.04$, where N is the number of cracks of radius a in volume V. qSP is polarized (P)arallel and qSR at (R)ight angles to the plane through the crack normals.
(a) Velocities in a quadrant of directions from perpendicular (0°) to parallel (90°) to the cracks.
(b) Equal-area projections (polar maps) over a hemisphere of directions showing the three-dimensional variations of split shear-waves through EDA-cracks striking east-west: upper diagram, polarizations of the faster split shear-wave; lower diagram, delays between the split shear-waves normalized to ms/km, with a north-south section to the left. The inner circle marks the critical angle within which shear waves are undistorted at the free surface.
(c) Schematic illustration of the effective distribution of cracks with crack density CD = 0.04: note that the velocity anisotropy is independent of the dimensions of the cracks, as long as they are much smaller than the seismic wavelengths.

parallel liquid-filled cracks with a crack density of *CD* = 0.04. This gives shear-wave polarizations parallel to the direction of maximum horizontal compressional stress as now widely observed. The differential shear-wave anisotropy of about 4% in Fig. 4 is typical of observed shear-wave splitting

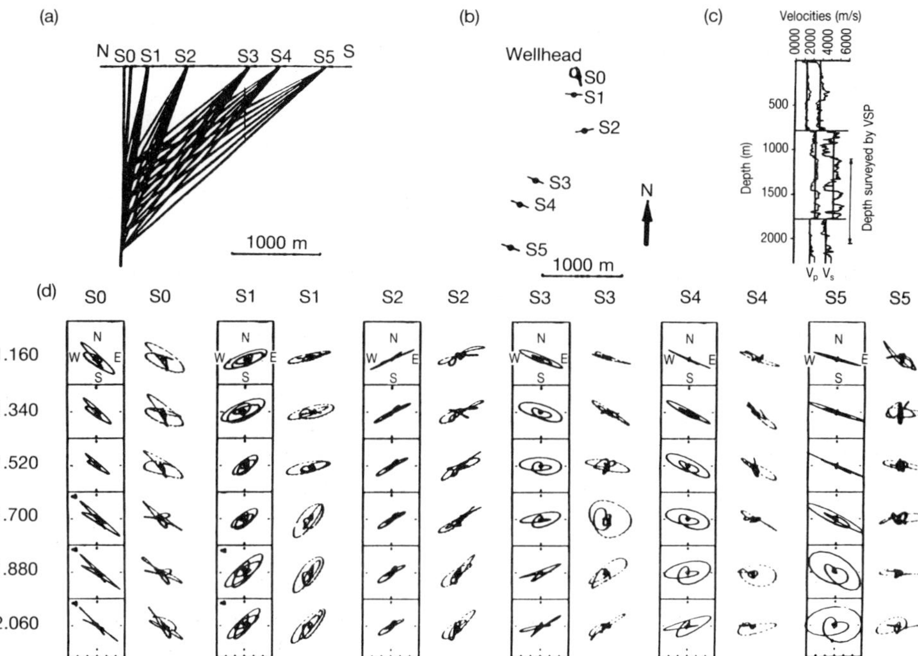

Fig. 5: Modelling multi-offset VSPs in the Paris Basin. (a) Cross-section of (straight-line) raypaths; (b) Plan of offsets; (c) Well log showing V_p and V_s velocity variations down the well; and (d) Polarization diagrams at 180m intervals down the well for six offsets, S0 to S5. Open columns are observed polarization diagrams, and boxed columns are polarization diagrams of synthetic seismograms calculated with the ANISEIS program package. (After Bush and Crampin, 1987, 1990).

in a wide variety of crustal rocks. Note that *P*-waves show very little anisotropy for propagation through thin parallel liquid-filled cracks (Fig. 4) which is the reason why *P*-waves in the Earth's crust can be successfully interpreted by assuming isotropic structures, whereas shear-waves display shear-wave splitting and require anisotropy for their interpretation.

5. Matching synthetic to observed seismograms

Although shear waves observed at the surface typically show the parallel polarizations expected of propagation through parallel vertical EDA-cracks and appropriate synthetic seismograms display similar patterns of polarization, the observed polarization patterns can seldom be modelled exactly. This is because of the complicated interactions of shear waves with the free surface (Booth and Crampin, 1985), where shear waves are also disturbed by near-surface low-velocity layers and topographic irregularities. Shear waves recorded subsurface in VSPs, however, can be matched very accurately. Note that most VSPs are in sedimentary basins, where there are two sources of anisotropy with different symmetry orientations: fine horizontal layering; and vertical EDA-cracks. These yield an orthorhombic symmetry system where the shear-wave polarizations are no longer wholly parallel to the strike of the EDA-cracks as they are in most other geological strata.

Fig. 5 shows a comparison of shear wave polarization patterns in a multi-offset VSP in the Paris Basin (Bush and Crampin, 1987, 1990). The synthetic patterns, in the boxed columns, are a very good match to the observed patterns, in the open columns. These patterns depend on the detailed amplitude and phase correlation of the three-dimensional waveforms. They are very sensitive to the internal structure along the raypath, and the quality of the match places very tight constraints on possible structures. In particular, it places constraints on the combination of the fine layering typical of sedimentary basins (Fig. 5c), the orientation of the cracks, parallel to the stress (N30°W), and the crack density and aspect ratio (separation of the faces) of the cracks. Such parameters are likely to be increasingly important for developing hydrocarbon reservoirs, and analyzing shear-wave splitting is a technique that can place tight constraints on the combination of parameters and in principle can give great accuracy.

Such techniques are likely to be increasingly important in the future, particularly for monitoring the progress of enhanced oil recovery (EOR). Present recovery techniques are generally able to extract less than 50% of the oil from a reservoir. There are several processes that may lead to higher recover and the ability to monitor such processes in deep inaccessible rocks by seismic techniques, of which monitoring shear-wave splitting is one, is currently exciting much interest.

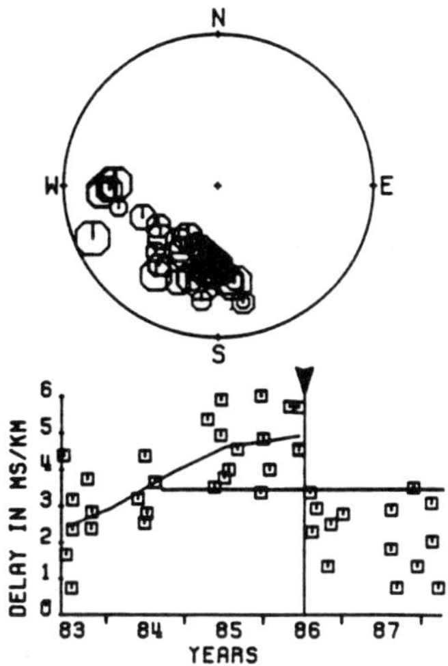

Fig. 6: Upper diagram: equal-area projection out to the critical angle of the delays between the split shear-waves from small earthquakes over five years to August 1988 in a range of directions to station KNW in the Anza network in Southern California. The diameter of the circles are proportional to the magnitude of the delay. Lower diagram: variation with time of the delay between the split shear-waves normalized to ms/km. The arrow marks the time of the M=6 North Palm Springs, 8th July 1986, 33km from KNW. (After Crampin et al, 1990).

6. Temporal changes in shear-wave splitting

The fluid-filled EDA-cracks are the most compliant elements of the rockmass, and shear waves are very sensitive to the details of the crack geometry (Crampin, 1987a). Consequently, when the conditions on the rockmass change in any way, either internally (as in EOR) or externally (as in the build up of stresses before earthquakes), the crack parameters and hence the behaviour of shear waves propagating through the cracked rock will be modified. Changes in splitting during EOR have not yet been identified (or at least not presented in the open literature), but changes in splitting before and after earthquakes, and before and after hydraulic pumping and other phenomena, have been reported.

Fig. 6 shows the variations in the delay between the split shear-waves over five years in a range of directions to a recording site near Anza, Southern California (Crampin et al, 1990). There is an abrupt change in the values of the delays near the time of the large earthquake, marked by arrow. Such changes can be simulated, to the first order, by an increase in the aspect ratio (elastic "bowing") of the EDA-cracks as the stress builds up, and then a subsequent decrease when the cracks relax after the stress is released by the earthquake (Crampin et al, 1990). Similar effects are observed in the laboratory when specimens are stressed until failure.

If this interpretation is correct, and the estimated effects in Fig. 6 are real, this would enable the changes of stress before earthquakes to be monitored in detail by analyzing shear-wave splitting. This would be an important step towards deterministic earthquake prediction.

7. Conclusions

It must be stated that neither the observations nor the interpretation of the phenomena reported above are yet wholly accepted by all the seismological community. It is clear, however, that shear waves contain much more information than *P*-waves and we are beginning to match synthetic to observed shear wave seismograms with a high degree of precision, and interpret them in terms of parameters of the rockmass that, in general, cannot be obtained in any other way (Crampin, 1987a). This development has come about at this time because recording and analysis technology has advanced enough for the phenomena to be observed, and the interpretation and understanding of the phenomena is based, and continues to depend, on numerical experiments with computer programs.

Shear-wave splitting depends on stress and cracks which are crucial factors whenever we penetrate the Earth or are interested in the dynamic response of the Earth. Consequently, there are many important applications, ranging from extracting hydrocarbons and exploiting geothermal heat to, possibly, earthquake prediction. We certainly do not know all the answers yet. These investigations have been a voyage through uncharted waters with treasure islands (and some reefs) that continue to surprise us along the way, but it is leading to what is probably the most fundamental advance in seismology for several decades (Crampin, 1987a).

8. Acknowledgements

This work was supported by the Natural Environment Research Council and is published with the approval of the Director of the British Geological Survey (NERC).

9. References

Alford, R.M. (1986). Shear data in the presence of azimuthal anisotropy: Dilley, Texas. Expanded Abstracts, 56th Annual International SEG Meeting, Houston, 1986, pp.476-479.

Armstrong, G.A. and Crampin, S. (1972). Piezoelectric surface-wave calculations in multilayered anisotropic media. Electronic Letters, 8:521-522.

Booth, D.C. and Crampin, S. (1985). Shear-wave polarizations on a curved wavefront at an isotropic free-surface. Geophysical Journal of the Royal Astronomical Society, 83:31-45.

Bush, I. and Crampin, S. (1987). Observations of EDA and PTL anisotropy in shear-wave VSPs. Expanded Abstracts, 57th Annual International SEG Meeting, New Orleans, 1987, pp.646-649.

Bush, I. and Crampin, S. (1990). Paris Basin VSPs: Case history establishing combinations of fine-layering and crack anisotropies for the first time. Geophysics (in preparation).

Crampin, S. (1970). The dispersion of surface waves in multilayered anisotropic media. Geophysical Journal of the Royal Astronomical Society, 21:387-402.

Crampin, S. (1978). Seismic wave propagation through a cracked solid: polarization as a possible dilatancy diagnostic, Geophysical Journal of the Royal Astronomical Society, 53:467-496.

Crampin, S. (1981). A review of wave motion in anisotropic and cracked elastic-media. Wave Motion, 3:343-391.

Crampin, S. (1984). Effective elastic-constants for wave propagation through cracked solids. Geophysical Journal of the Royal Astronomical Society, 76:135-145.

Crampin, S. (1987a). Geological and industrial implications of extensive-dilatancy anisotropy. Nature, 328:491-496.

Crampin, S. (1987b). The basis for earthquake prediction. Geophysical Journal of the Royal Astronomical Society, 91:331-347.

Crampin, S. (1990a). Modelling a cracked and anisotropic Earth. Bulletin of the Institute of Mathematics and its Applications, in press.

Crampin, S. (1990b). Effects of singularities on shear-wave propagation in sedimentary basins, Geophysical Journal International, submitted.

Crampin, S., Booth, D.C., Evans, R., Peacock, S. and Fletcher, J.B. (1990). Changes in shear-wave splitting near the time of the North Palm Springs earthquake. Journal of Geophysical Research, 95:11197-11212.

Crampin, S., Bush, I., Naville, C. and Taylor, D.B. (1986). Estimating the internal structure of reservoirs with shear-wave VSPs. The Leading Edge, 34(11):94-99.

Crampin, S., Evans, R. and Atkinson, B.K. (1984). Earthquake prediction: a new physical basis, Geophysical Journal of the Royal Astronomical Society, 76:147-156.

Crampin, S., Evans, R. and Ucer, S.B. (1985). Analysis of records of local earthquakes: the Turkish Dilatancy Projects. Geophysical Journal of the Royal Astronomical Society, 83:1-16.

Haskell, N.A. (1953). The dispersion of surface waves in multilayered media. Bulletin of the Seismological Society of America, 43:17-34.

Hudson, J.A. (1980). Overall properties of a cracked solid. Mathematical Proceedings of the Cambridge Philosophical Society, 88:371-384.

Hudson, J.A. (1981). Wave speeds and attenuation of elastic waves in material containing cracks. Geophysical Journal of the Royal Astronomical Society, 64:133-150.

Love, A.E.H. (1892). A Treatise on the mathematical theory of elasticity. Cambridge University Press.

Musgrave, M.J.P. (1970). Crystal acoustics. Holden Day.

Taylor, D.B. (1987). Double contour integration for transmissions from point sources through anisotropic layers as used in ANISEIS software. Geophysical Journal of the Royal Astronomical Society, 91:373-381.

Computer modelling of the geometry of outcrop sandstone

M.C. Spearing and G.P. Matthews

Polytechnic South West

1. Introduction

Although this paper is concerned with the modelling of a particular outcrop rock, namely Clashac sandstone, the ideas outlined could be applied to any rock or indeed any porous medium whose pore size distribution is known. This model will, we hope, be used for gas diffusion/permeability studies in reservoir rocks, but could be used to look at other systems which depend on porous media, for example the storage of toxic and nuclear waste, infiltration of inks in paper or aquifer pollution.

In this paper we describe the ways in which we have used computing resources in the course of our research into the geometry of Clashac sandstone. Section 2 reports the setting up of a 3D geometrical model together with the simulation of characteristic properties of sandstones using the model, and the results obtained. Section 3 reports our initial efforts in using commercial operational research software as a step in the prediction of the permeability of the sandstone from its geometry. Section 4 briefly explores the possibilities of using a fractal analysis in the modelling of the intricate random structure of a sandstone.

2. Main Program

Our main aim is to produce a porescale geometric model of a real porous sandstone which can simulate characteristic properties of a porous medium, i.e. its porosity, surface area, tortuosity, pore connectivity and eventually its diffusion and permeability properties. The meanings of these terms will be described later. To this end we have written a program in Fortran 77 to produce a three-dimensional network of pores and throats, a pore being a void space within the solid rock and a throat being the constricted connection between two pores. The program is run on a Prime mainframe computer.

The model consists of a 10x10x10 network of such pores as shown in Fig. 1. The distribution of pore sizes is initially estimated from inspection of the mercury porosimetry data of the actual sandstone, and later refined as we

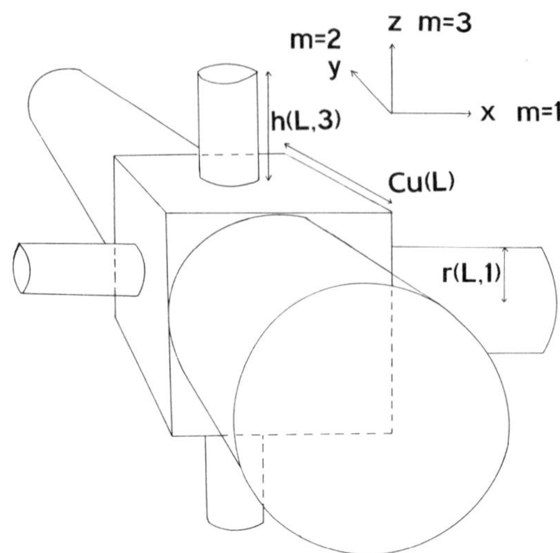

Fig. 1: One pore and its throats.

shall explain. Table 1 shows the pore size distributions (p.s.d.'s) used. These pore sizes are then distributed randomly within the network and assigned x, y and z coordinates. The effect of this is to produce a "cube" of rock which we call the unit cell. This is shown in Fig. 2 (p.s.d. 5). (This diagram was produced by SAMMIE - System for Aiding Man/Machine Interaction Evaluation - an interactive CAD system, combined with Prime MEDUSA). The largest pore in the unit cell is 180μm and the spacing between rows of pores is set greater than this so that no pores overlap. In Fig. 2, a pore row spacing of 200μm is used, so that the minimum throat length is 20μm. The size of the unit cell is therefore 10 x 200μm, or 2mm, in each direction. The unit cell is assumed to repeat infinitely in the x and y directions.

2.1 Blocked throats

Fig. 3 shows the top layer of the unit cell in 2D. It can be seen that not all the pores are connected. In fact in Fig. 3, 40% of the throats are omitted. By this method we can model the possible blockage of a throat by crystallisation of clay minerals or fines movement (Wojtanowicz et al, 1988). Any percentage blockage can be entered and this has the effect of altering the average connectivity of a pore and leads us to a method, described later, of calculating the average pore connectivity of the sample.

Table 1. *P.s.d's showing percentage of pores in each size range (total 1000 pores).*

throat diameter/microns	p.s.d. 5 / %	p.s.d. 12 / %	p.s.d. 13 / %	throat diameter/microns	p.s.d. 5/%	p.s.d. 12/%	p.s.d. 13/%
1	5	4	4	21	4	4	4
2	4	4	4	22	3	2	2
3	3	4	4	23	1	1	1
4	3	3	3	24	2	2	2
5	4	4	4	25	2	2	2
6	2	2	2	26	2	2	2
7	4	4	4	27	1	1	1
8	3	3	3	28	1	2	2
9	3	3	3	29	0	1	1
10	3	3	1	30	3	3	3
11	2	2	2	35	5	5	5
12	2	2	2	40	7	5	5
14	1	2	2	50	0	3	3
15	4	3	3	60	3	3	3
16	3	2	2	80	3	3	3
17	3	2	2	100	2	4	6
18	3	1	1	140	2	4	4
19	2	1	1	180	1	3	3
20	4	1	1	200	0	0	0

2.2 *Porosity and surface area*

The program then proceeds to calculate the porosity ϕ of the system, that is the ratio of void space in the sample to the overall volume of the sample. This correlates well with the experimentally measured porosity of 11% with the program run at a value of pore row spacing of 250μm with p.s.d. 12.

The internal surface area (S.A.) of the sample can also be predicted by the program but this turns out to be approximately 2 orders of magnitude less than the experimental value of 0.37 $m^2.g^{-1}$. Results of porosity and surface area are shown in Table 2. The reason for the disagreement in S.A. values is that the majority of the surface area of the sandstone is due to surface roughness and the surface area of clay minerals in the sample. Consequently the "smooth" surfaces of the pores and throats in the model lack any fine structure which makes up the bulk of the surface area.

Fig. 2: The unit cell.

We did try to model this fine structure by letting the surfaces of the pores and throats be of a zig-zag pattern, and by adjusting the ratio of the height (rl) to the width (rw) of the zig-zag, the experimental surface area could be matched. The ratio of rl/rw needed to match the experimental surface area is of the order of 50. However, the match has only been achieved by use of an empirical roughness factor, which though it has physical meaning, is of little further value. The way forward in the modelling of surface roughness is by using fractal analysis as described in Section 4.

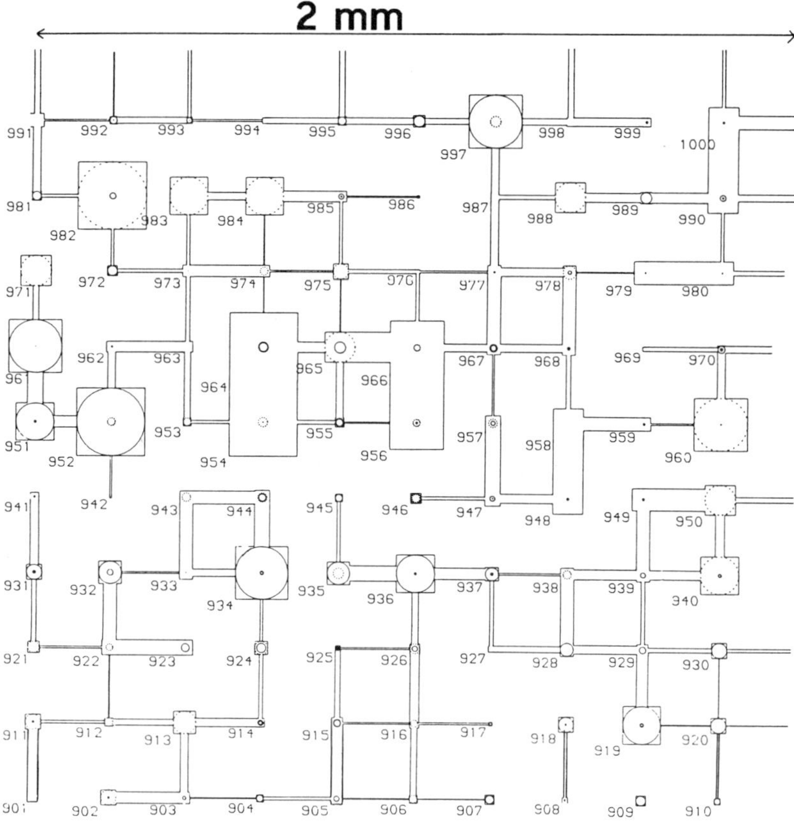

Fig. 3: 2D diagram of pores 901 - 1000. (Circles are throats: dashed denotes out of page).

2.3 *Tortuosity*

In the research of porous media, the one factor which has been used for half a century to fit data to predicted results is the tortuosity. Over the years it has been used to account for every possible rock heterogeneity and experimental failure (Van Brakel, 1975). We have given it a rigid definition based on an electrical conduction model so that the tortuosity, τ, is *the actual path length of charge carrying ion through medium divided by the length of straight path through medium.*

Using a random walk weighted by r^2/l (where r is the radius and l is the length of the throat pore system) theoretical conductivity experiments are carried out by the mainframe according to the flow diagram shown in Fig 4.

Table 2. Porosity and S.A. simulated values.

p.s.d. 12
pore row spacing = 250 microns

% B.T.	porosity/%	S.A./x10−3 sq.m/g
0	15.9	5.39
10	14.5	4.99
20	13.5	4.7
30	12.1	4.29
40	10.5	3.78
50	8.7	3.2
60	7.3	2.73
65	6.5	2.46
70	5.8	2.19

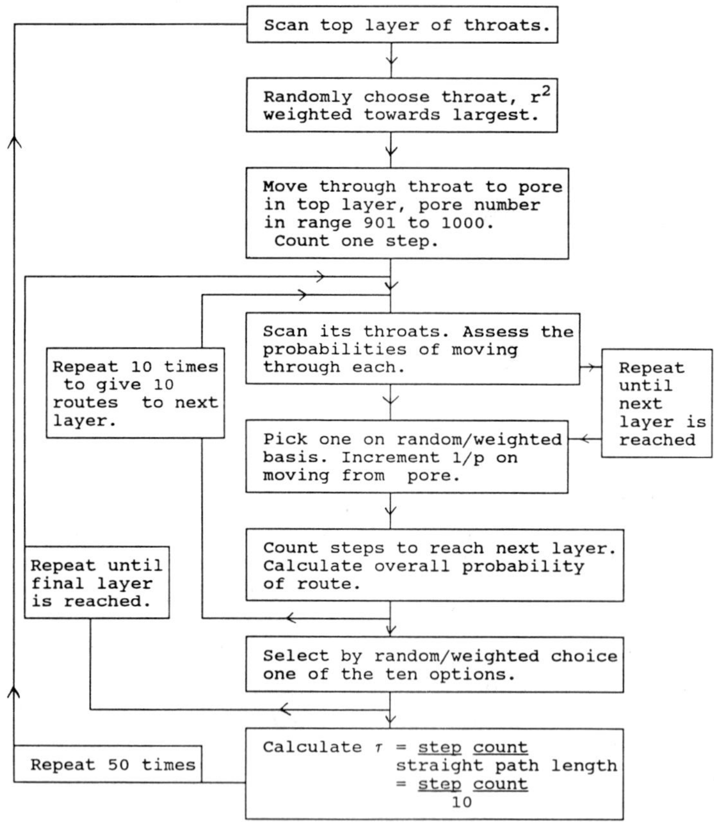

Fig.4: Flow diagram of tortuosity routine.

From 50 simulated values of tortuosity the actual tortuosity is defined as the median of the distribution with the interquartile range also quoted i.e. the values at 1/4 (Q1) and 3/4 (Q3) of the distribution. The tortuosity of the sandstone is found to vary with percentage blocked throats and these results are shown in Table 3. This then gives us a potential method of finding the average connectivity of a pore by matching an experimentally measured tortuosity to a simulated one. This should give us the same value as calculated from the mercury porosimetry simulation.

Table 3. Simulated tortuosity values.

p.s.d. 12
pore row spacing = 250 microns

%B.T.	median (tortuosity)	Q(1)	Q(3)
0	2.6	2.2	2.9
10	2.4	2.2	2.75
20	2.6	2.4	3.1
30	2.8	2.5	3.6
40	3	2.6	3.9
50	2.8	2.3	3.55
60	3.9	2.95	5.4
65	4.5	3.1	5.6
70	FINITE CLUSTERS		

2.4 *Application of the simulation to a model system*

The computer model can be set up to simulate the pores between an array of stacked solid spheres by setting all pore diameters to 2 units and the pore row spacing to 4.76 units. To test the tortuosity routine, it was first applied to this system of stacked spheres. The theoretical tortuosity of a simple cubic packed system of unconsolidated spheres is given by Chen (1973) as $\phi^{-1/4}$ and by Wyllie and Rose (1950) as $((3 - \phi)/2)^2$ which gives 1.2 and 1.6 respectively. The simulation gives a tortuosity of 1.6 with Q1=1.3 and Q3=1.9. Therefore it can be seen that there is agreement, giving confidence for more realistic simulations. Also for this system, the porosity is calculated as 31.6% against the theoretical value of 47.6%. This disagreement is a simply a geometric one; the pores are cubic and so only the corners of the cubes touch the surface of the assumed solid spheres. Therefore each simulated pore is of a smaller volume than the actual void between spheres. But the porosity of a real sandstone would be less than that of a simple cubic pack of spheres due to: (a) the sand particles being non spherical; (b) the sand particles not all being of equal diameter; and (c) consolidation of particles.

2.5 *Mercury porosimetry*

Mercury porosimetry is a common method of determining the p.s.d. of a sandstone and involves the forcing of mercury into the rock under increasing pressure. As the pressure increases, progressively smaller pores are invaded until all of the connected pathways have been filled. The volume of mercury injected is recorded as a function of pressure. The pressure readings are converted to throat diameters by the Laplace Equation

$$d = \frac{-4\ \sigma\ \cos\theta}{P} \tag{2.1}$$

where: σ is the surface tension of mercury; P is the capillary pressure; θ is the contact angle (140° for mercury on sandstone); and d is the throat diameter. By knowing the volume of sample and its effective porosity, the volume of mercury taken up by the sample is converted to a percentage pore volume. Thus a pore size distribution can be plotted.

The experimental curve for our sample of Clashac is shown in Fig. 5a. It is from this curve that a range of pore sizes was taken to give the initial estimate of the p.s.d. On the basis of this p.s.d., the model produces a porosimetry curve. If the p.s.d. is correct, then the same point of inflexion and shape of curve should be seen as on the experimental curve.

The point of inflexion is taken to be the "characteristic" pore size of the sample, although our model shows that this is not the most frequently occurring pore size, as is often assumed. At lower pressures below this point there may be large pores within the body of the sample which would be filled if exposed to the mercury but remain empty because they are shielded by smaller pores. At the point of inflexion the shielding effect of the characteristically sized pores is lost and so a large increase in pore volume occurs for a small increase in pressure. By carrying out the simulations at increasing percentage blocked throats we are mimicking the porosimetry of a sample of decreasing pore connectivity. The effect of this is to reduce the point of inflexion to smaller pore diameters showing an increasing shielding effect. This is shown in Fig 5b.

By adjusting the p.s.d. and percentage blocked throats until the simulated curve matches the experimental curve we can find the optimum p.s.d. and pore connectivity. Two such optimisations are shown in Fig 5c. This again gives a method of calculating the average connectivity.

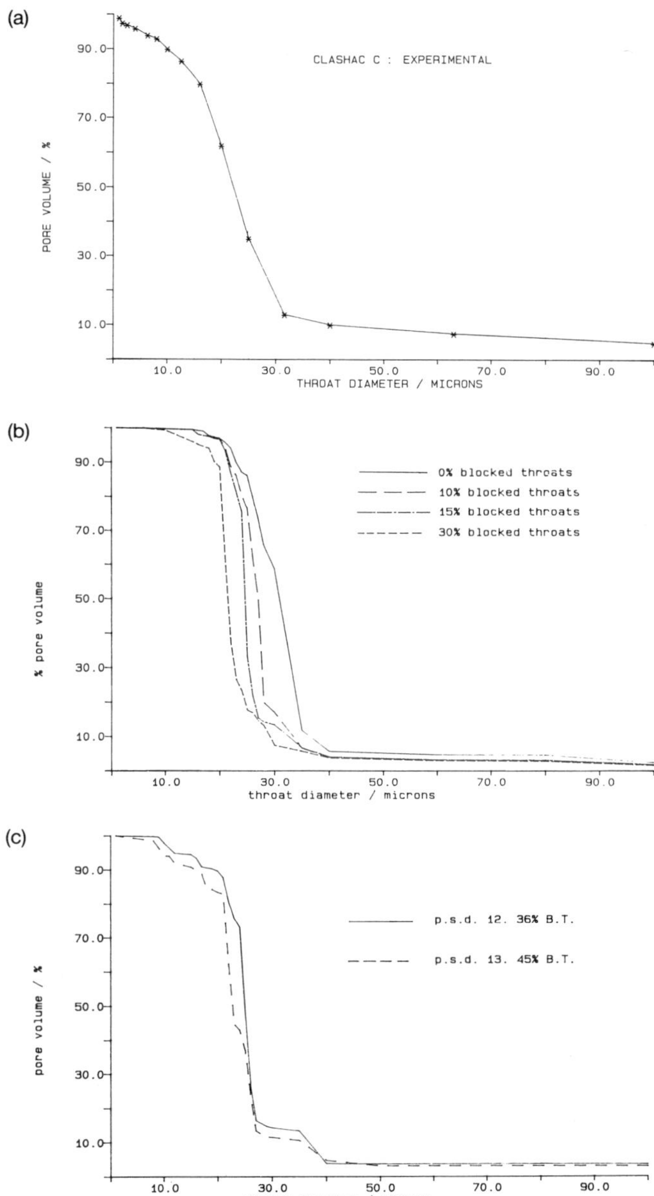

Fig.5: Mercury porosimetry curves: (a) experimental curve of Clashac sandstone; (b) simulated curves at decreasing pore connectivity; (c) optimised curves.

2.6 *Discussion*

The optimised mercury porosimetry simulation shows a percentage blocked throats of 36% is required with p.s.d. 12 and 45% with p.s.d. 13 to best fit the experimental curve. This indicates an average pore connectivity of 3.30 - 3.84. There is no experimental pore connectivity data available for Clashac sandstone, but this compares well to the value quoted by Koplik et al (1984) of 3.49 (= 42% blocked throats) for the average pore coordination for a Massillon Sandstone. The porosity of the unit cell matches the experimental porosity of 11% at 36% blocked throats with a pore row spacing of 250μm indicating a minimum throat length of 70μm. If we then simulate the tortuosity at 36% B.T. with the pore row spacing of 250μm we find that the predicted tortuosity is too large at a value of 2.9 compared to the experimental value of 2.42 from conductivity experiments and 2.46 from diffusion experiments (Spearing and Matthews, 1990). This indicates that the model of the tortuosity is not quite correct for real porous media. It could be said that this follows from the simulation of cubic packed spheres which was 25% greater than the theoretical value due to Chen. Further work is needed therefore to improve the model. The main result from this is the illustration that the tortuosity of a sandstone is not a defined value, as thought of in early literature on the subject of flow through porous media (Macmullin and Muccini, 1956), but is a distribution of values dependent on pore size distribution and connectivity.

3. Network analysis applied to the sandstone model

Network analysis is a method of analysis used in operational research to solve such problems as finding the optimum route for a postman so that he can deliver all his letters whilst travelling the least possible distance, or finding the shortest route between two cities in a road network.

The software used is produced by SAS and is run on an IBM-compatible personal computer. SAS is a powerful suite of programs used for data management, report writing and graphics, statistical and mathematical analysis, business planning and operational research. Network analysis is just one small part of SAS/OR. We have investigated network analysis in order to give us the maximum flow through the network of pores and throats.

Certain terminology is used in network analysis, Whitaker (1982). A connection between two points is termed an *arc* and crossing an arc will incur *cost*. An arc may have a capacity limitation. For example, with the analogy of a road between two cities, the road may have an *arc capacity* of 100 tons if there is a bridge with that restriction en route. Arcs can also be *directed* (can only be traversed in one direction) or *undirected* (can be traversed in either direction). The beginning and ending of an arc is marked by a *node*. Each network also has a *source node* from where paths originate and a *sink node* where paths end. Applying this to our model, an arc represents a throat and a pore a node. Unit cost is incurred on passing from one pore to the next. This is so that a minimum cost route through the medium is found. Without this stipulation the flows tend to swirl around pores within the same horizontal layer and so the true splitting of flows through the major pathways cannot be deduced. To model permeability, the capacity of the path is equal to an r^4/l term (derived from the Poiseuille equation for laminar flow) for each pore/throat system. Slip flow is appreciable for small pores ($<10\mu m$) and an allowance is made for this (Dawe, 1973).

The network analysis then calculates the theoretical flows that can pass through various branches of the network depending on the capacity of the route. There is an overall "conservation of flow" from source node to sink node so that what goes in, comes out with no build up through the network. The value then obtained for the maximum flow through the system is equivalent to an effective r^4/l term for the complete network. This value can then be combined with the Poiseuille Equation and Darcy's Equation or used with equations given by Koplik et al (1984) to make a prediction of permeability.

This problem proved to be more difficult than first thought. The full network of the unit cell consists of 5100 arcs: 5000 undirected within the network and 100 directed from the source node to the top layer. So far we have only managed to run the procedure to a conclusion for a small portion of the unit cell: 320 arcs which is 2 pores x 10 pores x 3 pores. Adding one more row of the unit cell to the problem - 3 pores x 10 pores x 3 pores - gives 480 arcs but for a problem of this size the computer can be left for many hours with no success.

Once this problem has been overcome a realistic permeability prediction should be possible purely from the pore geometry of the sample.

4. Fractal analysis

The geometry of a sandstone is random, but our model, although containing randomly allocated pore sizes, is nevertheless regular in the positioning of the pores. One way of further mimicking the random nature of the rock itself is by means of fractal analysis.

A fractal is a geometrical entity which retains its complexity, no matter how much it is magnified. Typically it has a dimension which is not an integer, making it intermediate in nature between the classical geometric concepts of line, surface and solid. We would expect a porous medium such as a sandstone to have a fractal dimension of less than three.

More rigorously, the fractal dimension D is calculated (Barnsley, 1988) from the equation

$$D = \lim(e \rightarrow 0)\left[\frac{\log N(e)}{\log(\frac{1}{e})}\right] \tag{4.1}$$

Experimentally D may be determined by constructing a grid of boxes covering the fractal object. One simply counts the number of boxes N(e) which contain some of the object, then changes the size e of the boxes, and counts again. A plot of log N(e) vs log (1/e) then has a slope equal to the fractal dimension D.

Although a mathematical object can be a fractal over an infinite range of magnifications, a real-world object will have a constant fractal dimension over a range limited by its physical structure. In the case of a sandstone this range is between the largest and smallest pores present, between about 0.1 and about 200 micrometers. The fractal dimension has been determined in this range by means of scanning electron microscopy. Katz and Thompson (1985) and Krohn and Thompson (1986) counted the number of geometrical features observed as a function of magnification and obtained D values between 2.55 and 2.85, depending on the particular type of sandstone.

We have developed a program to measure the fractal dimension of mathematical fractals and our own model, and to compare them with the required sandstone fractal dimension.

The program was tested by applying it to structures based on the "Menger sponge" (Plate 3, pp.14-15). The first stage of this structure is created by dividing a cube up into 27 smaller cubes, 3 on an edge. The centre cube is removed, together with the cubes at the centres of each of the faces. A total of 7 smaller cubes are removed, leaving 20 cubes. In the second stage, each of the 20 remaining cubes is partitioned in the same way. This process is repeated indefinitely, to create a structure whose fractal dimension is ln(20)/ln(3) = 2.727.

The program showed fractal dimensions of second and third stage Menger sponges as 2.94 and 2.91 respectively, showing a trend towards the correct value after an infinite number of stages. Application of the program to our model resulted in values of fractal dimension between 2.90 and 2.96, depending on the pore size distribution used. In general, the greater the percentage of large throats and pores present, the smaller the fractal dimension. Structures with much greater porosity are required in order to lower the fractal dimension below 2.90, but incorporation of a surface roughness microstructure within the model would lower its fractal dimension substantially.

5. Conclusion

We have presented a three-dimensional geometric model of a real porous medium, namely Clashac sandstone. The power of the model lies in the fact that it is the first network in which the three-dimensional geometry makes a full contribution to the predicted experimental properties. So far we have successfully modelled porosity, mercury porosimetry data, pore/throat size correlation, connectivity and tortuosity. We are continuing the development of the model by further examining permeability, fractal dimension and also two-phase hysteresis. Once these are incorporated, we believe we will have a powerful means of correlating and predicting pore-level properties of porous media.

6. Acknowledgements

We would like to thank Terry Mangles for useful discussions concerning statistics in various parts of the work, Dave Buckley for his effort in producing the 3D diagram of the unit cell and Professor Emerson Heald for his work on the fractal analysis.

7. References

Barnsley, M. (1988). Fractals Everywhere. Academic Press, New York.

Chen, L.C. (1973). PhD thesis. Univ. Michigan, Ann Arbor.

Dawe, R.A. (1973). The slip correction in accurate viscometry. Rev. Sci. Instrum., 44:1271-1273.

Katz, A.J. and Thompson, A.H. (1985). Fractal sandstone pores: implications for conductivity and pore formation. Phys. Rev. Lett., 54:1325-1328.

Koplik, J., Lin, C. and Vermette, M. (1984). Conductivity and permeability from microgeometry. J. Appl. Phys., 56:3127-3131.

Krohn, C.E. and Thompson,A.H. (1986). Fractal sandstone pores: Automated measurements using SEM images. Phys. Rev. B., 33:6366-6374.

Macmullin, R.B. and Muccini, G.A. (1956). Characteristics of porous beds and structures. AIChE Journal, 2:393-403.

Spearing, M.C. and Matthews, G.P. (1990). Modelling characteristic properties of sandstones. J. Transport in Porous Media, in press.

Van Brakel, J. (1975). Pore space models for transport phenomena in porous media. Powder Technology, 11:205-236.

Whitaker, D. (1984). OR on the Micro. Wiley and Sons, pages 43-65.

Wojtanowicz, A.K., Krilov,Z. and Langlinais, J.P. (1988). Experimental determination of formation damage pore blocking mechanisms. J. Energy Resources Technology, 110:34-42.

Wyllie, M.R.J. and Rose, W.D. (1950). Some theoretical considerations related to the quantitative evaluation of the physical characteristics of reservoir rock from electrical log data. Petroleum Trans. AIME, 189:105-118.

Using transputers for gravity and magnetic modelling in three dimensions

P.A.V. Young, S.B.M. Bell, D.C. Mason,
G. Wadge

*NERC Unit for Thematic Information Systems,
University of Reading*

1. Introduction

Magnetic and gravity surveys, in conjunction with other geophysical techniques, are often used to provide a model of subsurface geology. For instance, a granite body may cause a negative gravity anomaly if it is of lower density material than the country rock, or a basaltic intrusion may cause a magnetic anomaly, because it may contain more magnetic minerals aligned differently from those in the country rock. Although algorithms have been devised to calculate the gravity and magnetic response for 3D models, at present most modelling is performed in 2D, which is simplest to visualise and requires less computation than 3D.

The Regional Geophysics Research Group at the British Geological Survey (BGS) are developing an interactive 3D geophysical modelling system, workstation-based, in which models are built and altered on a graphical screen. The system aims to implement forward modelling of gravity and magnetic fields; a geological model is built by the operator, the magnetic and gravity response to that model computed and compared with observations, and then the model is progressively altered, usually a little at a time, to bring the calculated results closer to the observations. Although the (Apollo) workstation has full 3D graphics capability, it cannot perform 3D model calculations fast enough for anything like "interactive" responses. Currently, the calculations are done on a mainframe VAX or CONVEX vector processor, but this solution increases turnaround time and removes the calculations from the intended workstation environment.

NUTIS (the Natural Environment Research Council Unit for Thematic Information Systems) is collaborating with BGS to test the use of a transputer array to perform the magnetic and gravity calculations. A transputer is a very powerful computer-on-a-chip, capable of performances up to 10 million instructions per second (i.e. about equivalent to a small VAX). Transputers are usually linked into arrays, and then provide impressive computing power at much lower cost (about 1000 pounds sterling per transputer) than the conventional alternatives such as a CRAY. A large transputer array may be used as a mainframe, such as at Edinburgh University, or a smaller array can be used to enhance a PC or workstation (see Fig. 1). This latter approach appears useful for this collaboration with BGS.

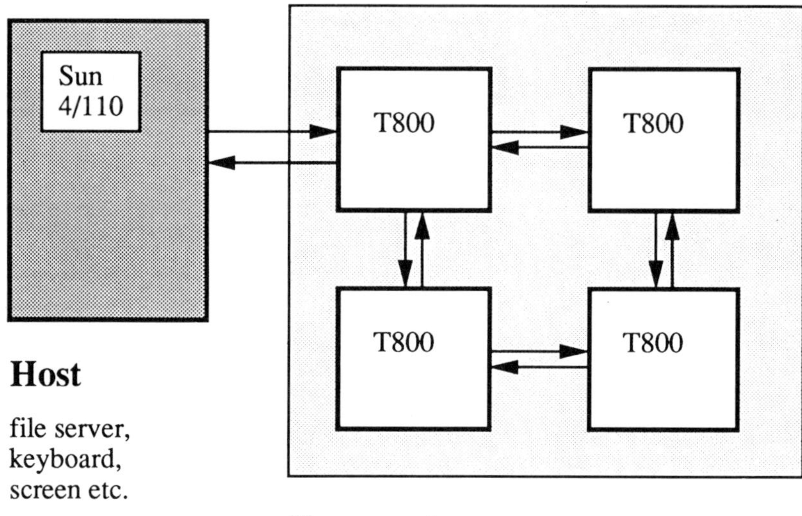

Fig. 1: The transputer array at NUTIS. The transputers are T800s, in a Meiko box, using a Sun workstation as the host.

To make use of a transputer array, the computing problem must be divided into a number of parts, one part for each transputer, which are then computed in parallel. Theoretically, an array of N transputers may perform up to N times faster than one transputer for any given problem. However, 100% efficiency is hardly ever achieved, because of the time taken in communication between transputers, and because parallel algorithms are often more complicated than the conventional sequential algorithms. High efficiency is only possible if the application requires intense computation with relatively little communication between transputers. Gravity and magnetic potential field calculations fall into this category.

2. The algorithm

An algorithm determined by Okabe (1979) allows the gravity and magnetic potential fields of a 3D body to be calculated for any point outside the body. The body's surface is described by triangular facets, and each facet's contribution to the gravity or magnetic potential is calculated independently for every observation point (see Fig. 2). The facets' potentials are then summed at each observation point. The calculation of the response at one observation point due to one facet of the model is completely independent of any other observation point or model facet. This offers a variety of ways of "parallelising" the problem.

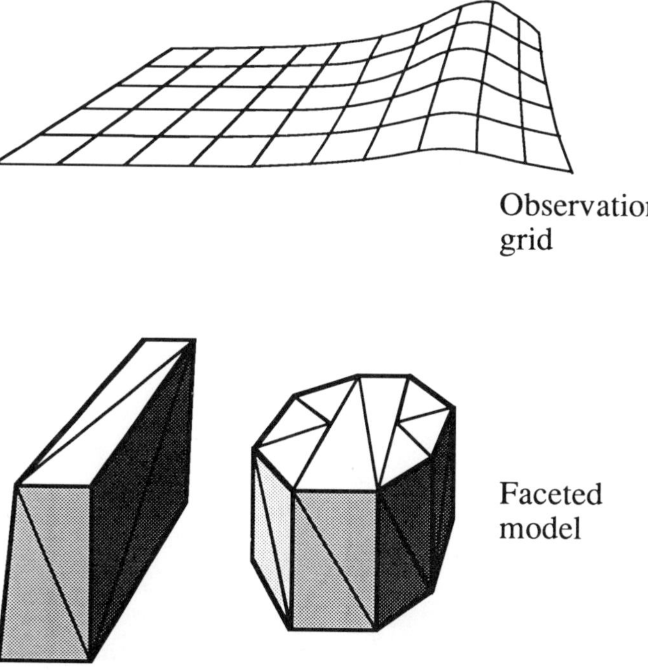

Fig. 2: A cartoon of a 3D geological model, defined by triangular faceted surfaces. Above is an observation grid (perhaps the ground surface), comprising observation points which each have an x, y and z position in space.

The transputer array at NUTIS is shown in Fig. 1. Each transputer may have up to 4 two-way links to other transputers, and these can be changed to create a variety of networks (see Fig. 3). Because the gravity and magnetic computations only require communication from the transputers to the host (rather than communication between neighbouring transputers), the most efficient network is expected to be the "tree", which ensures the minimum number of intermediate transputers between each transputer and host. All the results presented here use this configuration.

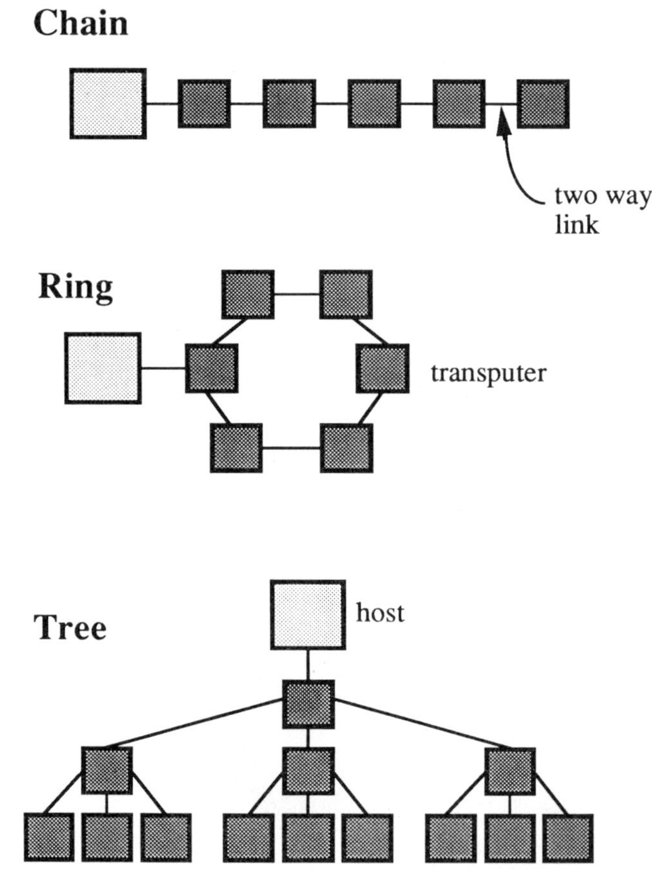

Fig. 3: Three possible ways to connect up transputer arrays. Each transputer has only 4 two-way links, and all input and output has to be channelled via one transputer to the host / file system. The "tree" is the most efficient for the application described in this paper.

A sequential Fortran program using Okabe's algorithm, written by K. Dabek at BGS was converted to the C language and run in parallel within an occam "harness". The problem was "parallelised" in three ways:-

A - Divide the model into an approximately equal number of facets for each transputer. The facets are sent out to the transputers one at a time, so that calculation starts immediately one facet has arrived. The results from the transputers then have to be added together.

B - Divide the observation grid into strips of approximately equal width, one for each transputer, and each transputer receives all the facets. All the facets are sent out together, calculation cannot start until all the facets are received.

C - Divide the observation grid into strips as in B, and each transputer receives all the facets, but the facets are sent out individually, as in A.

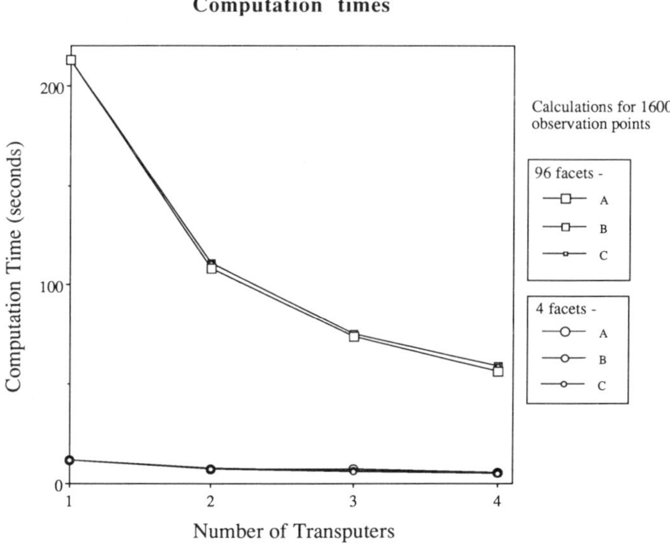

Fig. 4: Time taken for computation of 2 different models, for the 3 different approaches, A, B and C, described in the text. In comparison, the 96 facet computation took 456 seconds cpu time on an Apollo workstation, 114 seconds on a VAX 8530 and 654 seconds on a VAX 8350.

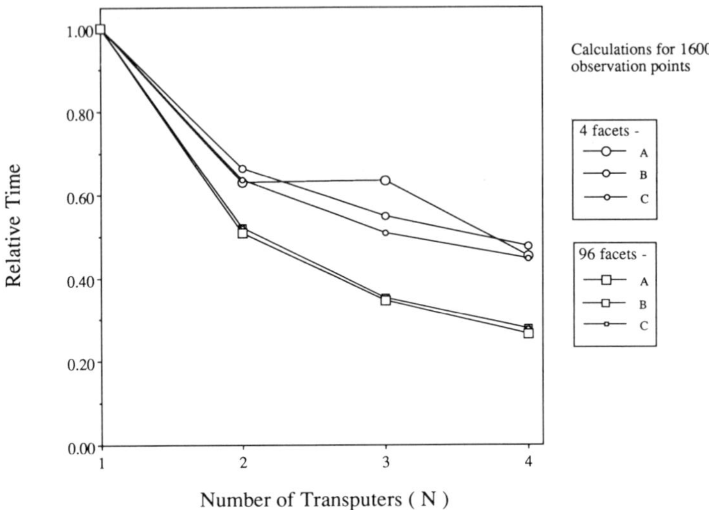

Fig. 5: Time taken for each computation, relative to the time taken by 1 transputer. The same models as for Fig. 4.

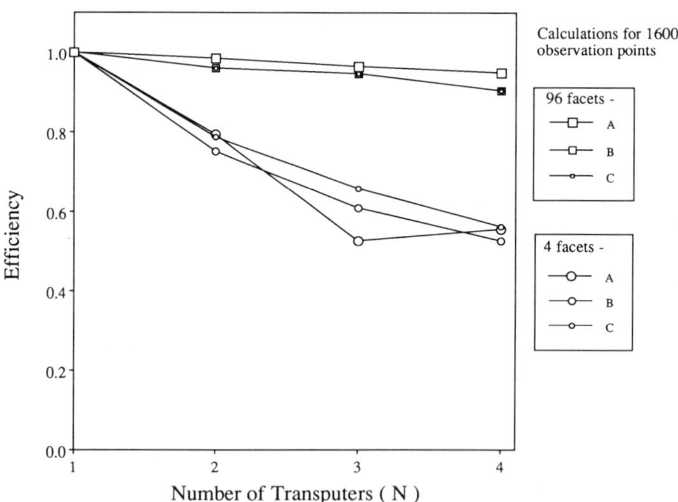

Fig. 6: Variation of efficiency with the number of transputers, compared with the performance of 1 transputer, for the same models as Fig. 4.

The time taken, relative times and corresponding efficiencies for two input models are shown in Figs. 4, 5 and 6. The calculations were performed for a 40 x 40 point observation grid (1600 points). The 96 facet computation represents a reasonably complex geological structure, which would be calculated at the beginning of a forward modelling session. The 4 facet computation represents a small adjustment to an existing model, which might be calculated as one step in the modelling procedure.

The equivalent sequential program was run on a VAX 8350, a VAX 8530 and the Apollo workstation, using the 96 facet model. One transputer alone performed 3 times faster than the VAX 8350, over twice as fast as an Apollo 4500, and a 2 transputer array was faster than a VAX 8530.

There is not much difference between the timings for approaches A, B and C, although A appears about 1% faster for the large model on 4 transputers. Approaches B and C provided very similar results for most of the models tested. The 4 facet model is calculated in a few seconds on 4 transputers, indicating that this would be useful in an interactive modelling environment, although the array is performing at only 60% efficiency (Fig. 6). It takes 4 transputers just over a minute to calculate the 96 facet model, which is an acceptable time to wait at the beginning of a modelling session. The relative timings and efficiency graphs (Figs. 5 and 6) indicate problems with approach A when there are few facets compared with the number of transputers - it takes 3 transputers longer to perform the 4 facet calculation than 2 transputers! However, approach A performs more efficiently than B and C for a large number of facets. In a final implementation, it might be useful to select the most efficient approach for the number of facets and observation points chosen relative to the number of available transputers.

3. Conclusions

These results indicate that a 4 transputer array provides gravity and magnetic computations that are acceptably fast for an interactive forward model environment. Efficiencies of 90 - 95% were achieved for a reasonably sized model, which is a good performance compared with other transputer applications (Freeman and Phillips 1990).

4. Acknowledgements

We would like to thank Konrad Dabek at BGS for providing the original program, and for running the comparative computation timings on VAX 8530 and Apollo, also Paul Williamson and Mick Lee for their co-operation. This work was partly supported under NERC Contract No. F60/G6/12.

5. References

Freeman, L., and Phillips, C. (editors) (1990). Applications of Transputers 1, Proceedings of the First International Conference on Applications of Transputers, 23-25th August 1989 - Liverpool, UK. I.O.S. Press: 332pp.

Okabe, M. (1979). Analytic expressions for gravity anomalies due to homogeneous polyhedral bodies and translations into magnetic anomalies. Geophysics, 44:730-744.

The rheological strength of the lithosphere following extensional and compressional tectonics

S.S. Egan

University of Keele

1. Introduction

The outer rigid layer of the Earth, the lithosphere, consists of both the crust and the top of the mantle. The crust and the mantle are divided by a compositional boundary known as the Moho, above which the rocks are composed dominantly of the mineral quartz, while below, the upper mantle rocks are rich in the mineral olivine. The base of the lithosphere occurs at about 125km depth in a continental setting and is defined by the 1333°C isotherm (McKenzie, 1978). This thermal boundary separates the rigid rocks of the lithosphere from the less rigid environment in the asthenosphere. Although it makes up a small proportion of the radius of the Earth, the extensional and compressional deformation of the lithosphere determines many of the large-scale geological structures seen at the surface. Extension or rifting of the lithosphere causes thinning of the crust by movement along fractures known as faults. The resulting surface hole fills with sediment over time to form sedimentary basins like those beneath the North Sea. The compression or shortening of the lithosphere, on the other hand, leads to a thickening of the crust by reverse movement along thrust faults. The Alpine mountain range is an example of the effects arising from this type of deformation. The study of the lithosphere is very important, therefore, in aiding our understanding of major structures seen in the geological record.

In this paper the factors affecting lithosphere strength are investigated. In particular, numerical modelling techniques have been used to simulate the evolution of lithosphere strength following both extensional and compressional tectonics. An effective representation of the overall strength of the lithosphere is the yield stress envelope, which defines either the tensile or compressive stresses that the lithosphere can sustain before it mechanically fails. In detail it differentiates the high strength, brittle regions from low strength, ductile environments (Fig. 1). The brittle yield stresses are controlled primarily by the pressure arising from rock overburden (lithostatic pressure) and, therefore, increase with depth. The lithosphere is not completely brittle, however, because the temperatures rise sufficiently in both the crust and mantle to induce a ductile deformational environment. The magnitude of the ductile yield stresses are controlled mainly by temperature such that the increase in temperature with depth causes a reduction in the ductile yield stress component.

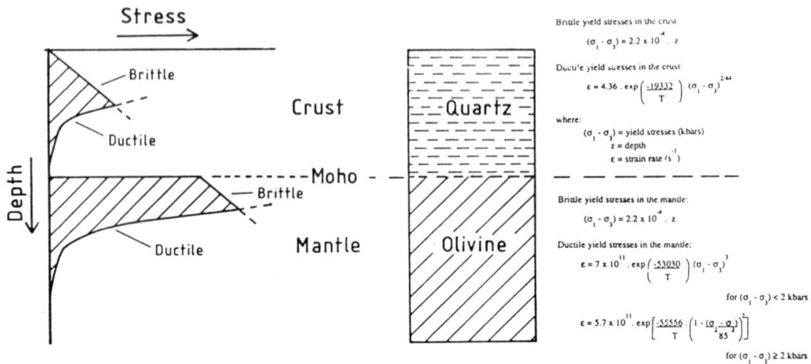

Fig. 1: Diagrammatic representation of the yield stress envelope for a lithosphere simplified compositionally into a crust of quartz and mantle of olivine. The equations used to calculate the brittle and ductile yield stresses are as follows:

Crust: *Brittle yield stresses*

$$(\sigma_1 - \sigma_3) = 2.2 \times 10^{-4} . z \tag{1.1}$$

Ductile yield stresses

$$\dot{\epsilon} = 4.36 . \exp\left(-\frac{19332}{T}\right) . (\sigma_1 - \sigma_3)^{2.44} \tag{1.2}$$

where: $(\sigma_1 - \sigma_3)$ *= yield stresses (kbars); z = depth; and* $\dot{\epsilon}$ *= strain rate* (s^{-1}).

Mantle: *Brittle yield stresses*

$$(\sigma_1 - \sigma_3) = 2.2 \times 10^{-4} . z \tag{1.3}$$

Ductile yield stresses

$$\begin{aligned} \dot{\epsilon} &= 7 \times 10^{11} . \exp\left(-\frac{53030}{T}\right) . (\sigma_1 - \sigma_3)^3 \quad for (\sigma_1 - \sigma_3) < 2 kbars \\ \dot{\epsilon} &= 5.7 \times 10^{11} . \exp\left[-\frac{55556}{T} . \left(1 - \frac{(\sigma_1 - \sigma_3)}{85}\right)^2\right] \quad for (\sigma_1 - \sigma_3) \geq 2 kbars \end{aligned} \tag{1.4}$$

Integration of the yield stress envelope over the thickness of the lithosphere provides a convenient "single-figure" definition of the strength of that particular section, which is useful for making regional comparisons.

2. Formulation of the thermo-rheological model

In this study a thermo-rheological model has been used to calculate yield stress envelopes associated with both extended and shortened lithosphere. For modelling purposes the lithosphere has been simplified compositionally into a crust (35km thick) assumed to be composed entirely of the mineral quartz, while the mantle lithosphere (90km thick) is represented by the mineral olivine. The equations used to calculate the brittle and ductile yield-stress values (σ_1-σ_3) for these minerals are summarised in Fig. 1 and have been derived by empirical work carried out by Post (1977), Goetze (1978), Koch et al (1980) and Bodine et al (1981). For the calculations an effective strain rate of 10^{-15} s^{-1} has been assumed.

A more detailed explanation of the thermo-rheological model used can be found in Kusznir and Park (1986 and 1987).

3. The factors controlling the strength of the lithosphere

The strength of the lithosphere is controlled by temperature, crustal thickness and composition (Kusznir and Park, 1987). In the sections below the influence of these parameters upon lithosphere strength is investigated.

3.1 The effect of temperature

In Fig. 2a a temperature-depth distribution has been defined for the lithosphere so as to generate a realistic surface heat flow of 45 $mW.m^{-2}$. The associated yield stress envelope has been calculated using the equations summarised in Fig. 1. A high strength, brittle region occurs in the upper crust, while the higher temperatures at mid- and lower crustal levels maintain a ductile environment, and the yield stresses drop sharply to negligible values. Within the top-most part of the mantle there is a strong increase in the yield stresses as the mineral olivine is strong enough to endure the high temperatures and deform by brittle deformation. The temperatures rise sufficiently in the lower lithosphere, however, to induce ductile behaviour and the strength again becomes negligible.

In Fig. 2b the temperature-depth distribution represents much warmer lithosphere with a surface heat flow of 75 $mW.m^{-2}$. As a consequence of the higher temperatures the depth ranges of the high strength, brittle deformational environment in both the crust and mantle is much reduced.

It can be concluded, therefore, that the hotter the lithosphere the lower will be its overall strength.

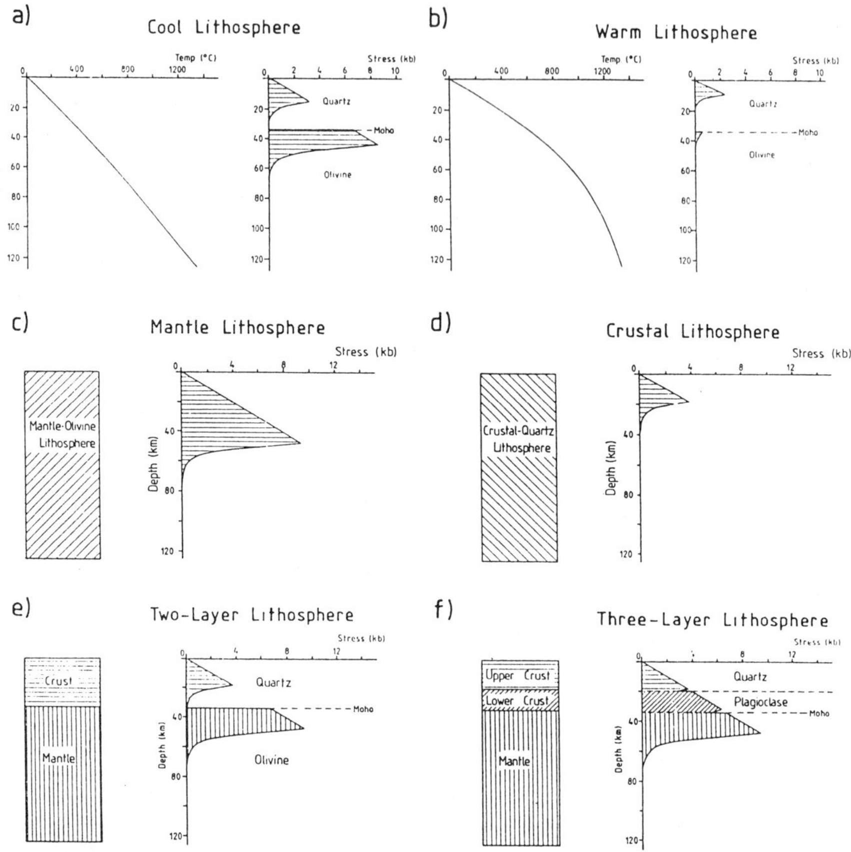

Fig. 2: The effects of temperature (2a and 2b), crustal thickness (2c and 2d) and composition (2e and 2f) upon lithosphere strength.

3.2 The effect of crustal thickness

In Figs. 2c and 2d yield stress envelopes are shown for lithosphere composed entirely of mantle and crust respectively, while the temperature-depth distribution has been kept constant for both examples. Each stress envelope consists of a single-layer vertical stack of brittle and ductile yield stress components because the lithosphere is represented compositionally by a single mineral. The model predictions show clearly that the crustal lithosphere is much weaker than the mantle lithosphere as the mineral quartz, which is abundant in crustal rock assemblages, is much weaker in terms of its resistance to high temperatures than the mineral olivine, dominating rocks within the mantle. The results imply, therefore, that the thicker the crust the weaker is the lithosphere.

3.3 The effect of lithosphere composition

In Fig. 2e, as in previous models, a highly simplified quartz-olivine compositional structure has been assumed to represent the lithosphere. Mineralogical studies reveal, however, that there is a dominance of the mineral plagioclase in the rocks of the lower crust, while quartz and olivine are concentrated in the upper crust and mantle respectively. The ductile yield stress component for the mineral plagioclase is predicted by the following equation and based on experimental work by Shelton and Tullis (1981)

$$\dot{\epsilon} = 8.2\times10^{2}.\exp(-\frac{28788}{T}).(\sigma_1-\sigma_3)^{3.2} \tag{3.1}$$

In Fig. 2f a diagrammatic representation of the lithosphere compositional structure is shown with plagioclase representing lower crustal depths from 20km to 35km. The associated yield stress envelope shows a three-layer stack of brittle-ductile yield stress components, resulting from the interaction of temperature upon the individual minerals composing the crust and mantle. The incorporation of plagioclase in the yield stress calculations has led to a much stronger crust and the lithosphere is strengthened overall.

4. The strength of the lithosphere following extensional tectonics

Lithosphere subjected to large-scale extension experiences a thinning of the crust and heating, as hotter material at depth is raised nearer to the surface. From the results in Section 3 these effects will induce strengthening and weakening components respectively. The question arises, therefore, as to whether the lithosphere is weakened or strengthened overall following extensional tectonics? In order to answer this question quantitative constraint is required for both the post extensional crustal thickness and temperature structure. These variables can then be included in the yield stress calculations described above.

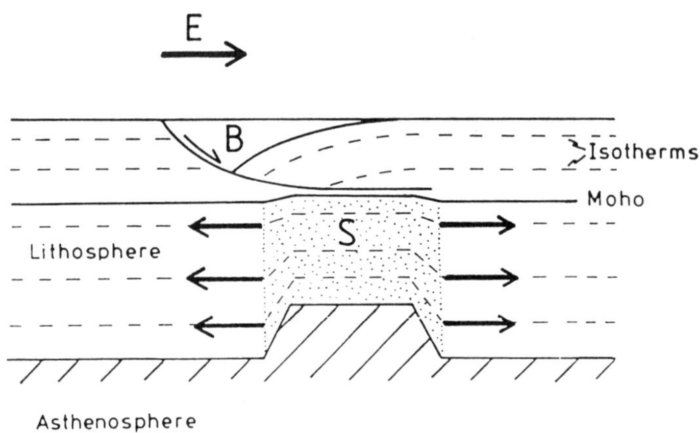

Fig. 3: Diagrammatic representation of the numerical model of lithosphere extension as used to quantify the thinning of the crust and temperature perturbations. The tectonic loading of the lithosphere is also taken into account using flexural isostasy (B = basin and S = stretching).

A two-dimensional numerical model of lithosphere extension (Fig. 3) has been used to define the post-extensional crustal thickness and temperature field. The model assumes that extension, and hence crustal thinning, in the upper brittle region of the lithosphere occurs by movement along a major fracture or fault, while extension in the lower ductile lithosphere is accommodated by stretching or pure shear (Jackson and McKenzie, 1983). The model also takes into account the perturbation of the temperature field. Faulting causes relative displacement between two blocks of crust and, in turn, strong lateral temperature gradients are generated. Stretching has the effect of raising the hotter material at depth closer to the surface. The resulting heating and thermal expansion leads to uplift of the overlying surface. These temperature perturbations, however, are temporary on a geological time-scale and following extension re-equilibration occurs as heat is transferred by conduction and, in turn, lost from the surface. This produces a thermal contraction effect and the overlying surface subsides to generate the thermal subsidence or sag phase exhibited by many sedimentary basins in the geological record. Thermal re-equilibration and associated subsidence is simulated to 100 million years after initial rifting.

The thinning of the crust and heating caused by extension imposes load forces upon the lithosphere. The dominant load is a negative or buoyancy load, arising due to crustal thinning. All tectonic loads are accommodated within the model by using the theory of flexural isostasy (Watts et al, 1982 and Kusznir and Egan, 1989), which assumes that the lithosphere responds to, loading by flexing. For example, the flexural isostatic response to crustal thinning is a regional upward flexing or rebound.

A more comprehensive account of the lithosphere extension model can be found in Kusznir et al (1987) and Kusznir and Egan (1989).

4.1 Model results

In Fig. 4a model predictions for 30km of extension along a fault, which flattens within the crust at a depth of 20km are shown. A surface hole and uplift of the Moho have been generated as a result of the thinning of the crust. Faulting in the upper lithosphere and stretching at depth have been positioned more or less laterally coincident in this model. The extensional deformation has been concentrated, therefore, into a relatively small region and a deep, narrow basin has been produced.

Both the post-extensional crustal thickness and temperature field predicted by the above model have been incorporated into the stress calculations described in Section 2. Yield stress envelopes have been calculated for selected positions across the model (Fig. 4b). Stress envelope A has been taken from unextended lithosphere for references purposes, while B, C and D represent sections of lithosphere that have experienced varying combinations of extension by faulting and stretching. It is shown that the strength of the crust has been reduced because faulting has led to the removal of a section of the high strength, brittle layer. The yield stresses are also reduced in the mantle due to the higher temperatures induced by stretching.

A two-dimensional strength distribution is obtained by integrating the yield stresses with respect to depth for each point across the model (Fig. 4c). The profile shows that the whole region of extension is weakened.

5. Geological implications of the model predictions

The above work suggests that extensional tectonics causes a reduction in lithosphere strength, indicating that the heating component, and the associated weakening it produces, dominates over the strengthening caused by the thinning of the crust. A further implication is that once a section of

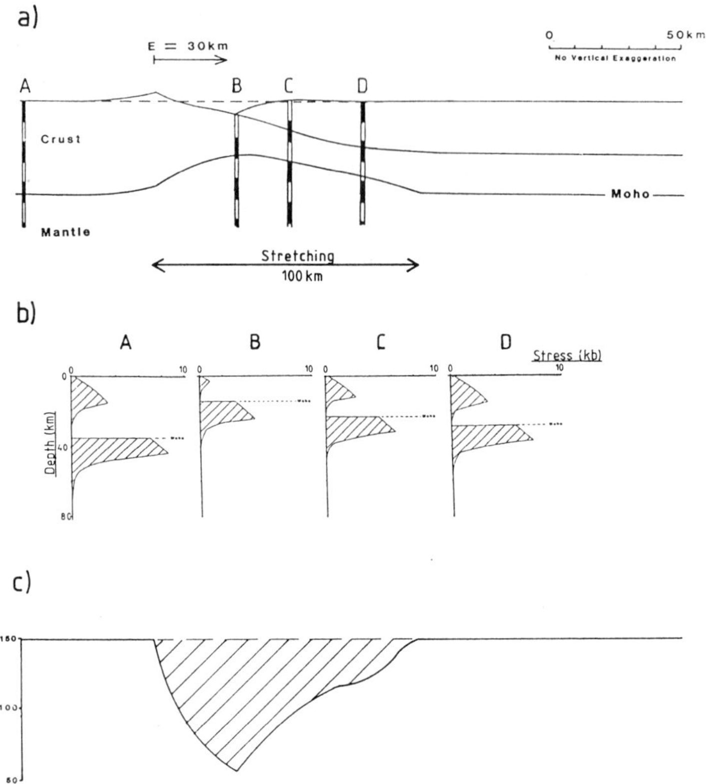

Fig. 4: (a) Model showing 30 km of fault-controlled extension in the upper, brittle region of the lithosphere, while extension in the deeper, ductile environment is accommodated by stretching (pure shear). Thinning of the crust is reflected by the generation of a basin some 7km deep and uplift of the Moho.

(b) The crustal thickness and temperature field predicted by the model have been used to constrain the yield stress calculations. Stress envelopes B, C, and D have been generated and show reductions in strength in both the crust and mantle when compared to the reference yield stress envelope A.

(c) Integration of the yield stresses with respect to depth for each point across the model produces a two-dimensional strength profile which shows clearly that the whole region of extension is relatively weak.

lithosphere extends it must continue to do so because it is weaker than surrounding unperturbed sections - the deformation will be spatially confined. Studies of regions of extension in the geological record, however, reveal that this is not always the case. For example, during the Late Jurassic (approximately 140 Ma) the continents of South America and Africa separated to form the South Atlantic (Fig. 5). Initially, the rifting took place close to the present Brazilian margin, exploiting an already existing crustal fracture. Some 20 million years later new extensional activity commenced in an offset distal region, where eventually the continental lithosphere failed completely (Ussami et al, 1986).

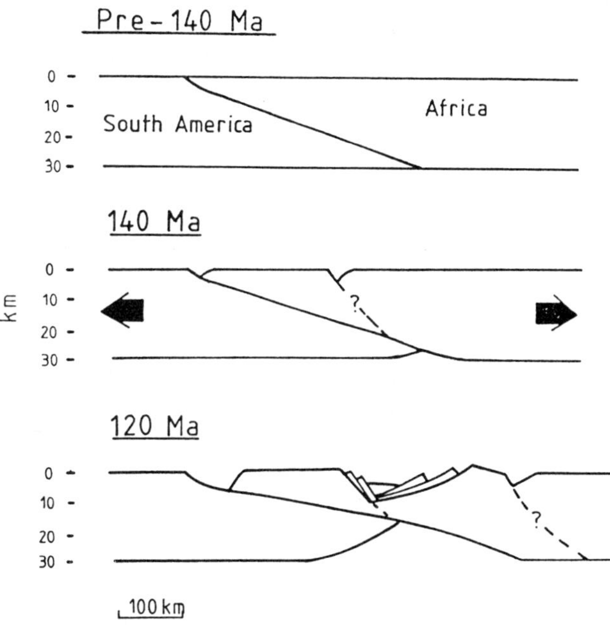

Fig. 5: The separation of the continents of South America and Africa began at the end of the Jurassic time period (140 Ma). Approximately 20 million years later renewed extensional activity began in an offset distal region.

The model predictions in Section 4 do not support a change in the position of extension over time and the question arises, therefore, as to what additional factors can be considered to explain why this occurred? An obvious factor, because of its time dependence, is the re-equilibration of the temperature field following extension. In addition, previous models have assumed rifting to be instantaneous, which is clearly a gross simplification, used as a modelling convenience. The effect of non-instantaneous extension or finite rifting should also be investigated.

5.1 *The effect of the post-rift re-equilibration of the lithosphere temperature field*

The perturbations caused to the lithosphere temperature field by extension gradually re-equilibrate over a time span of the order of 100 million years (McKenzie, 1978). The re-equilibration represents a cooling of the lithosphere back to its original state by both the lateral and vertical conduction of heat. The thermal subsidence (shaded ornament) caused by the cooling is illustrated in Fig. 6a. This has been induced by 30km of extension along a single fault balanced by a similar amount of stretching at depth. The strength profiles in Fig. 6b show the evolution of lithosphere strength at selected times during thermal re-equilibration. The lithosphere is weakened immediately following extension (0 Ma), whereas after rifting a gradual strengthening occurs. This is most rapid within the fault-induced basin because faulting causes temperature disturbances at shallow depths and as a result re-equilibration occurs relatively quickly by heat loss from the surface. In contrast, the weakening effects overlying the stretching are more prolonged due to the large perturbation to the temperature field this deformation creates within the mantle.

By 100 million years after rifting the temperature field has more or less completely stabilised, while a reduced crustal thickness, but enhanced proportion of strong olivine-rich mantle remain. The lithosphere is considerably strengthened as a result.

The implications of these results upon the evolution of extension across a region are quite profound because they explain how a region of extension, and initially weakness, strengthens over time. Thus the force profiles in Fig. 6b could be used to explain the distal transfer of extension that occurred during the initial stages of the evolution of the South Atlantic.

5.2 *The effect of non-instantaneous rifting*

The model predictions so far have made the assumption that extension is instantaneous, which has the effect of maximising the disturbance caused to the temperature field and also the associated weakening. Rifting over a finite time period, on the other hand, allows the temperature field to partially re-equilibrate while extension is taking place so that less heat remains in the lithosphere after rifting has ended. In Figs. 7a and 7b basin models and associated strength profiles are shown following instantaneous

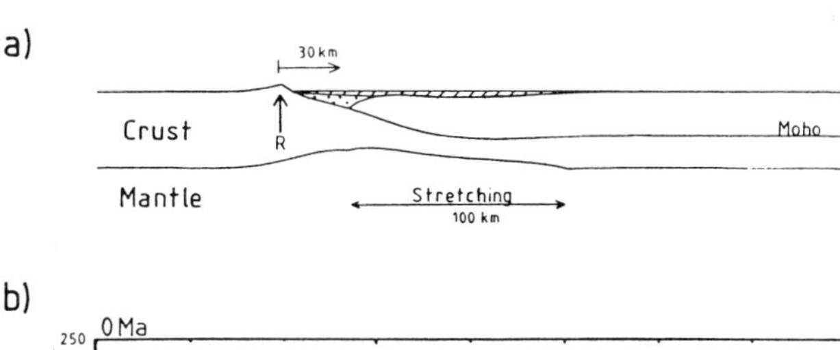

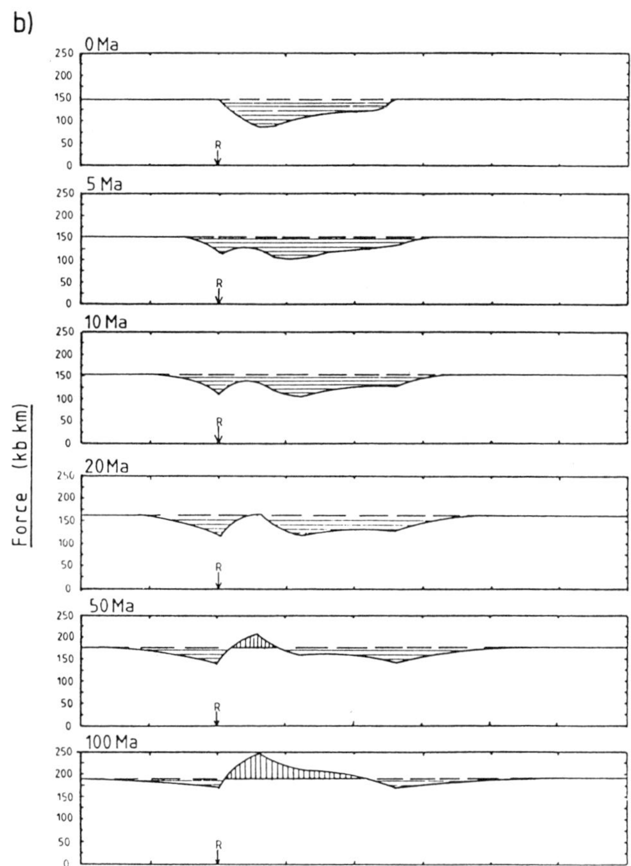

Fig. 6: (a) Predicted sedimentary basin geometry and crustal structure at 100 million years after rifting. During this time period the lithosphere temperature field re-equilibrates causing subsidence (shaded ornament). (b) Strength profiles shown at various stages from rifting (0 Ma) and during the re-equilibration of the temperature field. The region of extension gradually strengthens over the 100 million year time period, while flanking areas are weakened by the lateral conduction of heat.

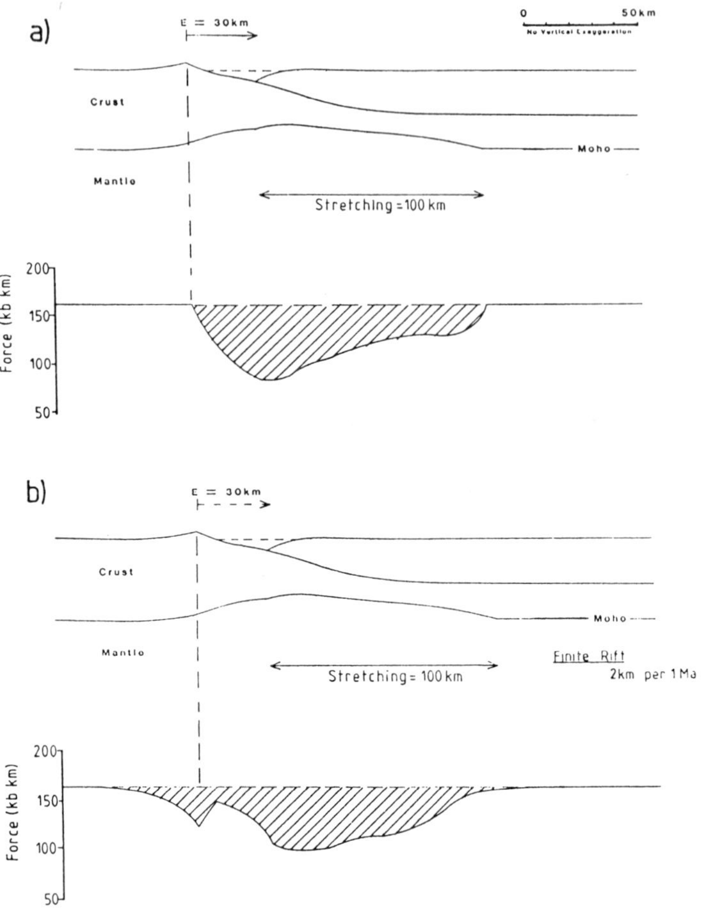

Fig. 7: Basin models and associated strength profiles for both instantaneous (a) and finite rifting (b). In the latter case lithosphere weakening is less pronounced because the temperature field re-equilibrates while extension is taking place.

and non-instantaneous extension respectively. Both models show 30km of rifting, but an extension rate of 2 km per one million years has been assumed for the calculations in Fig. 7b. The weakening effect is less enhanced in this model, which implies that for even slower rates of extension it should be possible for the lithosphere to be actually strengthened by the extensional deformation. This must surely encourage later extension to be accommodated in a new location as illustrated in Fig. 5.

6. Computing requirements

The modelling work described above takes into account crustal thinning, thermal effects and the flexural isostatic compensation of the tectonic loads associated with extensional tectonics. The calculation of these components has required the use of advanced mathematical techniques. For example, the heat conduction equation has been used to predict the cooling of the lithosphere temperature field over times of the order of 100 million years. Both vertical and lateral heat transfer has been taken into account and the equation has been solved numerically using the finite difference method. In addition, the flexing of the lithosphere in response to tectonic loading has been calculated using Fourier series. In order to carry out these calculations heavy demands have been made upon large "number crunching" computers, including the Cyber 205 supercomputer and Amdahl mainframes at Manchester Computer Centre, and the Sequent Symmetry mainframe at Keele University. The recent increased availability of powerful workstations has combined computing power with high resolution graphics, allowing results to be both generated and displayed in a relatively short time period.

The modelling software has been written in Fortran 77 and, as well, use has been made of commercial software such as compilation optimisers and the UNIRAS graphics package. In the near future it is intended to extend the modelling to three dimensions and, in addition, use will be made of transputer technology and the C programming language.

7. The strength of the lithosphere during compressional tectonics

Following compression or shortening of the lithosphere the crust is thickened and an overall cooling of the temperature field results. In Fig. 8a a computer generated model is shown in which 30km of shortening along a major thrust has been simulated. This thrust flattens within the crust at a depth of 20 km, below which shortening is achieved by a pure shear or "squashing". The contractional deformation produces an elevated topography adjacent to a large sag, which is known as a foreland or foredeep basin. These structures are generated by the flexing of the lithosphere in response to the loading imposed by crustal thickening (Beaumont, 1981).

The associated strength profile in Fig. 8b shows that following shortening the lithosphere is strengthened due to the cooling that it has experienced. After shortening, however, the temperature field re-equilibrates and the associated thermal expansion causes regional uplift of the surface (Fig. 9a). A gradual weakening of the region also occurs (Fig. 9b) in response to both the heating and existence of a thickened crust.

a)

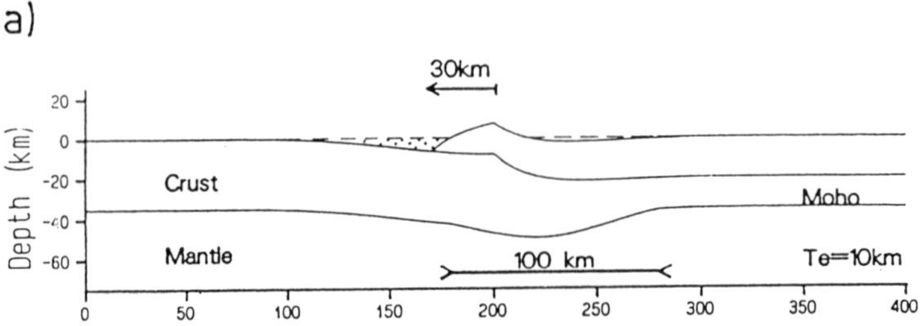

b)

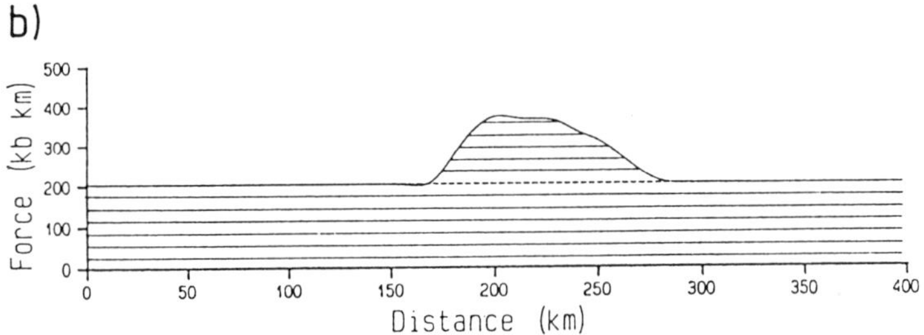

Fig. 8: (a) Model results for 30km of shortening along a major thrust fault. The lower ductile environment of the lithosphere accommodates the contractional deformation by a "squash" or pure shear mechanism. The effect of the shortening is to thicken the crust and cool the temperature field. The loading arising from the thickened crust flexes the lithosphere downwards and a foreland sedimentary is generated (dotted ornament). (b) Strength profile across the region of shortening shows that the lithosphere has been strengthened.

The model results are applicable to examples of mountain belt collapse within the geological record. In Fig. 10 two cross-sections are shown for an area to the north-east of Scotland. The section in Fig. 10a is a restoration and shows the area following the period of major contractional tectonics known as the Caledonian Earth Movements, which occurred at the end of the Silurian (approximately 400 Ma). The post compressional thickness of the crust has been estimated to have been about 60km. Some 20 million years later tensile forces experienced in the area caused an extensional reactivation of the Caledonian structures (Fig. 10b). This correlates with the model predictions in Fig. 9b, where the shortened lithosphere was also shown to weaken relative to surrounding areas some 20 million years following the termination of compressional deformation.

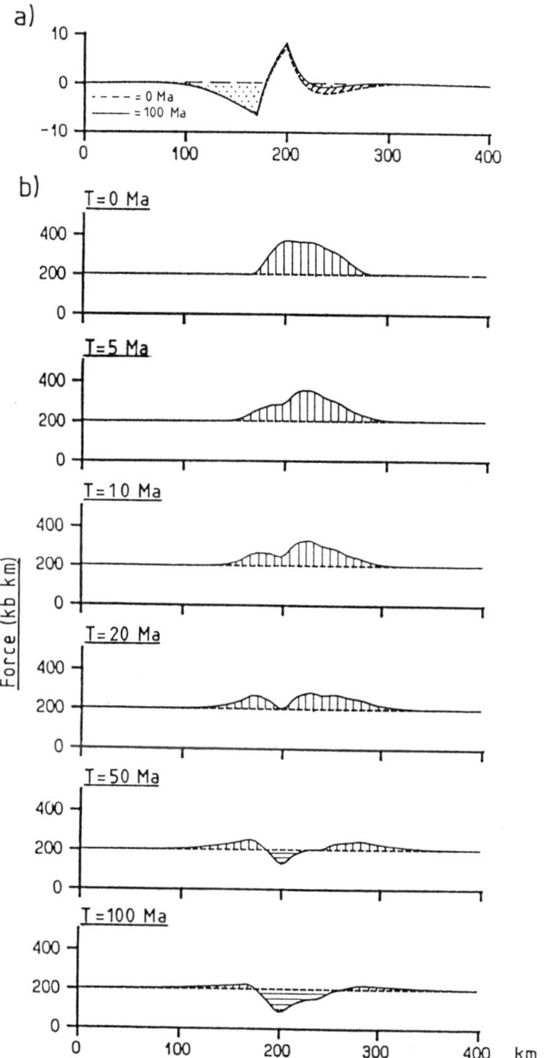

Fig. 9: (a) Following shortening the temperature field re-equilibrates and the associated heating of the lithosphere induces a regional uplift of the surface.
(b) Strength profiles for selected times during the temperature field re-equilibration. The shortened region gradually weakens over the 100 million year time interval and ultimately it becomes much weaker than surrounding undeformed sections.

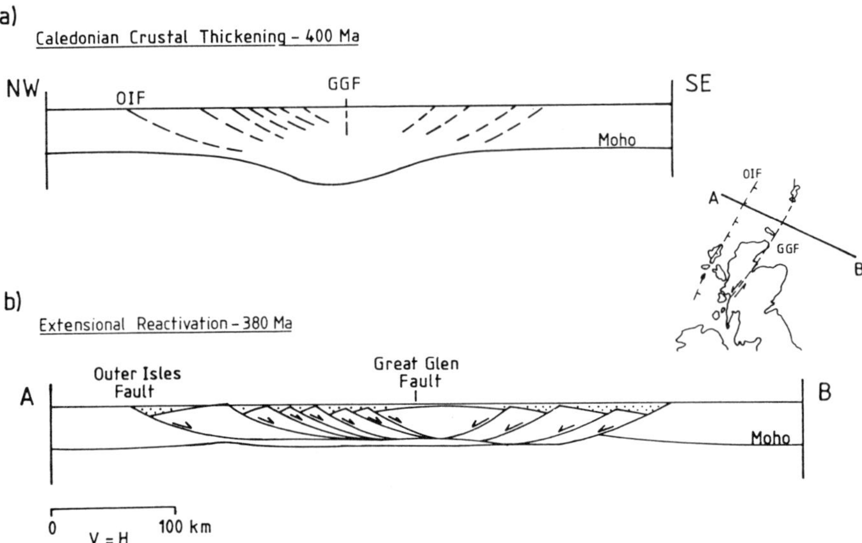

Fig. 10: Following the Caledonian earth movements some 400 Ma, the crust was thickened to an estimated 60km as shown in the section (10a). Tensile stresses were generated approximately 20 million years later, causing extensional reactivation of the Caledonian structures to form a sequence of extensional sedimentary basins as shown in Fig. 10b (M.L.F. Bamford, personal communication).

8. Conclusions

It has been shown that the strength of the lithosphere is strongly controlled by temperature and crustal thickness. Following extensional or contractional deformation there is a perturbation of both the temperature field and crust. The overall strength of the lithosphere is a result of the interaction of either the heating of the lithosphere and thinning of the crust following extension or the cooling of the lithosphere and thickening of the crust following shortening.

The relative weakness or strength of the lithosphere following tectonic deformation is, in fact, determined by the magnitude of the temperature disturbances caused by the earth movements. Model predictions indicate that temperature field perturbations are maximised for fast rates of extension and the resulting weakening effects are maintained in one position over a relatively long period of time. A spatial confinement of the extensional deformation over time will be encouraged. Slower rates of extension, on the other hand, produce less heating of the lithosphere and, in turn, the associated weakening will be less pronounced. A change in the position of extension over time is therefore likely.

Following a period of compressional tectonics the lithosphere is strengthened. Between 20 and 50 Ma after shortening, however, the strengthening component gradually disappears to leave a weakened lithosphere vulnerable to either mountain belt collapse or further compressional activity. Although the effects of finite shortening have not been investigated here, it is anticipated that following such an event weakening will occur in a relatively short time period and the region will be prone more quickly to further tectonic deformation.

9. Acknowledgements

I thank Nick Kusznir for useful discussion and guidance during the initial stages of this work. I am also grateful to Garry Karner and Maurice Bamford for pointing out suitable geological examples to which the model results could be applied.

10. References

Beaumont, C. (1981). Foreland basins. Geophys. J. R. Astr. Soc., 65:291-329.

Bodene, J.H., Steckler, M.S. and Watts, A.B. (1981). Observations of flexure and the rheology of the oceanic lithosphere. J. Geophys. Res., 86:3695-3707.

Goetze, C. (1978). The mechanisms of creep in olivine. Phil. Trans. R. Soc. A288:99-119.

Jackson, J.A. and McKenzie, D.P. (1983). The geometrical evolution of normal fault systems. J. Struct. Geol., 5:471-482.

Koch, P.S., Christie, J.M. and George, R.P. (1980). Flow law of "wet" quartzite in the α-quartz field. Eos. 61:376.

Kusznir, N.J. and Park, G. (1986). Continental lithosphere strength: the critical role of lower crustal deformation. In: The nature of the lower continental crust, J.B. Dawson, D.A. Carswell, J. Hall and K.H. Wedepohl, editors. Geol. Soc. Spec. Pub. No.24:79-93.

Kusznir, N.J. and Park, G. (1987). The extensional strength of the continental lithosphere: its dependence on geothermal gradient, crustal composition and thickness. In: Continental Extensional Tectonics, M.P. Coward, J.F. Dewey and P.L. Hancock, editors. Spec. Publ. Geol. Soc. Lond. No. 28:35-52.

Kusznir, N.J., Karner, G.D. and Egan, S.S. (1987). Geometric, thermal and isostatic consequences of detachments in continental lithsphere extension and basin formation. In: Sedimentary basins and basin forming mechanisms, C. Beaumontand A.J. Tankard, editors. Canad. Soc. Petrol. Geol. Mem. No. 12:185-203.

Kusznir, N.J. and Egan, S.S. (1989). Simple - shear and pure - shear models of extensional sedimentary basin formation: application to the Jeanne d'Arc basin, Grand Banks of Newfoundland. A.A.P.G. Mem.46:305-322.

McClay, K.R., Norton, M.G., Coney, P. and Davis, G.H. (1986). Collapse of the Caledonian orogen and the Old Red Sandstone. Nature, V. 323:147-149.

McKenzie, D.P. (1978). Some remarks on the development of sedimentary basins. Earth and Planet. Sci. Letts., 40:25-32.

Post R.L. (1977). High temperature of Mt. Burnet Dunite. Tectonophysics. 42:75-110.

Shelton, G. and Tullis, T. (1981). Experimental flow laws for crustal rocks. Eos, 62:396.

Ussami, N., Karner, G.D. and Bott, M.H.P. (1986). Crustal detachment during South Atlantic rifting and the formation of the Tucano-Gabon basin system. Nature, 322:629-632.

Watts, A.B., Karner, G.D. and Steckler, M.S. (1982). Lithospheric flexure and the evolution of sedimentary basins. In: The Evolution of Sedimentary Basins. P. Kent, M.H.P. Bott, D.P. McKenzie and C.A. Williams, editors. Phil. Trans. R. Soc., A305:249-281.

Ultra-low frequency variability in an atmospheric circulation model

I.N. James and P.M. James

University of Reading

1. Low frequency variability of the atmosphere

The atmospheric circulation varies on a wide variety of timescales. The typical lifetime of a depression system is a few days. Other systems, such as "blocking highs" may persist for 10 or 20 days. Similar timescales emerge when linearised analyses of wave propagation and fluid instability are carried out for typical atmospheric flows. On the longer timescale, the annual variation in the distribution of solar radiation over the globe strongly forces an annual cycle into the circulation. Even longer timescales are also observed, giving rise to the inter-annual variability of the circulation. For instance, Manley (1974) estimated ten year running means of the temperature in central England from historical sources from 1659 to 1973. His curves show variations in excess of 0.5K with periods of decades or longer.

The shorter timescale or "synoptic" variations dominate the fluctuations of weather, and their limited predictability is a well known and formidable problem in applied meteorology. These variations undoubtedly originate in the internal dynamics of the atmospheric flow, which is unstable and strongly nonlinear in character. But on timescales which substantially exceed the synoptic timescale, there is a tendency to assume that variations are inevitably driven by some external forcing. Such forcing might be the slow evolution of the ocean surface temperatures, variations of solar output or changes in atmospheric composition or aerosol load, either natural or man induced. Yet internal variability, due to the chaotic nature of atmospheric dynamics, is capable of giving surprisingly large fluctuations on long timescales. An example is afforded by the numerical experiments of James and Gray (1986), in which the effect of varying surface friction in a terrestrial type of atmosphere was investigated. Even with friction timescales as short as one day, the global circulation was highly variable throughout a 500 day run, such that it was quite difficult to be certain that transient effects had been eliminated from the model climatology. Yet the model included no variations in external forcing.

In studying climate, either using observations to identify impacts of perturbations on the atmospheric circulation, or running global circulation models (GCMs) to investigate climate sensitivity, it is most important to determine the relative contribution of internal variability and external forcing. This paper shows how experiments with models of intermediate complexity can be used to address such questions. A preliminary report was given by James and James (1989) and a fuller account is in preparation.

2. A simple global circulation model

In order to isolate the effects of internal nonlinear dynamics of the atmosphere, we have set up a model which includes a sophisticated representation of the internal dynamical processes, with a resolution at least comparable to that in many current GCMs. The model is driven by highly idealised heating while the growth of kinetic energy is restrained by an equally idealised friction. In this way, the model is much cheaper to run than a comparable GCM and very long runs can be attempted on the CRAY computers at the SERC Rutherford Appleton Lab (RAL) and the University of London Computer Centre (ULCC). We refer to such models as "simple global circulation models" or SGCMs.

The model is a dry, spectral primitive equation model of flow on the sphere. The momentum equation can be written

$$\frac{\partial \underline{v}}{\partial t} + N_M + L_M = -\frac{\underline{v}}{\tau_D} \tag{2.1}$$

while the thermodynamic equation can be written

$$\frac{\partial \theta}{\partial t} + N_T = \frac{(\theta_E - \theta)}{\tau_E} \tag{2.2}$$

Here, $\boldsymbol{N}_M$ and $\boldsymbol{N}_T$ represent the nonlinear self advection terms for momentum and potential temperature θ respectively. $\boldsymbol{L}_M$ represents the linear pressure gradient and Coriolis terms. The terms on the right hand side of the equations are the idealised friction and heating terms. Friction is given by a linear "Rayleigh friction". Most of this paper is based on a run in which the drag timescale τ_D was 1 day at the lowest model level and infinite at higher levels. Heating is given by "Newtonian cooling", with relaxation towards a state of "radiative equilibrium" θ_E. In the calculations reported here, τ_E was 30 days, and θ_E was chosen with slightly stable stratification, a pole to equator temperature difference of 80K and a temperature difference of 40K between the two poles, thus representing conditions at the solstice. We shall denote this integration "run 1".

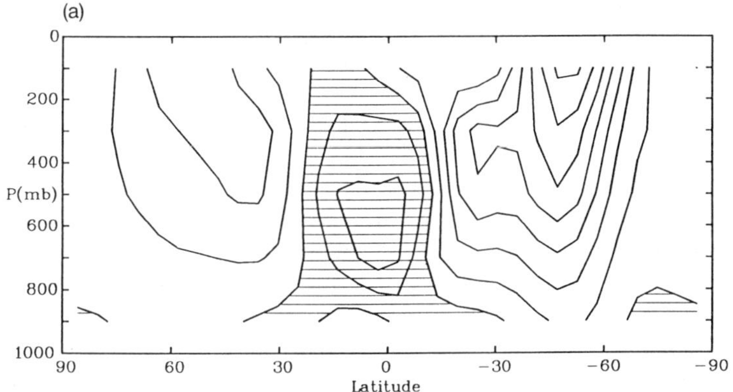

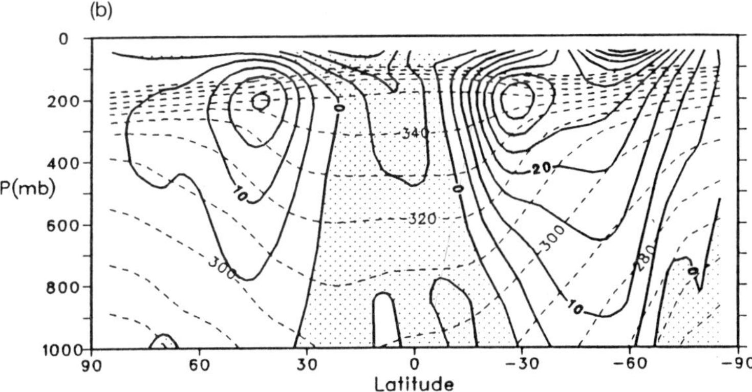

Fig 1: Latitude-pressure cross sections showing the time and zonal mean zonal wind. (a) for run 1; (b) observed, contour interval 5 $m.s^{-1}$, negative (easterly) values shaded.

The model was run with five levels in the vertical, and with triangular truncation at wavenumber 21 in the horizontal. This resolution is sufficient to model the development and decay of midlatitude depression systems quite reasonably, though the later stages of the life cycles, when the structures become highly distorted and sheared are less satisfactory. It was possible to run the model for very extended periods; run 1 was 100 years.

The model climatology is surprisingly realistic. In response to the heating which forces the radiative equilibrium thermal structure, the atmospheric circulation transports heat upwards and polewards, especially in the winter hemisphere. The pole-equator temperature gradient is reduced and the static stability increased. In the tropics, these transports are accomplished by axisymmetric overturning or "Hadley" circulations. In the midlatitudes, transient, baroclinically unstable eddies dominate. Fig. 1 shows the zonal mean westerly wind as a function of latitude and height, compared with observations given by Hoskins et al (1989). The main features, namely, tropical easterlies, a strong winter subtropical jet at upper levels and a midlatitude jet extending through a deeper layer in both hemispheres, are reproduced. The distribution of eddy activity, with a maximum in the midlatitudes, is also realistic. It is worth remarking that such a comparatively simple model provides an admirable pedagogical tool for studies of the global circulation.

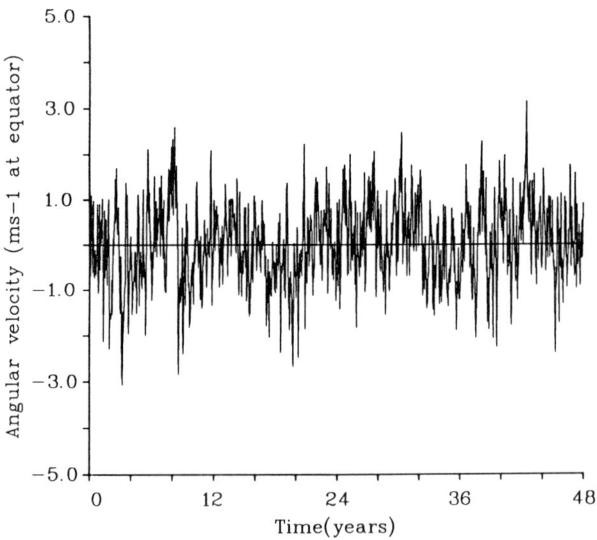

Fig 2: Time series of the global mean atmospheric angular velocity for the first 48 years of run 1. The time average has been removed, and the values of angular velocity are presented in terms of zonal wind at the equator.

The model includes disturbances which are transient on the synoptic timescale of a few days. Like the atmosphere, the flow is chaotic, with systems developing and collapsing in a quasi-random fashion throughout the midlatitudes. During their life cycle, these transient eddies interact with the zonal mean flow and modify it. The time variations of the model are illustrated by Fig. 2. This shows a single measure of the global circulation, namely the global mean angular velocity of the atmosphere relative to the solid earth, as a function of time for part of the run. Clearly, the angular velocity fluctuates rapidly in response to the individual weather systems. But superimposed upon this variation, there are more persistent changes which last for several years. For example, there is a general increase in angular velocity from year 20 to year 32, followed by a more rapid collapse. Other similar examples can be seen throughout the record.

A Fourier analysis of the data confirms the subjective impression of the importance of low frequencies. The time spectrum of angular velocity shows large amplitudes at the lowest frequencies contained in our time series. The largest amplitudes are at decadal time scales, although these peaks are probably not statistically significant, and would not be exactly reproduced in different realisations of the system. We shall refer to variability on such timescales as "ultra-low frequency" variability (ULFV), as distinct from the "low frequency variability" with timescales of 30-60 days discussed by other authors (e.g. Wallace and Blackmon, 1983).

3. Structure of low frequency variability

We have examined a number of measures of the flow and generated time spectra for them. It is clear that in this model, the ultra-low frequency variability is associated with the zonal mean state, and not with the eddies. The eddies have much shorter timescales of less than a year, although they have maximum variability on timescales which are considerably longer than the synoptic timescales we might expect. So instead of looking at the zonal mean flow arbitrarily averaged over the globe, we now study the spatial structure of the variability. The most compact means of describing such spatial structure is to use the method of "empirical orthogonal functions" (EOFs). A concise introduction to the method is given in Mo and Ghil (1987).

The EOF analysis was applied to the zonal mean zonal wind. The first EOF accounts for 31% of the total variance of the field, and it has a smooth large scale structure confined to the winter hemisphere, illustrated in Fig. 3. The remaining EOFs represent much smaller fractions of the variance and have an increasingly noisy and disorganised structure. Beyond the first few,

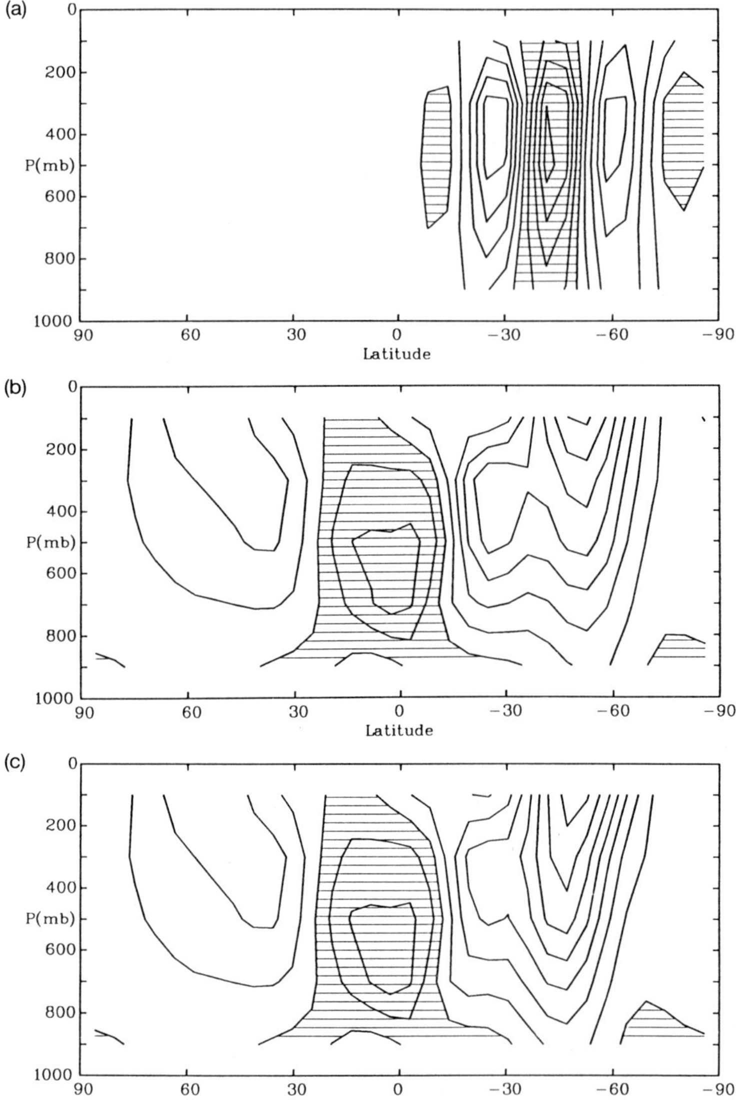

Fig. 3: (a) Showing the structure of EOF1 for run 1. Contour interval is arbitrary, negative values are shaded. (b) Showing the zonal wind in the positive phase of EOF1. (c) showing the zonal wind in the negative phase of EOF1. Contour interval in (b) and (c) is 5 $m.s^{-1}$, negative values shaded.

they probably have little physical significance. But the fluctuations of EOF1 have an interesting interpretation, also shown in Fig. 3. During the negative phase of EOF1, the subtropical jet is weakened and the midlatitude jet is shifted equatorwards. During the positive phase of EOF1, the subtropical jet is strengthened and the midlatitude jet is displaced polewards. EOF2 accounts for only 12% of the variance, and its structure corresponds to an alternate separation and coalescence of the subtropical and midlatitude jets.

The fluctuation of EOF1 is largely associated with the ULFV. Its timeseries has been Fourier analysed. Its spectrum is red and is generally similar to that of the mean angular momentum. The maximum amplitudes are at ultra-low frequencies. The remaining EOFs have white or only much more weakly red spectra than does EOF1.

Along with this ULFV of the zonal wind, there are associated changes in the structure of the temperature field and the eddy activity. These are illustrated by constructing composites, built up by averaging the fields for the 10% of the cases when the amplitude of EOF1 is most positive. ULFV of the temperature is concentrated in the winter midlatitudes, with cooling in subtropical and polar latitudes and warming in the storm track latitude. The magnitudes are around 1K, which is comparable with the temperature fluctuations shown by Manley (1974) for central England. The eddies are generally weakened during the positive phase of EOF1, and the largest heat and momentum fluxes are shifted polewards.

4. Sensitivity of ULFV

A number of other runs have been carried out to investigate the sensitivity of the ULFV to changes in the imposed parameters. For present purposes, we shall mention just one further run, denoted run 2, which shows a considerable increase in ULFV. The most unrealistic aspect of runs at low resolution is the small momentum fluxes. The poleward momentum fluxes are associated with the shearing and rotation of decaying vortices by the larger scale flow, resulting a cascade of energy to larger zonal scales and eventually to the zonal mean flow (Rhines, 1975). If the resolution is too coarse, this cascade cannot proceed very far. So run 2 employed lower dissipation on the eddy part of the flow. The model used the same equations as before, the only difference being that the Newtonian cooling was only applied to the zonal mean state, not to the eddies. The value of τ_D and the pattern of θ_E was chosen so that the levels of dissipation and eddy temperature flux were comparable in both runs. As a result, the thermal damping of the flow was concentrated in the zonal mean part of the flow and the cascade of energy to the zonal state was enhanced.

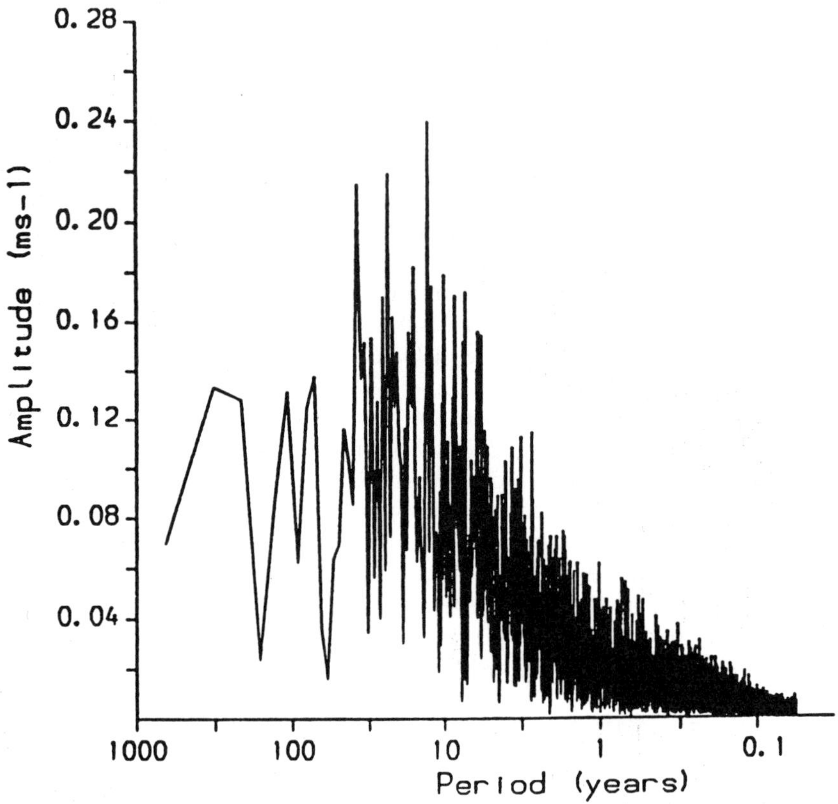

Fig. 4: Spectrum of the variations of atmospheric angular velocity for 648 years of run 2.

The ULFV was considerably larger than in run 1. Eventually we integrated run 2 much further than run 1, enabling the shape of the spectrum to be determined more accurately at low frequencies. The time spectrum of the atmospheric angular velocity is shown in Fig. 4. The maximum amplitudes occurred for periods between 10 and 30 years. It is remarkable that such a long timescale emerges from the model. Recall that the timescales associated with drag and heating are 1 and 15 days respectively, while the frequency and growth rates of the linear normal modes of the system suggest linear dynamical timescales no longer than this. It is the nonlinearity of the flow itself which leads to its chaotic evolution and hence to its very large low frequency variability.

5. Concluding remarks

The implications of this work for climate research are, it seems to us, very important. The existence of trends in certain kinds of climate data over periods as long as decades would be no evidence for systematic climate change caused by external forcing. The appearance of strong spectral peaks in comparatively short sequences of data (of order a few decades or less) does not necessarily indicate any periodic external forcing. For example, Labitzke and Van Loon (1988) suggest that they can detect the influence of the 11 year solar cycle in atmospheric circulation statistics. It is ironic that our preliminary analysis of the first 100 years of run 2 revealed the largest peak in the spectrum of angular velocity with a period of 10-12 years! Again, attempting to use GCMs, integrated over a few annual cycles, to predict circulation changes due to (for example) carbon dioxide increases may be quite misleading. Our model, with no such effects, shows substantial changes of eddy activity and shifts of the storm track latitudes which are comparable to those ascribed to greenhouse warming in many GCM studies.

In a conference of this nature, it is appropriate to conclude by pointing out that the use of SGCMs like ours is a very efficient use of computer resource. Our model required some 0.2 hours of CPU time on the RAL CRAY X-MP to carry out 1 year of integration. It is a small job, only requiring some 350kwords of memory, enabling other jobs to run easily on the other processors. Our 100 year runs therefore needed about 20 hours of computer time. A run of the UGAMP global circulation model at similar resolution would require around 3 hours of computer time per year and would also demand about 1.2mwords of memory. The difference is the massive extra computing needed to carry out radiative transfer calculations, calculations of surface exchanges of heat, moisture and momentum and parametrizations of moist processes. Our 1000 year integration of run 2 would be inconceivable with a GCM; with our SGCM it is merely extravagant.

6. Acknowledgements

This work was carried out as part of the programme of the UK Universities' Global Atmospheric Modelling Project, UGAMP. We thank our colleagues in UGAMP for encouragement and discussions. We are grateful to NERC for affording us the necessary computing time at RAL and ULCC. PMJ also acknowledges the support of an NERC Advanced Studentship during this research.

7. References

Hoskins, B.J., Hsu, H.H, James, I.N., Masutani, M., Sardeshmukh, P.D., and White, G.H. (1989). Diagnostics of the global circulation based on ECMWF analyses 1979-1989.

James, I.N. and Gray, L.J. (1986). Concerning the effect of surface drag on the circulation of a baroclinic planetary atmosphere. Q.J. Roy. Met. Soc., 112:1231-1250.

James, I.N. and James, P.M. (1989). Ultra-low frequency variability in a simple atmospheric circulation model. Nature, 342:53-55.

Labitzke, K. and Van Loon, H. (1988). Association between the 11-year solar cycle, the QBO, and the atmosphere. Part I: the troposphere and stratosphere in the Northern Hemisphere in winter. J.Atmos.Terrestrial Physics, 50:197-208.

Manley, G. (1974). Central England temperatures: monthly means 1659 to 1973. Q.J. Roy. Met. Soc., 100:389-405.

Mo, K.C. and Ghil, M. (1987). Statistics and dynamics of persistent anomalies. J. Atmos. Sci., 44:877-901.

Rhines, P.B. (1975). Waves and turbulence on a beta plane. J. Fluid Mech., 69:417-443.

Wallace, J.M. and Blackmon, M.L. (1983). Observations of low frequency atmospheric variability. In: Large scale dynamical processes in the atmosphere, B.J. Hoskins and R.P. Pearce, editors. Academic Press, London.

Towards an improved person-machine interface in atmospheric modelling

S.P. Cooper and W.A. Norton

University of Cambridge

1. Introduction

The aim of large-scale numerical modelling of the atmosphere is to increase our understanding of how the atmosphere works. Increased understanding will lead to more accurate weather forecasts and will enable us to quantify the effects of man-made pollutants on the climate and ozone layer.

With current supercomputers, it is feasible to run high-resolution three-dimensional numerical models of the atmosphere, e.g. the UGAMP (UK Universities' Global Atmospheric Modelling Project) model. Such models can produce vast amounts of data, typically hundreds of megabytes for a 100 day integration. Analysing this data and comparing it with the results from other models and with observations of the real atmosphere is often far from straightforward. One might wish to follow the evolution of a particular feature, e.g. a deepening cyclone, or to see how two features interact. This requires some means of viewing scalar and/or vector fields as they vary in space and time. It might be desirable to view two or more fields simultaneously, e.g. the wind and pressure fields for a cyclone. Furthermore fields of advected quantities, such as moisture and chemical mixing ratios, can become topologically complicated and vary on a wide range of spatial scales. Conventional presentation techniques rely heavily on the use of contour plots. However these rapidly become illegible as the complexity of the field increases - contours can crowd together making it difficult to distinguish between high and low values.

Recently, the problem of the "data glut" produced by supercomputers and the analysis of large numerical datasets has started to receive attention, e.g. Bitz and Zabusky (1990) with the DAVID system. New interactive analysis techniques have been made practical by the availability of high speed networks and fast window-based workstations, used more traditionally in CAD/CAM (Computer Aided Design/Manufacture) environments. Using "photographic" types of visual presentation, i.e. colour or greyscale (*note*: the term colour is used throughout the remainder of this paper to mean colour or greyscale), qualitative patterns can be revealed much more efficiently than with contours, and in particular the problem of contour crowding is avoided. Combining this form of presentation with animation makes full use of the ability of the human visual system to assimilate information. This is especially true when viewing advected fields, an immediate impression of fluid motion and the timescales associated with dynamical processes, such as mixing, is obtained. Animation is also an ideal way to quickly scan and locate time intervals of an integration that are of particular interest which can then be further studied.

New diagnostics that are derived from the basic model fields will also be important in the clear presentation of results from numerical models. Traditionally in analysing atmospheric data, the fields of winds, temperature and pressure have been used. However over the last few years there has been increasing interest in the scalar quantity *potential vorticity* (which we will refer to as PV), e.g. Hoskins et al (1985). This is because PV has two special properties. In the absence of diabatic heating, such as radiation or cumulus convection, it is advected like a chemical tracer. Hence PV acts as an air parcel marker. In addition, the distribution of PV can be considered, to a very large degree, to control the dynamical evolution. That is, from the PV field one can rederive almost exactly the wind, temperature and pressure fields and their time tendencies (see McIntyre and Norton, 1990). In particular high PV indicates cyclonic winds, low PV indicates anticyclonic winds. These properties mean that PV combined with colour animation is a particularly powerful and concise way of following the dynamical evolution of an atmospheric model. We present an example of a PV field in Fig. 1.

This paper presents a tool currently being developed, which we have named VIDI, that uses the above ideas to move towards an improved person machine interface in atmospheric modelling. VIDI could also be useful for analysing data from many other numerical experiments.

2. VIDI - Visual Interactive Data Interface

VIDI is a system for analysing of large numerical datasets. The basic data structure of VIDI is the *movie*, which is a sequence of rectangular *frames* of byte data. Files containing movies are routinely produced from atmospheric models run on the CRAY computers at ULCC (University of London Computer Centre) and RAL (SERC Rutherford Appleton Laboratory). They are transferred over JANET to a local workstation where they are viewed and analysed (there is more about the movie file format and data transfer in Section 3).

With a movie loaded into VIDI various interactive tasks can be performed. Care has been taken to provide ergonomically efficient controls.

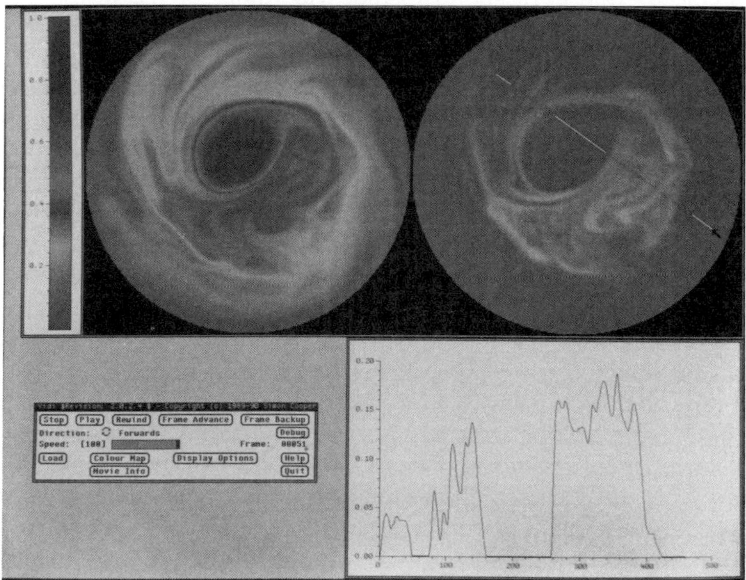

Fig. 1: An example of VIDI in use (see Plate 2, pp.14-15, for colour version). The top window shows frames of the potential vorticity field (left) and a tracer field (right) from a movie produced from an integration of a high resolution single layer atmospheric model (see text). The colour bar on the left hand side shows the colour map. The bottom left window is the control panel with the basic VIDI commands indicated by buttons and the animation speed by the slide control. A line has been drawn on the right hand frame with the mouse and the resulting cross-section is in the bottom right window.

Developing these constitutes a substantial part of the programming effort. The basic user interface is designed around the WIMP (Windows, Icons, Mouse and Pull down menus) environment. Much of the pioneering work in this area was done by Rank Xerox in the 1970s, e.g. Goldberg and Robsin (1983), and it is now used in many window-based applications. At the simplest level VIDI has the controls similar to those found on a video recorder. By selecting options in the control panel window with the mouse, a user can play, stop, frame advance or backup the movie. The playing speed can be altered by changing a slide control. A specific frame can be selected by entering its number on the keyboard. There are also a number of more advanced features which we shall now describe.

The colours that correspond to the byte values of the displayed frame are given by the *colour map*. A colour bar showing the colour map is positioned to the left of the displayed frame. The colour map can be changed by selecting the "colour map" option in the control panel window. This opens a new window which contains line graphs showing how the byte values, 0-247 along the horizontal axis, map to an intensity of colour indicated by the vertical axis (byte values 248-255 are reserved for use by VIDI). A colour bar along the bottom shows the resulting colour from the addition of the red, green and blue intensities. A greyscale is achieved by having the three intensities equal at each byte value. The colour map can be changed in a variety of ways. One can select predefined colour maps for a linear greyscale and, on colour displays, a rainbow. Alternatively a colour map can be selected from those included within a movie file. There are several ways to design a new colour map. Using the mouse, the line graphs which define the colour map can be manipulated - any combination of straight lines, cubic splines and freehand curves can be composed, and if necessary smoothed. The choice of colour map can make a significant difference to the visual information gained from a movie. A single colour map will normally be insufficient to clearly see all features of a given model run. Different features of a movie can be highlighted by suitably changing the colour map.

Cross-sections can be taken by drawing a line with the mouse on the displayed frame, a new window opens to show the result. Multiple sections can also be taken. The cross-sections can be made to automatically update upon frame advance or backup. Spot values of the data in the displayed frame can be taken. Although colour representation is the normal method of displaying a frame, contour plots and 3D mesh plots can also be selected.

It may be desirable to analyse more than one field from a numerical experiment. A movie file is structured so that it can contain movies of several fields. VIDI can selectively display any combination of the different movies. Movies from different movie files can also be loaded so that the results from different numerical experiments can be compared.

VIDI can accommodate frames of any size (provided they fit on the workstation display) and there is an option to expand the size of the frames when they are displayed. For example, a movie might consist of data in 256 x 256 frames, images of this size are rather small so VIDI normally expands frames of this size to 512 x 512 pixels. Better visibility is thus obtained and, furthermore the frames need be stored only in their unexpanded form. Efficient use of memory storage is important because for fast animation the whole movie must be resident in the host computers memory.

To enable users to quickly learn the operation of VIDI, on-line interactive help information can be displayed. With the help option selected on the control panel, a help window opens which explains the options currently available in the window pointed to by the mouse. For example, if the mouse is placed in the window of the displayed frame, help is given on how to draw cross-sections and obtain spot values. If help is no longer required then it may be disabled by closing the help window.

Displayed frames, cross-sections and other output can be converted to PostScript and subsequently be sent to a laser printer. Currently VIDI only supports black and white PostScript devices using "half toning" to achieve greyscale. Acceptable output can be produced from an Apple LaserWriter; higher quality output has been obtained from the Linotronic at the ULCC phototypesetting service. Colour output will be supported when colour PostScript printers become widely available.

3. Movie file format and transfer

Unfortunately different computers do not always represent numbers and characters in the same way. There are different representations of floating point numbers, ordering of bytes in words, and character sets (ASCII and EBCDIC). This makes the transfer of data between different computers somewhat difficult and hazardous.

The movie file format has been designed to overcome these problems as well as being flexible and easily extensible. By reading a standard initial string, VIDI can immediately recognize the data representation used to write the file. The necessary translations to the data representation of the host computer can then be made. The rest of the file is constructed using *tags*. A tag describes the type of the next data object. For example there are different tags for frames of a movie, scaling information, and comments (such as the parameters of the model integration). This structure allows the movie file to consist of any mixture of formatted and unformatted data and to be constructed in an almost ad hoc way.

Another problem is that the transfer of large datasets over networks can take a considerable time (perhaps many hours). VIDI supports two techniques to reduce dataset size.

The frames displayed by VIDI are rectangular but the actual model domain, or way the results are presented, might be some other geometry. This is often the case in global atmospheric modelling, for instance the frame displayed in Fig.1 uses a polar stereographic projection looking down on the north pole. By stencilling the frames, i.e. storing only the part of frames which contains data, savings are made in the size of datasets. In particular it means that more frames of a movie can fit into the memory of the host computer. With a circular mask, as in Fig.1, the saving in storage is approximately 21%.

The second technique is Lempel-Welch-Ziv coding (Welch, 1984), as used by the Unix "compress" utility. This is a data compression algorithm which extracts common bit patterns from data. It reduces the size of movie files by typically 70% and hence significantly reduces the time to transfer large movie files. The movie files are "uncompressed" as they are read into VIDI - this operation being too computationally expensive to perform when the movies are being displayed (although with dedicated hardware this could become possible).

4. Computer requirements

The VIDI system currently runs under the Sun Microsystems "SunView" windowing environment. It has been used on Sparcstation 1 and Sun 3/60 workstations. Maximum animation speed on a Sparcstation for a 256 x 256 frame movie expanded to an image of 512 x 512 pixels is approximately 12 frames per second. A workstation with 16 megabytes memory can store in memory a 256 x 256 frame movie of maximum length of approximately 160 frames. Movies of greater length require swapping to disk and so only animate at about 1 frame per second.

5. Example of the use of VIDI

VIDI has been used to analyse data from a variety of atmospheric models. Because of the complex nature of the atmosphere, some processes must be simplified or parameterised, to a varying degree, in all models. By using a hierarchy of models, where in different models some processes are accurately represented while others are simplified, progress in understanding many different atmospheric phenomena can be made. For example one model used has only a single horizontal layer but consequently can be run at very high horizontal resolution. Another model has typically 21 vertical levels but these are confined to the stratosphere (see Haynes, 1990). Both models have simplified representations of forcing from the troposphere. The most sophisticated model used by researchers in UGAMP is a general circulation model. VIDI has been useful in comparing the results from all these models. In particular, it has helped us to see what features are reproduced between models and also the inadequacies of each model.

Fig.1 shows an example of VIDI in use. The top window shows two frames of different fields from a movie. Note the colour bar on the left hand side which shows the colour map. The bottom left window is the control panel with the basic VIDI commands indicated by buttons and the animation speed by the slide control. A line has been drawn on the right hand frame with the mouse and the resulting cross-section is shown in the bottom right window.

The left hand frame is a potential vorticity (PV) field (see Section 1) from an integration with a single-layer high-resolution atmospheric model (PV in this model is defined as the Coriolis parameter plus the curl of the velocity divided by the thickness of the layer, and is approximately a "material invariant" or advected quantity). The data is presented in a polar stereographic projection looking down on the north pole. The equator is at edge of the circle. VIDI has expanded the 256 x 256 frame to an image of 512 x 512 pixels. This model is being used to study the dynamics of the stratospheric polar vortex and in particular to understand aspects of the "ozone hole" (e.g. McIntyre 1989). The dark (blue) region near the centre of the frame is the polar vortex consisting of high-PV air. There is a strong jet circulating anticlockwise around the edge of the polar vortex. The right hand frame is a passive tracer field from the same integration of the model. The initial distribution of the tracer field in the integration was a single blob situated half way between the north pole and the equator (i.e. outside the polar vortex). The dynamics of the model has mixed the tracer field during the integration. However as indicated by the cross-section, there is no tracer inside the polar vortex. Therefore this experiment indicates that the edge of the polar vortex acts as a barrier to air parcels and so the air inside the polar vortex is isolated from the mid-latitude and tropical air. This is thought to be an important requirement for the formation of the ozone hole.

6. Conclusions and future directions

VIDI has already proved a powerful tool in analysing data from atmospheric models. We believe the big challenge in the future is to develop powerful visualization techniques for 3D fields. At present VIDI can only display a 3D field as a series of horizontal sections. Other techniques such as the construction of "solid/translucent" images or shading a given isosurface in the 3D domain are being investigated. These algorithms can be computationally expensive, but could be precalculated on a supercomputer or parallel architecture machine for which they are ideally suited. One problem might be that small-scale features or numerically generated noise might hide the underlying large-scale structure.

A further extension to VIDI would be the inclusion of a command language so that more complicated manipulation and post-processing of the data could be performed. This could lead to a system where the initial conditions of numerical experiment could be interactively manipulated and "what if ..." type sensitivity experiments could easily be performed (perhaps with applications to numerical weather prediction: see McIntyre, 1988).

The future holds the promise of a fast portable graphics standard running on high performance workstations. We can also look forward to faster networks (such as JANET II) which would allow greater interaction with numerical experiments running on remote supercomputers. VIDI is perhaps a first step towards a new powerful person-machine interface in numerical modelling that takes full advantage of workstation technology. Such tools are surely needed if advances in supercomputing technology are to be most profitably exploited in the understanding of our natural environment.

7. Acknowledgements

We would like to thank P.H. Haynes and M.E. McIntyre for the contributions they have made to this project. We acknowledge support from the Natural Environment Research Council through the UK Universities' Global Atmospheric Modelling Project (UGAMP) and the British Antarctic Survey, from the Science and Engineering Research Council and from the US Office of Naval Research.

8. References

Bitz, F.J. and Zabusky, N.J. (1990). DAVID and "Visiometrics". Visualizing, diagnosing and quantifying evolving amorphous objects.
To appear in Computers in Physics.

Goldberg, A. and Robsin, D. (1983). Smalltalk-80: The Language and its Implementation. Addison-Wesley: 705pp.

Haynes, P.H. (1990). Environmental research - a user's experience. ULCC News, 241:15-17.

Hoskins, B.J., McIntyre, M.E., and Robertson, A.W. (1985). On the use and significance of isentropic potential-vorticity maps. Q. J. Roy. Meteorol. Soc., 111:877-946. Also 113:402-404.

McIntyre, M.E. (1988). Numerical weather prediction: a vision of the future. Weather, 43:294-298.

McIntyre, M.E. (1989). On the Antarctic ozone hole. J. Atmos. Terrest. Phys., 51:29-43.

McIntyre, M.E. and Norton, W.A. (1990). Potential vorticity inversion on a hemisphere. Submitted to J. Atmos. Sci.

Welch, T. A. (1984). A Technique for High Performance Data Compression. IEEE Computer, 17, no. 6.

Fast models on small computers of turbulent flows in the environment for non-expert users

D.J. Carruthers

Cambridge Environmental Research Consultants Ltd

J.C.R. Hunt, R.E. Britter, R.J. Perkins,
P.F. Linden, S. Dalziel

University of Cambridge

1. Introduction

Recently there have been many developments of new analytical formulations for airflow, turbulence and diffusion in the atmospheric boundary layer over complex terrain including changes in both surface elevation and roughness. These formulations have enabled the development of models for non-experts which run on personal computers. The models are mainly based on rather simple equations (generally differential equations of first order).

Analytical solutions to the governing equations can be derived for the mean atmospheric flow, turbulence and diffusion over hills and roughness changes given certain assumptions, namely:

(i) the slopes of the hills are small (typically less than about 1/4, although useful result are obtained for slopes as great as 1/2);

(ii) the profile of potential temperature $\theta(Z)$ in the atmosphere can be approximated into layers which each have single analytic forms. In our model we consider five basic forms;

(iii) the upwind velocity profile increases from ground upwards and does not have a strong elevated shear layer;

(iv) the upwind conditions are varying slowly on a time scale comparable with that taken by a fluid particle to cross the flow region under consideration. For a 20km long region and a wind of 10 $m.s^{-1}$, this means slow changes over half an hour;

(v) rapid cooling or heating of the surface of the hillside is absent; this can induce significant motions which are not considered here;

(vi) the turbulent structure near the surface can be approximated by the "mixing length" relation between the shear stress and velocity gradient in a region near the surface and by rapid distortion theory in an outer region away from the surface;

(vii) the stratification is not so strong that the flow is blocked by the hills.

Using these assumptions Hunt, Leibovich and Richards (1989), Xu and Hunt (1990), and Hunt, Richards and Brighton (1989), have derived formulae for the Fourier transform of the perturbation (or changes) in the mean velocity distribution over the terrain. To evaluate the actual flow the Fourier transform has to be inverted, usually numerically. These papers are based on earlier studies by Jackson and Hunt (1975) and Walmsley et al (1986) for neutral flow over hills; Walmsley et al (1982) for roughness changes; and Smith (1988), Carruthers and Choularton (1982), and Hunt and Richards (1984) for stratified flows over hills of low slope. The use of the Fourier transform technique to calculate flows over arbitrary terrain was pioneered by Walmsley et al (1982) but it was restricted to *neutral* flow and was based on the earlier and inaccurate model of Jackson and Hunt.

To understand the mean flow model, it is helpful to explain why it is different to models based on computing full equations of motion (e.g. PHOENICS, FLUENT etc.). In the full computations, it is necessary to compute five or six differential equations for six variables at each point in the flow domain and so the computer storage required is an order of magnitude greater than that needed by FLOWSTAR. In addition, FLOWSTAR requires much less computing time; for instance it uses about 5 minutes on an IBM PC AT to calculate the flow at one level on a 32 x 32 grid.

The method of calculation of the mean flow is to compute the Fourier transform of the velocity field; then the transform is inverted to calculate the actual flow variables at a point. There is, of course, no iteration involved and no doubt about the solution once the algorithm and its assumptions have been agreed. This is why the Fourier transform method is quite appropriate for use on small personal computers.

Similar procedures are used to calculate the shear stresses and from them and the mean flow, the turbulence velocities and length scales. These in turn are used to calculate the diffusion of pollutant from a continuous source (FLOWSTAR-D).

2. Model descriptions and analysis

2.1 Mean flow

In the solution for airflow over hills used to produce the algorithms, the turbulent boundary of the lower atmosphere is divided into three layers (Fig.1): the inner, middle and upper layers. The velocity is $\mathbf{u} = (U + u,v,w)$, where U is the upwind mean velocity and u,v,w are the turbulent wind components which vary with x,y and Z, the height above the ground.

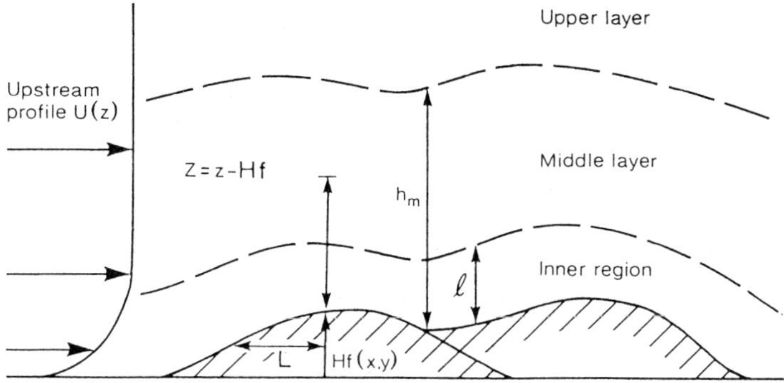

Fig. 1: Pictorial representation of flow regimes.

The inner layer is adjacent to the ground and so the perturbation shear stresses are important. In this layer the equations for the mean flow are solved by using a mixing-length closure for the turbulence. They are

$$
\begin{aligned}
U\frac{\partial u}{\partial x} + w\frac{\partial U}{\partial Z} &= -\frac{1}{\rho}\frac{\partial p}{\partial x} + \frac{\partial \tau_{xz}}{\partial Z} \\
U\frac{\partial v}{\partial y} &= -\frac{1}{\rho}\frac{\partial p}{\partial y} + \frac{\partial \tau_{yz}}{\partial Z} \\
\frac{\partial u}{\partial x} + \frac{\partial v}{\partial y} + \frac{\partial w}{\partial Z} &= 0 \\
\tau_{xz} &= 2\kappa u_* Z\frac{\partial u}{\partial Z}
\end{aligned}
\tag{2.1}
$$

where: u_* is the friction velocity; p and ρ are the pressure and density; and κ von Karman's constant.

The middle layer, whose height is h_m, is sufficiently far above the ground that shear stresses are unimportant; however, the effects of shear are important. Here the governing equation is

$$
\frac{\partial^2 w}{\partial x^2} - \frac{d^2U/dZ^2}{U}w = 0 \tag{2.2}
$$

The outer layer contains the outer part of the turbulent boundary layer and may also include part of the free, non-turbulent atmosphere (e.g. in case S_2, see Section 2.2, the air above the inversion is generally non-turbulent but does not form part of the boundary layer). Stratification now has an important effect but shear and perturbation stress are unimportant.

In this layer we solve the equations for inviscid stratified flow, i.e.

$$\frac{\partial^2 w}{\partial x^2} + \frac{\partial^2 w}{\partial Z^2} + \left(\frac{N^2}{U^2} - \frac{U''}{U}\right)w = 0 \qquad (2.3)$$

Note that to first order it is the pressure field developed in the outer layer at $Z=h_m$ which drives the flow in the two lower layers. This pressure field is strongly affected by stratification and so by this means flow in the lower layer is affected by stratification in the upper layer.

2.2 *Effects of stratification*

We consider five profiles of stratification. In this section we briefly discuss the physical effects of different stratification types whilst in Section 4.5 we define their forms precisely.

In case S_0, there is no stratification and the perturbation in the upper layer decays with height above the hill surface.

In case S_1 where there is uniform stratification (i.e. the buoyancy frequency N is constant with height), the equations show that horizontal waves with wavenumber $k<N/U_0$ can propagate, whilst waves with wavenumber $k>N/U_0$ are evanescent and thus decay away from the surface (U_0 is the speed of the upwind flow). A radiation boundary condition has to be used for low frequency/low wavenumber limits, namely that energy propagates upwards. This gives a downward phase velocity and the familiar asymmetrical flow pattern with the strongest velocities downwind of the summit of the hill. The amount of flow moving in horizontal planes around the hill increases with stratification.

In the second kind of stratification considered here, S_2, there is an inversion at height z_i. Above z_i the air is stably stratified. Waves with wavenumber $k>N/U_0$ can propagate *within* and along the inversion layer but not in the layers above and below z_i. It is found that energy may be trapped giving large-amplitude perturbations which, in the atmosphere, appear downwind of the hill. In the linearised model we have to make an assumption about the amplitude of these resonant waves. For the time being, we adopt a plausible value until such time as work in progress provides a definitive solution.

Two further stratification types are also accounted for: S_3, a uniformly stratified boundary layer capped by a strong inversion and S_4, a boundary layer of decreasing stratification with uniform stratification above. Both these forms exhibit strong wave activity.

2.3 *Shear stresses and turbulence*

These were calculated following the linear model of Hunt, Leibovich and Richards (1989). The perturbation shear stress is taken as

$$\tau = 2\kappa Z u_* \frac{\partial u}{\partial Z} \qquad (2.4)$$

This expression for the stress is accurate in the lower part of the inner region. However, where the perturbation velocity gradient reverses in the upper part of the inner region, the expression leads to an overestimate in the decrease of the stress perturbation.

For the calculation of the turbulence components, estimates using the calculated shear stress and an assumption of local equilibrium are made in the lower part of inner region, while rapid distortion theory is used in the upper part of the middle and outer regions since the structure is mainly determined by the distortion of the upwind turbulence structure and not by local nonlinear interactions. Between these two regions there is a layer where there are a number of effects determining the structure of the turbulence including advection, distortion, curvature and local non-linear effects (Finnigan, 1988). It is not possible to calculate these effects with an approach suitable for use in conjunction with the other calculations in FLOWSTAR, since such calculations would require too much computer time. So, since we know from observations that the turbulent velocities do not attain locally large values in this region, we use blending functions to match the solutions in the inner and outer regions across this layer (see Carruthers and Hunt, 1990). The longitudinal component of turbulent velocity is referenced as σ^2_u, the transverse component as σ^2_v, and the vertical, σ^2_w, but only the latter two are important for diffusion calculations.

The blending functions are such that the solutions are continuous at $Z=\ell$. A factor $(1\text{-}0.8Z/h)^2$ has been introduced to allow for the decrease with height in the turbulent energy upwind of the hill; h is the height of the boundary layer which is specified by the user. It is assumed that mechanically driven turbulence dominates convectively produced turbulence. However, in moderate or strongly convective conditions this condition is no longer held. In these situations, which occur in stratification Cases 0 and 2 when the Monin Obukhov Length $L_{MO}<0$, we include a contribution to the turbulent velocities due to convection; this contribution is not affected by the flow over the hills.

Turbulence length scales: These are required for the diffusion calculation. Following Weng, Richards and Carruthers (1988), we assume a vertical length scale $L_x^{(w)}$ which includes the effects of the surface blocking effect, local shear and boundary layer depth.

$$L_x^{(w)} = \left[\frac{A}{Z} + \frac{BdU/dZ}{\sigma_w} + \frac{4}{h}\right]^{-1} \qquad (2.5)$$

where the constants are A=0.6, B=1.0. The transverse length scale ($L_x^{(v)}$) includes contributions from scales of the same order as the height of the boundary layer and local scales.

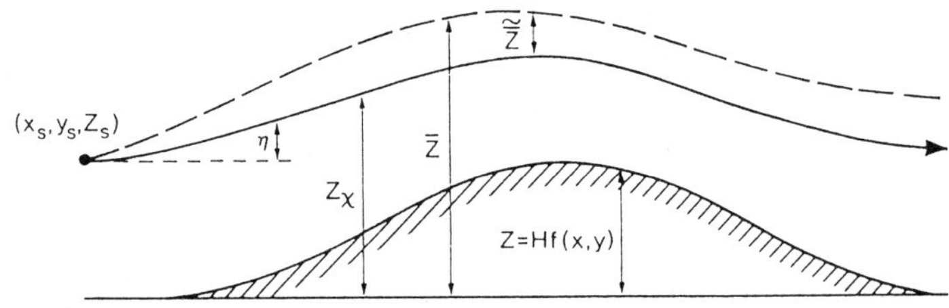

Fig. 2: Streamline (solid line) and plume centreline (dashed) downwind of a source at (x_s, y_s, Z_s).

2.4 *Streamlines and diffusion*

The first stage in dispersion calculations is to determine the height Z_χ of the mean streamline through the source at position (x_s, y_s, Z_s); then we calculate the position of the centre of the plume relative to the mean streamline (see Fig. 2). We use a simple non-linear algorithm for this which is an improvement on a strictly linear approach. Letting the vertical deflection of the mean streamline be $\eta(x,y;Z_s,y_s)$, so that $Z_\chi = Z_s + \eta(x,y;Z_s,y_s)$, then

$$(U+u)\frac{d\eta}{dx} = w(x,y,Z_s+\eta) \qquad (2.6)$$

This equation satisfies identically the condition that the streamlines are parallel to the hill surface. Because of vertical gradients in the vertical component of turbulence and length scales, the mean height of the plume does not follow exactly the streamline through the source.

Concentration distribution: Following the approach of Hunt (1985), we calculate the transverse and vertical mean square displacement of concentration about the mean position of the plume. The thin plume approximation with allowance for reflection at ground and at the inversion layer is used to calculate the concentration field, so that

$$C(x,y,z) = \frac{q}{2\pi\sigma_y\sigma_z U(\bar{Z})} \Big[\exp\Big(-\Big[\frac{(Z-\bar{Z})^2}{2\sigma_z^2} + \frac{(y-\bar{y})^2}{2\sigma_y^2}\Big]\Big) + \exp\Big(-\Big[\frac{(Z-\bar{Z}_i)^2}{2\sigma_z^2} + \frac{(y-\bar{y}_i)^2}{2\sigma_y^2}\Big]\Big) + \exp\Big(-\Big[\frac{(Z-\bar{Z}_h)^2}{2\sigma_z^2} + \frac{(y-\bar{y}_h)^2}{2\sigma_y^2}\Big]\Big) \Big] \tag{2.7}$$

Z_i, y_i and Z_h, y_h are the coordinates of the image streamlines below the ground (suffix i) and above any inversion if it exists (suffix h). The expressions for the change in plume width and depth σ_y and σ_z are derived from statistical theory (Hunt, 1985).

3. Computer requirements and main elements in the computational model

There are six main elements in the program and software to implement the model:

(i) receiving and storing the data about the terrain height and roughness, and meteorology (or upwind flow). The program specifies the total number of grid points on which the terrain data can be defined which at present is 64 x 64 (but not all these need be used). Although the data must be specified within a square grid, the spacing of points need not be at regular intervals. The meteorology is assumed to have certain standard forms (which are defined in Section 4.3) but the program operator decides which form is most appropriate for the atmospheric conditions under consideration and what are the relevant numerical factors;

(ii) the input data is then processed to enable the flow computations to proceed. Chiefly, this stage involves taking fast Fourier transforms of the terrain height and of the logarithm of the roughness length over the flow region. It also involves defining the depths h_m and ℓ of the middle layer and inner region over the terrain;

(iii) computing the distribution of the mean flow at locations supplied by the user;

(iv) computing the shear stress and turbulence;

(v) computing the diffusion from a specified source (FLOWSTAR-D);

(vi) outputting the results of the data in a form also specified by the user. This may be either numerical data or a graphical display.

Each element of the program has been designed to run under MS-DOS on an IBM PC AT or equivalent with at least 640kb of random access memory (RAM).

4. Procedures for computing mean flow over hills

4.1 Terrain height and lateral scale

Terrain is specified within a domain D which is a square of side $2L_*$ (typically $L_* \sim 10$km). The height of the terrain $z_T(x)$ or $z_T(x,y)$ is specified on a mesh of scale δL_* (i.e. z_T is specified at points x_i, y_j). In general, this might be an irregular mesh, but the program will put this into a regular mesh. (Typically, if $\delta L_* = 400$m, and $L_* = 10$km, the number of points will be $(2\times 10^4/400)^2 = 2500$). Any mean slope in the terrain can be subtracted out so that the computation will provide air flow over a surface. This is done by taking the mean height of the terrain along the two edges perpendicular to the upwind direction and then subtracting out the mean slope between those edges.

4.2 Roughness length z_0 over the terrain

z_0 is specified at every grid point to enable the model to treat changes in surface roughness. Only the first-order solution is included in the current algorithm. (Belcher et al, 1990 show that when second-order effects are included the roughness changes lead to significant horizontal divergence which is absent from the first-order solution.) Advice on the choice of z_0 for different fetches of actual terrain (country, sea, suburbs etc.) is given, for example, by Wieringa (1976).

4.3 Input profile of wind for different directions and meteorological conditions

For each wind direction defined by θ_i and each stratification by S_i (i=0 is neutral, i=1-4 are different stratifications), the wind profile U(z) is defined in algebraic form. We recommend either the log-linear or power-law forms for the stably stratified layer, where the user provides the input data.

(i) The log-linear form is

$$U(z) = \frac{u_*}{\kappa}(\ell n(\frac{Z}{z_0}) + \frac{\alpha Z}{L_{MO}}) \quad (\alpha=5;\ \kappa=0.4) \tag{4.1}$$

where u_* is the surface shear stress, L_{MO} is Monin-Obukhov length and z_0 is the average roughness length along the upwind edge of D. (This form is recommended only for $Z \leq 100$m).

(ii) The power-law form is

$$\begin{aligned} U(Z) &= U_{10}(\frac{Z}{10})^p \quad Z<h_m \\ U(Z) &= U_0 = constant\ for\ Z>h_m \end{aligned} \tag{4.2}$$

where U_{10} is the upwind speed at 10m above the ground.

(Note that if u_* is not supplied, it must be calculated from (i) and data for U_{10}).

4.4 Turbulence

The upstream vertical profile of the turbulent components must be specified.

4.5 Types of stratification

The type of stratification is defined only in the outer layer in all conditions considered in these models; the middle and inner layers are effectively neutral, although flow in these layers is strongly affected by the outer layer stratification.

S_0: neutral

S_1: uniform potential density (ρ) or potential temperature gradient (θ_p) and uniform wind speed U_0 (N/U_0 constant with height, where N is the buoyancy frequency defined by
$N^2 = (-g(\partial\rho/\partial Z)/\rho(Z=0)) = g(\partial\theta_p/\partial Z)/\theta(Z=0))$.

S_2: zero density gradient below inversion ($Z<h_i$),
density discontinuity at the inversion ($Z=h_i$),
and uniform density gradient above inversion ($Z>h_i$).

S_3: uniform density gradient below inversion ($Z<h_i$), density discontinuity at the inversion.

S_4: decreasing density gradient below boundary layer top ($Z=h_i$); $N^2=N_0^2e^{-\alpha Z}$; uniform density gradient above boundary layer top ($Z>h_i$); $N=N^2_0e^{-\alpha hi}$.

4.6 *Source position*

This must be specified by the user for the diffusion calculation.

5. Model output and comparisons

Output from FLOWSTAR and FLOWSTAR-D is presented for four different flows:

(i) The calculated and observed profile of the mean flow in neutral conditions over the summit of a low hill on Askervein, an island in the Hebrides (Zeman and Jensen, 1987). These show good agreement, as Fig. 3 demonstrates.

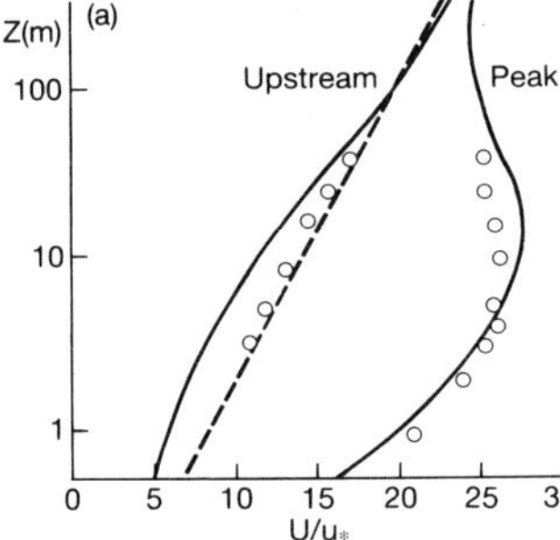

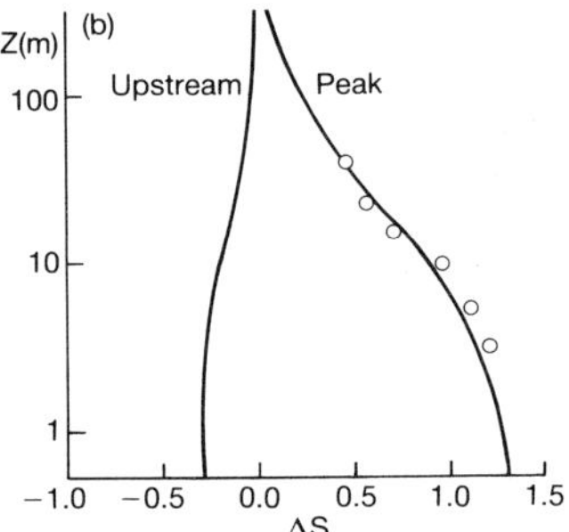

Fig.3: Calculated (using FLOWSTAR) and measured profiles of the mean flow over Askervein in neutral flow. z_0=0.03m, u_*=0.45 m.s^{-1}, h=115m, L=225m. (a) $U(z)/u_*$; (b) speed-up ΔS.

(ii) The second set of calculations relate to flow over Brent Knoll (Fig. 4). In neutral flow, the speed-up (ΔS) at the summit was calculated to be 1.4 which compares well with the observed value of ΔS=1.3 (Mason and Sykes, 1979). The cross-section of the flow shows that over the summit there is a low-level maximum at about 10m above the surface, but away from it there is a monotonic increase with height.

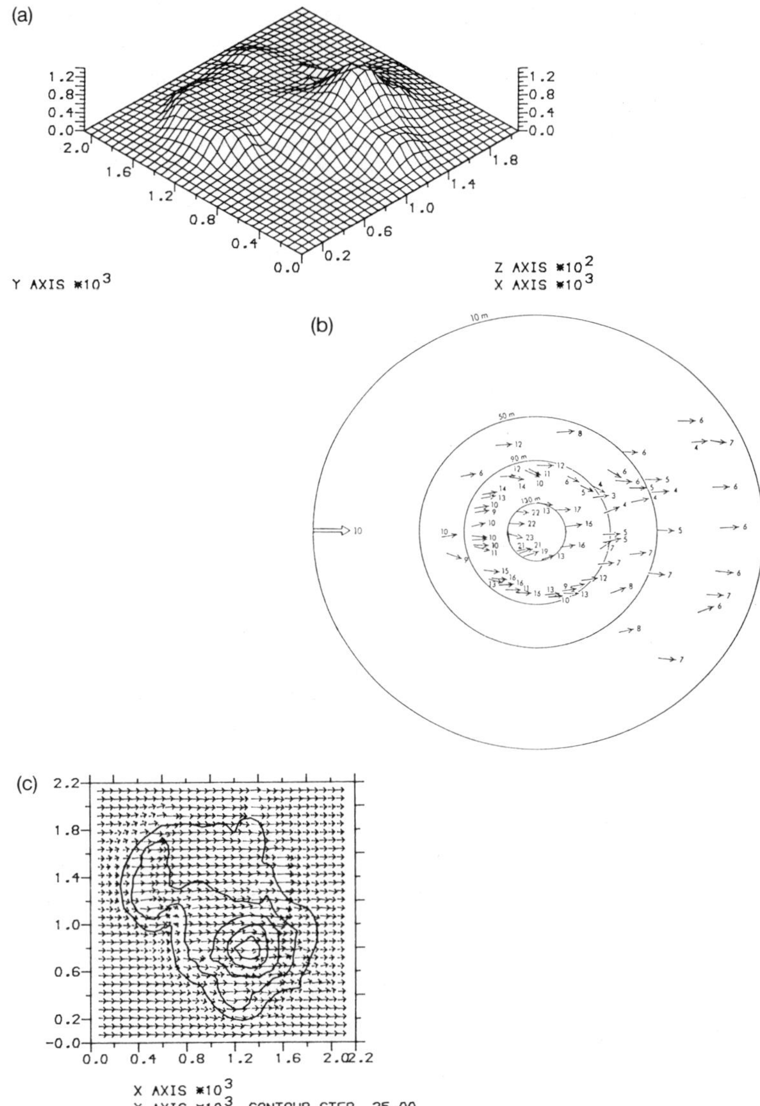

Fig.4: Flow over Brent Knoll. (a) Isometric projection of Brent Knoll topography; (b) Measurements of Mason and Sykes (1979) at 2m; (c) FLOWSTAR calculations at 2m for neutral flow. Upwind speed at 2m is 10 m.s^{-1}, L=290m, ℓ=14m, h=100m, z_0=0.02m. The maximum speed of 24 m.s^{-1} compares well with the observed value of 23 m.s^{-1}.

(iii) Turbulence profiles (Fig. 5) have been calculated for flow over the idealised hill f(x)=h(1+cos2nx/L) with L=1000, h=200m. These show that the longitudinal component decreases monotonically with height, the values in the rapid distorted region (z>h) being smaller than their upwind values, while there is an elevated maximum in the vertical component, as has been observed by Bradley (1981) and predicted by the numerical model of Newley (1985).

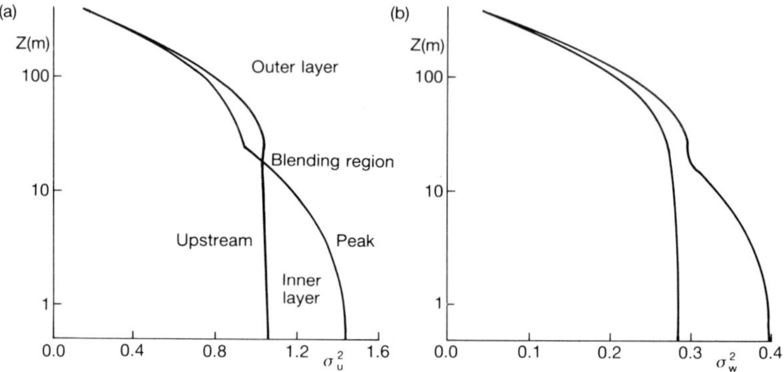

Fig.5: Turbulence profiles calculated using FLOWSTAR in neutral flow over an idealised hill $S(x)=h/(1+\cos 2\pi x/L)$, L=1000m, h=200m. The height of the inner region ℓ=45m. The blending functions between the inner and outer regions smooth the solution in the range $z_0<z<100m$. (a) σ_u^2; (b) σ_w^2.

(iv) Pollutant concentrations (Fig. 6) were calculated using FLOWSTAR-D over a digitised representation of Cinder Cone Butte (Lavery et al, 1981) for a moderately stable flow, N=0.0423 s^{-1}, L_{MO}=87m, u_*=0.354 $m.s^{-1}$. The calculations are compared with field measurements and other models for the concentration developed at the United States EPA (Lavery et al, 1981).

6. Other models

CERC has also developed codes for GASeous Transport from Accidental Releases (GASTAR) and Liquid Pool Spills (LiPS). Both models have also been designed to run on IBM PC ATs or compatible computers.

GASTAR has a modular structure that allows straightforward updating of the various parts. A basic and central dispersion model for neutral and dense releases, isothermal or non-isothermal with instantaneous, continuous or time-varying source conditions. This central module is serviced by a

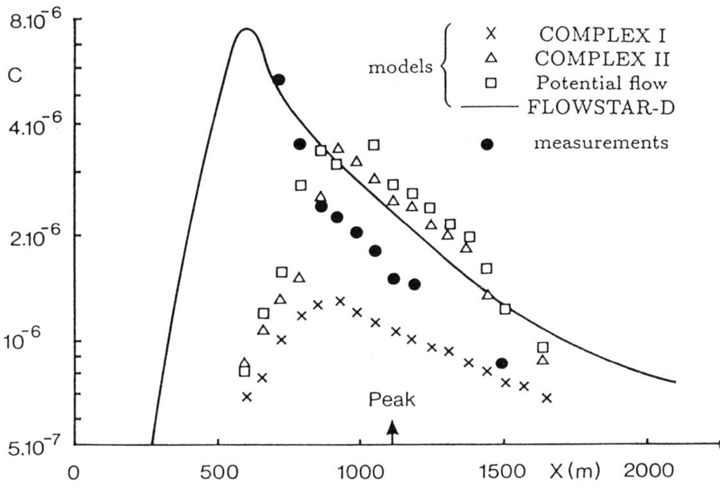

Fig.6: Concentrations calculated by FLOWSTAR-D at ground level over Cinder Cone Butte (h=112m, horizontal length scale L≈400m), $u_*=0.35\ m.s^{-1}$, $N=0.0423\ s^{-1}$, $L_{MO}=87m$, U(source)=7.5 $m.s^{-1}$. Height of source is 40m (for details of hill, see Lavery et al, 1981). (Measurements denoted by solid circles; models: FLOWSTAR-D, line; Complex I, crosses; Complex II, triangles; potential flow model, squares).

library of various source modules, e.g. cryogenic pool spill, catastrophic release, multi-phase jet etc.

There are then various output and peripheral modules and these address aspects such as: (i) the influence of slopes and general topography; (ii) the influence of obstacles, buildings, plant structure etc; (iii) assessment of hazard areas based on weighted toxic dose calculations; (iv) mass and area covered by released material between the UFL and LFL; (v) customised models for incorporating chemical reactions; (vi) incorporation of empirical data on concentration fluctuations and/or variations between members of an ensemble.

The LiPS program has been developed by CERC to model spills of multicomponent cryogens (e.g. liquified natural gas) on water using the time-dependent shallow-water equations and thermodynamic properties of the cryogen, the spread, mass loss and compositional changes of the boiling pool may be determined for both planar and axisymmetric flows. Drag, viscous effects and surface tension are included in the model which may be applied to a wide range of source conditions and cryogen types. Problem specification and graphical output production are achieved through a user

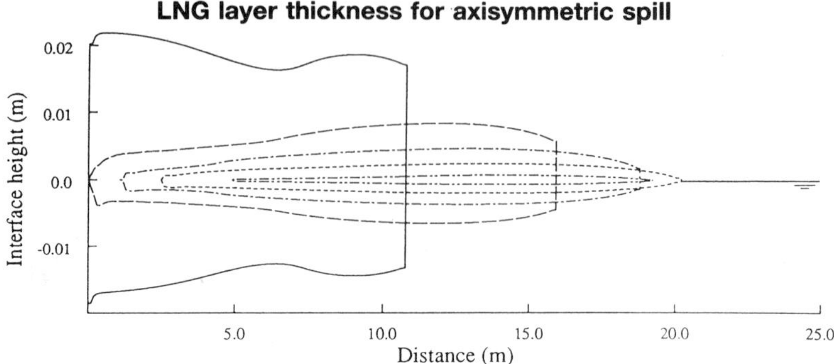

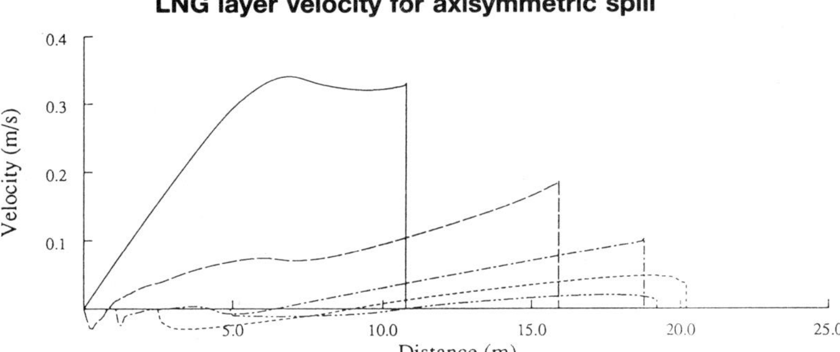

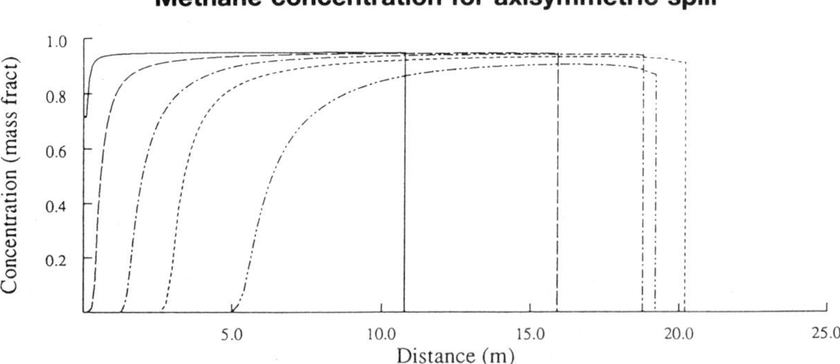

Fig.7: The evolution of a 12m^3 spill of liquified natural gas (LNG). The LNG is initially confined to a barge 4.8m in radius. The barge sinks over a period of 10s, releasing the LNG. Plots are given at 20s intervals for the interface locations, LNG velocity and methane concentration.

friendly, forms-oriented, menu system. Both on-screen and hard copy plots may be readily produced. Running the simulation itself is batch-oriented to facilitate unattended running. The system is written in carefully documented ANSI standard Fortran to facilitate using the model on systems other than MS-DOS. The modular structure of LiPS ensures the maximum flexibility in handling a wide range of real flows with sources ranging from continuous to finite volume releases such as dam break or a sinking barge; a sample output is shown in Fig. 7.

7. References

Belcher, S.E., Xu, D-P. and Hunt, J.C.R. (1990). The response of the turbulent boundary layer to arbitrarily distributed surface roughness. Q.J. Roy. Met. Soc., 116:611-635.

Bradley, E.F. (1980). An experimental study of the flow of the profiles of wind speed, shearing stress and turbulence at the crest of a large hill. Q.J. Roy. Met. Soc., 94:361-379.

Britter, R.E. and McQuaid, J. (1988). The development of a workbook on the dispersion of dense gases. In: Internal Conference on Vapor Cloud Modelling. Boston, Center for Chemical Process Safety, A.I.Ch.E.

Carruthers, D.J. and Choularton, T.W. (1982). Airflow over hills of moderate slope. Q.J. Roy. Met. Soc., 108:603-624.

Carruthers, D.J. and Hunt, J.C.R. (1990). Fluid mechanics of airflow over hills: turbulence fluxes and waves in the boundary layer. Meteorological Monographs, 23(45):83-107.

Finnigan, J.J. (1988). Airflow over complex terrain. In: Flow and Transport in the Natural Environment: Advances and Applications, W.K. Steffen and O.T. Denmead, editors. Springer Verlag.

Hunt, J.C.R. (1985). Turbulent diffusion from sources in complex flows. Ann. Rev. Fluid Mech., 17:447-485.

Hunt, J.C.R., Leibovich, S. and Richards, K.J. (1989). Stratified shear flow over low hills. I: Effect of wind shear. Q.J. Roy. Met. Soc., 114:1435-1470.

Hunt, J.C.R. and Richards, K.J. (1984). Stratified airflow over low hills. Boundary Layer Met., 30:223-265.

Hunt, J.C.R., Richards, K.J. and Brighton, P.W.M. (1989). Stratified shear flow over low hills. II: Stratification effects in the outer flow region. Q.J. Roy. Met. Soc., 114.

Jackson, P.S. and Hunt, J.C.R. (1975). Turbulent wind flow over low hills. Q.J. Roy. Met. Soc., 101:929-955.

Lavery, T.F., Bass, A., Strimaitis, D.G., Venkatram, A., Green, B.R., Drivas, P.J. and Egan, B.A. (1981). EPA Complex Model Development. First Milestone Report, No. EPA-600/3-82-036.

Mason, P.J. and Sykes, R.I. (1979). Flow over an isolated hill of moderate slope. Q.J. Roy. Met. Soc., 105:383-395.

Newley, T.M.J. (1985). Turbulent Airflow over Hills. PhD dissertation, University of Cambridge, Cambridge, UK.

Smith, R.B. (1980). Linear theory of stratified hydrostatic flow past an isolated mountain. Tellus, 32:348-364.

Walmsley, J.L., Salmon, J.R. and Taylor, P.A. (1982). On the application of a model of boundary layer flow over low hills to real terrain. Boundary Layer Met., 23:17-46.

Walmsley, J.L., Taylor, P.A. and Keith, T. (1986) A simple model of neutrally stratified boundary-layer flow over complex terrain with surface roughness modulations (MS3 DJH/3R). Boundary Layer Met., 36:157-186.

Weng, W.-S., Richards, K.J. and Carruthers, D.J. (1988). Some numerical studies of turbulent waves over hills. Proc. 2nd European Turbulence Conf., Berlin, H.H. Fernholz, H.E. Fiedler, editors.

Wieringa, J. (1976). An objective exposure correction method for average wind speeds measured at a sheltered location. Q.J. Roy. Met. Soc., 102:241-253.

Zeman, O. and Jensen, N.O. (1987). Modifications to turbulence characteristics in flow over hills. Q.J. Roy. Met. Soc., 113:55-80.

Massively parallel computers and their applications to numerical weather prediction

S.F.B. Tett

University of Edinburgh

1. Introduction

L.F. Richardson in his 1922 book described the first scheme for numerical weather forecasting. In the same book he imagined using a massively parallel computer, each element being a human being.

Parallel computation may offer the present generation of environmental scientists a cost effective way to obtain large amounts of computing power. However, there is a whole menagerie of parallel computers available. These range from computers having small numbers of highly complex processors (such as the CRAY Y-MP and X-MP) through to computers with large numbers of extremely simply processors (such as the ICL/AMT DAP).

Parallel computers are conveniently divided into two classes. Single Instruction Multiple Data (SIMD) computers, in which each processor performs the same instruction (or does nothing) on its data. The DAP, which is well described in Hockney and Jesshope (1988) and Connection Machine described in Hillis (1984) are examples of SIMD machines. These architectures are best suited for problems which are homogeneous. The second class is Multiple Instruction Multiple Data (MIMD) computers, in which each processor can do different things to different pieces of data. The Hypercube, all multi-processor vector machines (CRAYs etc) and transputer-based machines, such as the Meiko Computing Surface are all MIMD machines.

MIMD computers further divide into two classes - computers with a global memory which all processors can access and computers with no global memory, each processor having its own local, private memory. In the second case processors will communicate with each other via (message passing) "links". Transputer-based machines and Hypercubes are examples of the second class, while the CRAY computers are examples of the first. It is considerably easier to implement models on computers with shared memory; the programmer does not to have to worry about communications between processors or with the processor topology. However as the number of processors increases it will become increasingly difficult to apply these computers effectively because of memory locking and bandwidth problems.

Various parallel computers have been used for atmospheric modelling. Dent (1988) has implemented a multitasking version of the European Centre for Medium Range Weather Forecasting model on the CRAY X-MP. A spectral model has been implemented on the DAP by Carver (1988), and Tuccillo (1990) has implemented a local area grid point model on the Connection Machine. This paper will describe an implementation of the UK Meteorological Office's (UKMO) global grid point model on the Edinburgh Concurrent Supercomputer (ECS). When a parallel computer is referred to henceforth, we shall mean a processor with local memory which has to communicate with its neighbours. Most of what is said is based on experiences with transputers and the ECS but some should be of relevance to other machines.

A description of transputers and the ECS is given followed by a brief summary of the UKMO grid point model. Following this the various parallel strategies are described before the details of the model implementation are described.

2. Transputers at Edinburgh

2.1 The transputer

Generically, a transputer consists of a central processing unit, on-chip memory, an external memory interface and four communication links, or external links, which provide bi-directional communications with other transputers.

There are several varieties of transputer; at Edinburgh we use the T800. On chip it has a floating-point unit, in addition to the integer processor, and 4kbytes of memory. Each of our transputers has 4mbytes of external memory. This computing resource is capable of sustaining approximately 0.8 x 10^6 floating-point calculations per second per transputer.

To get this kind of performance the data must be stored in the on-chip memory rather than the slower external memory. Each external link is capable of communicating at a rate of 0.8 mbytes per second each way.

Each processor can simulate the running of several parallel processes by time-slicing them. If a process is idle - for example, if it is waiting for a communication to proceed - then it takes up only a very small part of the processor's time. It is also possible to do communications and calculations concurrently. This is real concurrency, the communications being handled by link controllers leaving the arithmetic unit to proceed with calculations, but it does consume some of the available memory bandwidth.

2.2 The Edinburgh Concurrent Supercomputer (ECS)

The ECS is an electronically reconfigurable array of transputers built by Meiko Ltd. Its computational engine currently comprises approximately 400 T800 transputers. The machine is divided into several "domains". Each domain is a single-user resource ranging in size from 1 to 131 processors. The processors within each domain can be reconfigured by the user into most two-dimensional topologies; reconfiguration of the domains is determined by the system manager.

The computer is fileserved from a microVAX running VMS and three 800mbyte discs which run a very simple Unix-like operating system. The machine can be programmed in any one, or mixtures of three languages - occam, C and Fortran 77.

3. UKMO global grid point model

The UKMO model is a three-dimensional finite difference grid point model. It uses latitude, longitude (λ,θ) coordinates for horizontal discretisation and σ coordinates for vertical discretisation, but the implementation uses hybrid coordinates as described in Simmons and Strüfing (1981). A detailed description of the model can be found in Bell and Dickinson (1987).

The dynamics part of the model uses a split explicit integration scheme. Its main features are that it decomposes the forward integration step into three "Forward-Backward" adjustment steps followed by one advection step. Near the poles the meridians converge and in order to allow a reasonable equatorial timestep it is necessary to Fourier damp the model near the polar regions. The way that Fourier damping works is that for a linearised version of the model, the growth rate of each Fourier mode is computed. This growth rate is multiplied by a factor (presently 1.1) to compensate for non-linear effects. We transform the data into Fourier space and then each mode is then damped by this growth rate factor. See Gadd (1978a,b;1980) for details. Processes which are not explicitly resolved on the model scale are parameterised in terms of the bulk variables.

4. Parallel Strategies

This section will first outline three paradigms for programming a massively parallel computer. More details of these as well as a detailed description of the transputer and the ECS are in Bowler et al (1987). After this it will examine what a model requires for efficient parallel implementation.

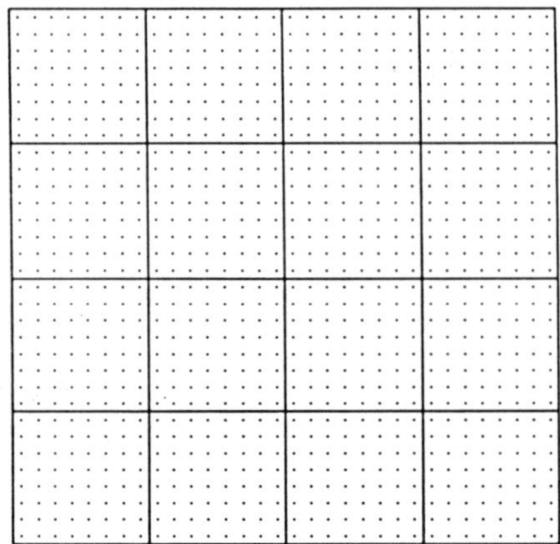

Fig. 1: Processors with grid overlayed. Processors are shown as thick boxes and grid points by dots. Each of the 16 processors has 8x8 grid points.

For atmospheric models the most obvious paradigm is geometrical decomposition. In this method the space is decomposed over the processors such that neighbouring processors have neighbouring physical regions. Fig. 1 shows an example of this. Another paradigm that is useful is task farming. Here tasks are considered to be atomic, i.e. they do not require any outside information. One way that this method has been used is to have one processor generate many tasks and pass them to the rest of the processors where they are computed and then sent back. The final paradigm that we consider is Algebraic Parallelism. Here a complex algebraic expression is decomposed into several parts, each part is computed and then everything is brought back together. This paradigm will suffer from load balancing problems in that each independent part may have very different computational requirements.

Before implementing a model on a parallel computer we must do some analysis. We must find out what parts of the model are serial and what parts are parallel. For parts of a model to run in parallel they must be independent, if there are some dependencies between them then they must run sequentially.

If we examine the UKMO model we find that the serial part of the model is twofold. First the time-stepping part of the algorithm. Secondly the calculation of the geopotential and most of the parameterisation schemes are serial. Thus we should not decompose our problem vertically or in time. The parallelism in the model arises from the calculations of differences and means, all of which are independent for each gridpoint.

5. Implementing UKMO model on ECS

5.1 Dynamics

For reasons discussed earlier, a two dimensional geometrical decomposition was used. Fig. 2 is a diagram of the processor topology used to map a sphere using 16 processors.

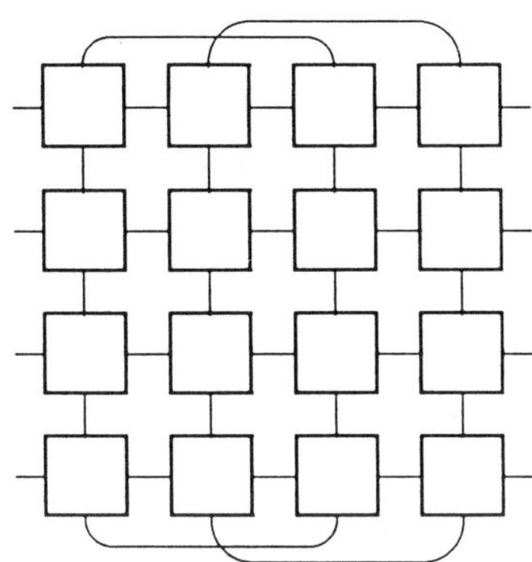

Fig. 2: Processor topology used to map a sphere onto, in this case, four by four processors. The links shown at the east and west sides wrap around. The links shown at the north and south edges are connected to processors 2 away, to provide the "over the pole" communications.

For the bulk of the dynamics code each processor needs to communicate only with its neighbours. It does this in order to compute the finite difference approximations to differentials. As we are using a staggered grid, computing the differentials usually involves computing a mean in one direction and a difference in the other. The first of these will take one operation while the second will take two operations.

The grid on each processor will split into two regions: an inner region which does not require communication with neighbouring processors and an outer region where communication with neighbouring processors is required. Fig. 3 shows a general example of this.

In the case of the transputer, which can overlap calculations and communications, we can send edge regions to neighbouring processors while simultaneously computing interior regions. When these have *both* finished then we compute the outer region. We should be careful not to consume much of the memory bandwidth doing communications. In the analysis that follows we assume that we can ignore the effects of limited memory bandwidth.

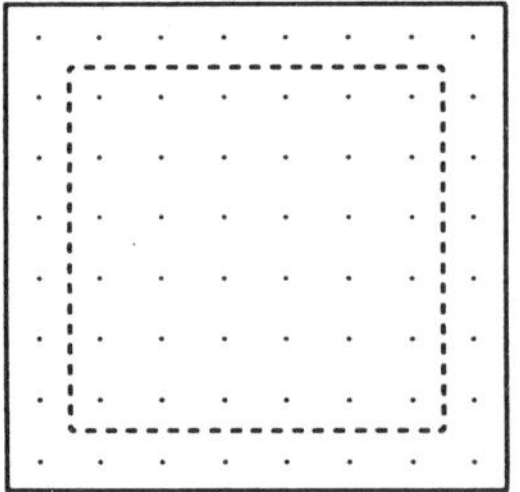

Fig. 3: How the grid points on a processor split up into two regions: an outer region (requiring communication then computation) and an inner region (requiring computation only).

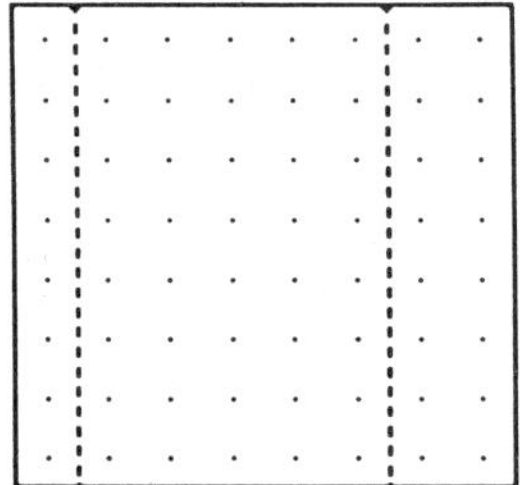

Fig. 4: A more typical example of how the gridpoints split up into two regions.

At this point it is helpful to define the *interaction range* (I_R). This is the largest distance (in grid points) over which any point interacts. A more formal definition is given in Tett et al (1990). The total amount of computational work that each processor needs to do is simply given by cn^2, where c is the amount of work per gridpoint and the total number of grid points on the processor is n^2. We now wish to define a critical value n_c of n which is such that if $n>=n_c$ then the processor never has to wait for communication to complete. This can be achieved if the time to send all edge regions is less than the time to compute on interior regions.

For the centred finite difference schemes (as used by UKMO) a total number of points given by $n(2I_R\text{-}1)$ need to be transferred, one transfer of nI_R and another of $n(I_R\text{-}1)$. Fig. 4 shows how the grid on a processor splits up for computing an east-west (or λ) difference with $I_R=2$. As stated earlier, for grid point models c will on the whole be 1 or 2.

As all edge regions can be sent simultaneously, the total time spent communicating is given by $rnI_R\tau$, where r is the number of floating point calculations that can be performed in the time it takes to transfer one word and τ is the time it takes to do one floating point operation. The time to do these calculations on the interior region is $c(n^2-(2I_R-1)n)$. Thus we have for n_c

$$(c(n_c^2-(2I_R-1)n_c))\tau \geq n_c I_R \tau r \tag{5.1}$$

or after some rearrangement

$$n_c \geq I_R(\frac{r}{c}+2)-1 \tag{5.2}$$

In most of the UKMO model, c is one or two and I_R is one, except for some parts of the advection step where I_R is two. This relationship gives a minimum patch size on each processor if we want optimal efficiency. We could, of course, reduce the patch size, in which case our efficiency would reduce to $\approx n/n_c$. We should be aware that the larger c is, the smaller the patch size need be. If we want to have the whole model run at 100% efficiency then we should look at the worst case. This occurs when $I_R=2$ and $c=1$. For the transputer τ is four but this value could be higher in practice because of software overheads. Substituting these into Eq. 5.2 we find that n_c is 11. Only a small part of the model has an I_R of 2; for most of the model I_R is 1 and in this case n_c is 5.

Current forecasting models have approximately 200 x 100 points in them, so they could use approximately 200 processors optimally and 800 processors with reasonable efficiency. As forecast models get larger they could use more processors. In the case of low resolution climate models with approximately 64 x 64 points, we could only use approximately 36 processors at full efficiency and approximately 170 with reasonable efficiency. Massive parallelism requires massive models. We find that for our model implementation, the adjustment and advection steps have a computational speed of approximately 0.5mflops.

5.2 Load Balancing the Fourier Transforms

For an explanation of the Fast Fourier Transform see Press et al (1986; Section 12). As we use geometrical decomposition in both λ and θ, to compute the Fourier transform we have to communicate data between processors in the same row.

For the adjustment it is only necessary to carry out Fourier Damping North or South of 60°N and 60°S respectively. In the case of the advection step, where Fourier Damping is required depends on the model velocities, though usually near the winter pole more damping is required than near the equator. These both pose load balancing problems since processors which have to carry out computations for near polar regions will have to carry out Fourier Transforms while those processors which do computations for equatorial regions will not be doing any Fourier Transforms.

In order to make full use of all the processors, tasks (in this case Fourier Damping) should be shared around all the processors. This can be achieved, to some extent, by task diffusion. In task diffusion each processor finds out how many jobs neighbouring processors have and then either gives to or receives from each processor some fraction of the difference. The number of jobs sent over the link is limited by some maximum value. In addition one job is always retained so the processor always has some work to do.

Since the data for the Fourier transforms are distributed over a processor row then each processor in the same row must make the same decision about which jobs to give to its two neighbouring processors. This is done by each processor telling both of its north and south neighbours how many tasks they have. After this stage all processors know how many tasks each of their neighbours have. Each processor starts doing a Fourier transform (if it has any work to do) and simultaneously sends to or receives from each neighbour a number of tasks t, given by 80% of the difference in the number of tasks between the processor and its neighbour. In the event of the processor having to send more jobs than it has, the number of jobs sent to each processor is scaled by the total number of jobs on the processor. The fraction 80% was determined from simulations: a more advanced version would allow the program to determine such value itself. This technique gave us a roughly 30% speed increase over the non-balanced case even despite the considerable overhead for the load balancer.

6. Conclusion

We believe that massively parallel computers may offer an economic alternative to vector computers. For simple grid problems both SIMD and MIMD machines are suitable. SIMD machines have an advantage due to their relative ease of programming, but if there is much inhomogeneity in the problem, as is the case in most atmospheric models then MIMD machines are to be preferred.

For optimal efficiency in atmospheric models, there is a minimum data patch size that each processor should have. Therefore models with large numbers of gridpoints could be successfully implemented on large numbers of processors but models with small numbers of gridpoints will only be able to use small numbers of processors.

7. Acknowledgements

I would like to thank R.S. Harwood for helpful criticism, and the UKMO and SERC for financial support.

8. References

Bell, R., and Dickinson, A. (1987). The meteorological office operational numerical prediction system. Scientific Paper 41, UKMO, HMSO.

Bowler, K., Kenway, R., Pawley, G. and Roweth, D. (1987). An introduction to occam 2 programming. Chartwell-Bratt, Sweden.

Carver, G. (1988). A spectral meteorological model on the ICL DAP. Parallel Computing, 8:121-126.

Dent, D. (1988). The multitasking spectral model at ECMWF. In: Multiprocessing in meteorological models, G.-R. Hoffmann and D. Snelling, editors. Springer-Verlag, Berlin.

Gadd, A. (1978a). A numerical advection scheme with small phase speed errors. Q. J. Roy. Met. Soc., 104:583-592.

Gadd, A. (1978b). A split explicit integration scheme for numerical weather prediction. Q. J. Roy. Met. Soc., 104:569-582.

Gadd, A. (1980). Two refinements of the split explicit integration scheme. Q. J. Roy. Met. Soc., 106:215-220.

Hillis, D. (1984). The Connection Machine. M.I.T. Press, Cambridge, Mass., USA.

Hockney, R.W. and Jesshope, C. (1988). Parallel Computers 2. Adam Hilger, Bristol.

Press, W.H.B., Flannery, B.P., Teukolsky, S.A. and Vetterling, W.T. (1986). Numerical Recipes: the Art of Scientific Computing. Cambridge University Press.

Richardson, L.F. (1922). Weather Prediction by Numerical Process (1965 reprint). Dover, New York.

Simmons, A. and Strüfing, R. (1981). An energy and angular momentum conserving finite difference scheme, hybrid coordinates and medium-range weather prediction. Technical Report 28, European Centre for Medium Range Weather Forecasting (ECMWF), Reading.

Tett, S.F.B., Harwood, R.S. and Kenway, R.D. (1990). Implementation of atmospheric models on large multiprocessor surfaces. In: The Dawn of Massively Parallel Processing in Meteorology, G.-R. Hoffman and D.K. Maretis, editors. Springer-Verlag.

Tuccillo, J.J. (1990). Numerical solution of the primitive equations on the Connection Machine. In: The dawn of massively parallel processing in meteorology, G.-R. Hoffman and D.K. Maretis, editors. Springer Verlag.

The application of the ICL DAP to finite element meteorological modelling

G. Carver

University of Edinburgh

1. Introduction

The ICL Distributed Array Processor (DAP) (Reddaway, 1973) is a parallel computer, comprising a 64 by 64 array of single bit processing elements (PEs). Each processor is connected to 4kbits of local memory and to its 4 nearest neighbours so that data can be moved across processors on the array. All the PEs execute the same instructions at the same time, but on the different data in their own memory. The DAP is therefore classed as a Single Instruction/Multiple Data stream (SIMD) machine. The original DAP service at Edinburgh (Brown, 1986) closed in 1987 but a new generation of DAP technology is now produced by Active Memory Technology (AMT).

Finite difference, spectral and finite element methods are all used in meteorological modelling. The first two have been applied to the DAP by Reddaway et al (1976), Fishbourne (1980) and Carver (1988). New parallel algorithms had to be developed in both cases to obtain an efficient utilization of the processors.

The finite element (FE) method is used infrequently for meteorological modelling, although its use is growing. It is mainly employed for limited area models. Studies using the FE method for engineering applications on the DAP have shown it to be well suited to the array processor architecture (e.g. Lai and Liddell, 1987a). The FE method applied to a limited area meteorological model with a rectangular domain would therefore appear to be suited to the DAP architecture. The key aspects of implementing such a model are described in the following sections. The model is based on that of Orlanski and Ross (1977) and described in detail by Carver (1990).

2. Computing requirements

The ICL DAPs at Edinburgh offered a local high-speed computing facility to study the suitability of this type of architecture to meteorological modelling. In implementing any application efficiently on the DAP, it is important that as many processors as possible are utilized for the duration of the program and the available memory (2mbytes) is sufficient. Both these requirements are met by the FE model considered here. To obtain the best performance on the SIMD architecture, a different approach to the design of algorithms is often required.

Whilst the ICL DAPs at Edinburgh were hosted by an ICL mainframe computer, the latest generation DAP, from Active Memory Technology (AMT), can be attached to a Sun or VAX workstation. The processing speed for 32-bit floating point arithmetic on the ICL DAP is 25mflops for addition and 14mflops for multiplication. The AMT 64 x 64 DAP is 10 times faster.

3. Solution method

3.1 Choice of elements

Staniforth (1987) has shown that linear, bilinear or trilinear elements are superior to triangular or higher order elements for meteorological problems in regular domains. First, fourth order accuracy can be obtained for the solution at the nodes of a regular grid, as super-convergence occurs. Second, the resulting system of equations can be solved as sets of one-dimensional tridiagonal matrix problems, rather than the more usual higher bandwidth problems. Last, for a first order derivative, a two-dimensional calculation reduces to one dimension. Since economic and efficient algorithms for the solution of tridiagonal systems exist for the DAP, the use of rectangular bilinear elements is made more attractive.

3.2 Approximation of variables

To make the best use of the DAP architecture, a nodal grid of 64 by 64 points was used. Any variable, f, can be expanded as

$$f(x,z,t) = \sum_{i=0}^{63} \sum_{j=0}^{63} f_{ij}(t)\phi_i(x)\phi_j(z) \tag{3.1}$$

where a FE node is identified by the integers i and j. The product $\phi_i(x)\phi_j(z)$ is the bilinear piecewise basis function where ϕ_n is the one-dimensional Chapeau function defined by

$\phi_n(\eta) = 0$ for $\eta<\eta_{n-1}$ and $\eta>\eta_{n+1}$
$\phi_n = (\eta-\eta_{n-1})/\Delta$ for $\eta_{n-1}<=\eta<=\eta_n$
$\phi_n = (\eta_{n+1}-\eta)/\Delta$ for $\eta_n<=\eta<=\eta_{n+1}$

where η may be x or z and Δ is the spacing between nodes. The model has constant spacing in the x and z directions, although it is possible to use a variable spacing in the vertical.

3.3 *Data mapping*

How data are mapped onto the DAP PEs is important because it determines their utilization efficiency. This is perhaps the major factor in determining the suitability of the DAP to certain applications.

Two methods have been proposed to map FE grids onto the DAP. Lai and Liddell (1988) use the "long-vector" mode of the DAP, where elements are stored consecutively according to their element numbering and columns of PEs on the array can be treated as a single vector of 4096 elements. Dixon and Singh (1984) used an "upper-leftmost" storage arrangement in which the axes of their domain were mapped along rows and columns of the DAP array with the origin in the top-left corner of the array. The long-vector storage format is advantageous for arbitrary element shapes and irregular boundaries. However, the upper-leftmost format is appropriate for this model because the equations are solved successively in each direction i.e. along rows and columns of the DAP PE array. Also, the FE matrices, arising from the approximation of the terms in the model equations, can be efficiently computed by shifting data between processors across the FE nodes, since the PE array has the same structure as the nodal arrangement for the model.

3.4 *Procedure*

The solution procedure uses the Galerkin FE method; the terms of the equations are multiplied by the basis functions, integrated over the domain and the variables are approximated according to Eq. 3.1. Each term results in a full matrix that must be computed. Typically in engineering problems, a single matrix for all of the terms on the RHS of the equation is computed, the solution is obtained by premultiplying by the inverse of the "mass matrix" on the LHS. However, as Staniforth (1987) has pointed out, using bilinear elements on a regular grid, a first order derivative and a product of two functions can be computed to fourth order accuracy (as opposed to second order if all the terms are considered together). Using these basic operations of differentiation and multiplication, it is possible to compute the FE matrix for every term in the equations and solve the equations to fourth order accuracy, assuming a constant spacing between nodes.

The calculation of the derivative and product terms can be identified as "kernel" subroutines to the model with which more complex terms can be computed. The diffusion terms of this model were also approximated by the FE method to save CPU time, rather than calculated by differentiation and multiplication operations. These terms were therefore only calculated to second order accuracy.

4. Efficient calculation of the finite element matrices

In this section, the algorithms used to calculate the FE matrices arising from the approximation of the derivative, product and diffusion terms are briefly described. As such calculations require information from neighbouring nodes, how the nodes are mapped onto the PE array determines the amount of data movement required between processors. Moving data between processors can be regarded as an overhead in implementing the model on a parallel computer of this class.

4.1 Derivative term

The FE solution to a derivative of the form $w=\partial u/\partial x$ over the model domain is given by, $\boldsymbol{w}=\boldsymbol{M}^{-1}\boldsymbol{P}_x\boldsymbol{u}$ where $\boldsymbol{w}$ and $\boldsymbol{u}$ are matrices whose elements are the nodal values of w and u respectively. $\boldsymbol{M}$ is the tridiagonal mass matrix for the x direction whilst $\boldsymbol{P}_x$ is known as the projection matrix of the differentiation operator. The value of the product matrix $\boldsymbol{P}_x\boldsymbol{u}$ for each node is given by $(\boldsymbol{P}_x\boldsymbol{u})_{i,j}=(u_{i+1,j}-u_{i-1,j})/2$ and can be computed simultaneously for each node by shifting data between processors. However, the values at the boundary nodes must be computed separately. This matrix can be computed at a rate of 11.3mflops.

4.2 Product term

Staniforth and Beaudoin (1986) showed that it is more efficient to compute the integrals resulting from applying the FE method to the product of two functions , w(x,z)=u(x,z)v(x,z), by Simpson quadrature rather than following the formal FE approach. The integrals are given by

$$F_{ij} = \int_{x_{i-1}}^{x_{i+1}} \int_{z_{j-1}}^{z_{j+1}} uv\phi_i\phi_j dx dz \tag{4.1}$$

which, after applying Simpson quadrature to the integral over z becomes

$$\begin{aligned} F_{ij} = {} & \frac{1}{3}\Delta z_{j-1}\int_{x_{i-1}}^{x_{i+1}}[uv]_{j-0.5}\phi_i dx + \frac{1}{6}(\Delta z_{j-1} + \Delta z_j)\int_{x_{i-1}}^{x_{i+1}}[uv]_j\phi_i dx \\ & + \frac{1}{3}\Delta z_j\int_{x_{i-1}}^{x_{i+1}}[uv]_{j+0.5}\phi_i dx \end{aligned} \tag{4.2}$$

where $\Delta z_j = z_{j+1} - z_j$ constant for all j. If the first integral in Eq. 4.2 is evaluated for all nodes, the result can be used for the third integral. On the DAP this is done by shifting the values on the PE array. Therefore, the DAP algorithm computes the one-dimensional integrals along the levels z_j and $z_{j-0.5}$ since the results at $z_{j-0.5}$ give those for $z_{j+0.5}$.

The calculation at each node requires data from its 9 nearest neighbours. The DAP algorithm to compute $\boldsymbol{F}$ in Eq. 4.2 computes the values at all nodes simultaneously. In particular, the values at the boundaries can be computed concurrently with those of the interior by making use of the hardware boundary conditions to shift zeros into empty processors when data is moved across the PEs and by setting values at unused PEs to zero. This contrasts the equivalent algorithm on a conventional serial architecture where boundary values would need to be computed by separate code. This is not an overhead for the DAP algorithm. The matrix $\boldsymbol{F}$ can be computed on the DAP at a rate of 15.2mflops and on average 98.4% of the PEs are utilized during the calculation. The solution, $\boldsymbol{w}$, is then obtained from $\boldsymbol{w} = \boldsymbol{M}^{-1}\boldsymbol{N}^{-1}\boldsymbol{F}$, where $\boldsymbol{N}$ is the tridiagonal vertical mass matrix.

4.3 Diffusion term

The model contains diffusion terms of the form $\partial/\partial x\{\kappa(x,z)\partial\phi/\partial x\}$ where κ is a diffusion coefficient and ϕ is any of the 3 model prognostic variables. Whilst this term can be evaluated using the kernel operations of differentiation and multiplication, it is computationally cheaper to compute the term by applying the Galerkin FE scheme directly. The matrix arising from the application of the FE scheme can be evaluated using Simpson quadrature as before, to reduce the operation count below that if the formal FE approach was used. The calculation proceeds at 16.9mflops and 99% of the DAP PE array is utilized during the calculation.

5. Solution of matrix equations

The solution of the matrix equations requires the solution of 64 sets of 64 tridiagonal simultaneous equations, of the form $\boldsymbol{Ax} = \boldsymbol{b}$. The matrix $\boldsymbol{A}$ is diagonally dominant for most cases with a ratio of 4:1 between the main and off diagonals. For the remainder of cases, the main diagonal is the sum of the other two, so the matrix is not diagonally dominant which restricts the choice of algorithm for solving the equations.

Several efficient and economic algorithms for the solution of tridiagonal systems on the DAP already exist. The parallel cyclic reduction algorithm is considered by Hockney and Jesshope (1981). This is a direct method in that it always completes in $\log_2 n$ steps for a n x n system and is therefore suitable for systems that are not diagonally dominant. The parallel Jacobi iterative algorithm, considered by Bowgen (1981), uses an initial pass of the cyclic reduction algorithm to increase the diagonal dominance and therefore accelerate convergence. The iterative conjugate gradient algorithm or preconditioned conjugate gradient algorithm has been favoured recently for FE applications on the DAP (Lai and Liddell, 1987b).

The conjugate gradient algorithm is usually applied to problems with a greater bandwidth than the tridiagonal equations considered here. Each iteration of this algorithm is more expensive than that of the cyclic reduction or Jacobi algorithms. To reduce the iterations required, the matrix equation $\boldsymbol{Ax} = \boldsymbol{b}$ was preconditioned using two passes of the cyclic reduction algorithm and two passes of the Jacobi algorithm to reduce the condition number of $\boldsymbol{A}$ and improve the convergence such that only one iteration of the conjugate gradient algorithm was required. However, the resulting algorithm was specific to tridiagonal, diagonally dominant systems and tuned to the requirements of the model. Although the execution time of this algorithm was the same as the Jacobi algorithm (which is slightly faster than the cyclic reduction algorithm), the Jacobi and cyclic reduction algorithms were preferred for their simplicity for solving the matrix equations in the model. The performance rate of the cyclic reduction algorithm was 9.7mflops, whilst the Jacobi algorithm's was 12.9mflops.

6. Conclusions

The FE method applied to meteorological modelling in rectangular domains is well suited to the processor array architecture. The matrix notation of the equations allows their easy translation to DAP FORTRAN (the parallel programming language for the DAP). As the computation of the FE matrices for each node requires information only from neighbouring processors, the model has a low data routing overhead of 4% of the CPU time per model step. Also, on average, 90% of the processors were kept busy doing useful work each timestep and the average performance for the model was 13.1mflops. Therefore, efficient use is made of the DAP processors. The model also fitted into the DAP memory, occupying about 30% of the available store.

There are several disadvantages to the use of the DAP however. First is the constraint that the model grid should ideally be a multiple of the PE array size to obtain the best performance. Second is the lack of support for variable size arrays (the first two dimensions of any DAP array must be 64 by 64). It is often useful to test models at higher resolutions than they are normally run, something that is difficult to do on the ICL DAP.

7. Acknowledgements

This paper forms a part of the author's PhD thesis for which financial support from SERC is acknowledged. Thanks are due to Dr Charles Duncan at Edinburgh for his support for this work.

8. References

Bowgen, G. (1981). Iterative solution of tridiagonal linear equations. DAP Support Unit, Technical Doc. No.2-9, Queen Mary College, London.

Brown, M.W. (1986). Integrating Distributed Array Processing into EMAS 2900. Software - practice and experience, 16:517-529.

Carver, G.D. (1988). A spectral meteorological model on the ICL DAP. Parallel Computing, 8:121-126.

Carver, G.D. (1990). PhD thesis. Department of Meteorology, Edinburgh University, 378pp.

Dixon, L.C.W. and Singh, P. (1984). The finite element method for Navier-Stokes equations: A least squares approach. Hatfield Polytechnic, Numerical Optimisation Centre, Technical Report No.142.

Fishbourne, J. (1980). Unfinished PhD thesis. Computer Science Dept., Reading University.

Hockney, R.W. and Jesshope, C.R. (1981). Parallel computers: architecture, programming and algorithms. Adam Hilger Ltd, Bristol: 423pp.

Lai, C.H. and Liddell, H.M. (1987a). A review of finite element methods on the DAP. Appl. Math. Modelling, 11:330-340.

Lai, C.H. and Liddell, H.M. (1987b). Preconditioned conjugate gradient methods on the DAP. Math. of Finite Elements and Appl., IV:145-156, Academic Press.

Lai, C.H. and Liddell, H.M. (1988). Finite elements using long vectors of the DAP. Parallel Computing, 8:351-362.

Orlanski, I. and Ross, B.B. (1977). The circulation associated with a cold front. Part I: dry case. J. Atmos. Sci., 34:1619-1633.

Reddaway, S.F. (1973). DAP - a distributed array processor. Proc. 1st IEEE/ACM Annual Symposium on Computer Architecture.

Reddaway, S.F., Hunt, D.J. and Parkinson, D. (1976). Study of a meteorological operational suite. ICL Research and Advanced Development Centre, Doc. No. CM52:15pp.

Staniforth, A. (1987). Formulating efficient finite element codes for flows in regular domains. Int. J. Num. Meth. Fluids, 7:1-16.

Staniforth, A. and Beaudoin, C. (1986). On the efficient evaluation of certain integrals in the Galerkin finite element method. Int. J. Num. Meth. in Fluids, 6:317-324.

Tracer transport model for the study of sources and sinks of trace gases important to climatic change

J.A. Taylor

Australian National University

1. Introduction

Understanding of the physics and chemistry of the global atmosphere is currently changing at a rapid rate. Perhaps the most obvious manifestation of this revolution is the perceived need to improve our understanding of the greenhouse effect and the potential for global climate change and the destruction of the ozone layer. The high probability that major changes in the earth's climate currently being predicted by climate models shall actually eventuate has lead to an urgent need to develop our understanding of the biogeochemical cycles of carbon, nitrogen, sulphur and phosphorus. With the exception of the chlorofluorocarbons, the atmospheric concentrations of the key greenhouse gases, carbon dioxide, methane and nitrous oxide, are controlled through a combination of anthropogenic and biospheric sources and biospheric sinks. Consequently, if we wish to develop effective and efficient control strategies for these trace gases and accurately predict their future concentrations we must improve our understanding of biogeochemical cycles.

2. A global atmospheric tracer transport model

While comprehensive and precise measurements for a few chemical species have been available for some time, detailed measurements, with global coverage, of the thousands of trace chemical species present in the troposphere remain beyond the reach of modern technology. Accordingly, theoretical studies requiring the development of high-resolution global chemical transport models able to represent tropospheric chemistry and global atmospheric circulation have commenced. These models must also include a realistic treatment of small scale processes responsible for producing significant perturbations at the larger scale if an accurate understanding of global tropospheric chemistry and the cycles of chemical species which pass through the atmosphere are to be developed.

Ideally, it would be advantageous to perform studies of the biogeochemical cycles of trace gases within the framework of an existing climate model. However, the computational cost of such an approach is currently too great for detailed calculations to be performed. In addition, the benefits of performing such expensive calculations when considerable uncertainty in the mechanisms of biogeochemical cycles confounds precise calculation, will be very limited. This problem arises when studying the sources and sinks of greenhouse gases, models of atmospheric chemistry and models of biosphere, as a large range of alternative formulations would need to be investigated. A wide range of sensitivity studies would need to be performed before any confidence could be placed in the model output. As a consequence, the approach adopted here is to first develop a parameterized three-dimensional model of atmospheric transport. This allows the transport component of the model to be run very efficiently rather than dominating the calculations.

As a first step in the process of developing a global tropospheric chemistry and transport model, a high resolution transport model must be developed. A stochastic global three-dimensional Lagrangian tracer transport model has recently been developed for the purpose of studying the sources and sinks of trace gases important to climatic change (Taylor, 1989). Model advection terms are derived from the ECMWF analysed grids. Source and sink terms for carbon dioxide, methane, CFC-11, radon and methyl-chloroform have been incorporated within the model. Model predictions are compared with atmospheric observations of the respective trace gases when appropriate data is available (Taylor, 1989).

The Lagrangian equation for atmospheric dispersion of a chemical species is as follows (Zanetti, 1989)

$$<c(r,t)> = \int_0^t \int_{-\infty}^{\infty} p(r,t/r^1,t^1)\ s(r^1,t^1)\ dr^1\ dt^1 \qquad (2.1)$$

where: $<c(r,t)>$ represents the ensemble average concentration at r at time t; $s(r^1,t^1)$ is the source of the chemical species ($g.m^{-3}.sec^{-1}$); $p(r,t/r^1,t^1)$ is the probability density function (m^{-3}) that an air parcel moves from r^1 at t^1 to r at t, where for any given r^1 and $t>t^1$ the following must hold (Zannetti, 1989)

$$\int_{-\infty}^{\infty} p(r,t/r^1,t^1)\ dr = 1 \qquad (2.2)$$

in order to satisfy the conservation of mass of the chemical species. If physical or chemical removal processes are included, the integral in (2.2) is set to a value less than one. The term $s(r^1,t^1)>0$ only at the point of release and where chemical transformation of secondary species within the atmosphere occurs.

As it is not always possible to evaluate the release of the chemical species for $-\infty \leq t^1 \leq t$, (2.1) can be rewritten as (Zannetti, 1989)

$$<c(r,t)> = \int_{-\infty}^{\infty} p(r,t/r^1,t^0)\ c<(r^1,t^0)>\ dr^1 + \int_{t^0}^{t}\int_{-\infty}^{\infty} p(r,t/r^1,t^1)\ s(r^1,t^1)\ dr^1\ dt^1 \tag{2.3}$$

where only the trace gas emissions during $t^0 \leq t^1 < t$ are included as the first term represents the release of the trace gas prior to t^0. The first integral term is approximated by the average concentration $<c>$ at time t^0 for the simulated atmospheric domain.

In applying (2.1) and (2.3), Taylor (1989) uses a stochastic Lagrangian advection scheme to move 100,000 equal mass air parcels containing a known mass of tracer in air according to the ECMWF analysed wind field grids. At each time step an air parcel displacement is computed taking into account the transport due to the mean fluid flow and the random turbulent flows, for each wind speed component u, v and w separately, as follows (Taylor, 1989)

$$D_{u,v,w} = \Delta t\ (m + \sigma N[0,1]) \tag{2.4}$$

where: D is the air parcel displacement; Δt is the time step; m is the appropriate bi-monthly mean wind speed and σ the standard deviation of the wind speed averaged over a bi-monthly time period; and N[0,1] represents a sample from the standard normal distribution.

3. Comparison with Eulerian approaches

Previous attempts at modelling atmospheric tracer transport and the study of sources and sinks of trace gases were based on the Eulerian modelling approach incorporating eddy diffusion. The majority of these models are two-dimensional zonally averaged representations of atmospheric circulation with relatively low resolution. Eddy diffusion coefficients are themselves very uncertain and should strictly be evaluated separately for each chemical species under study. A further difficulty with Eulerian models is the introduction of artificial advection arising from the numerical solution of the finite difference form of the advection equations which leads to a failure to conserve tracer mass. This problem is currently resolved through the introduction of numerical fixes. The Eulerian approach also requires that the transport of chemical species be independently evaluated for each chemical species included in a model simulation. Finally, the computational costs of Eulerian models applied at high resolution and for simulations over a number of years are so considerable that no such studies have been undertaken (Rood and Kaye, 1989).

As an example of the current limitations of the application of Eulerian-based models, Rood and Kaye (1989) estimate that for a one year model simulation, a stratospheric general circulation model with chemistry using a 4° latitude by 5° longitude grid with 19 vertical levels would require approximately 76 days on a dedicated CDC Cyber 205 supercomputer, assuming that the model could efficiently fit into memory. Clearly, even if such resources could be made available the results of such a model run would be of limited use due to the difficulties of performing sensitivity studies. Other Eulerian three-dimensional tracer transport models have been developed where a general circulation model is run independently of the chemistry sub-model (Prather et al, 1987). The model transport is stored as coefficients and used as a parameter to the chemistry model. Such model simulation runs are reported as requiring about 20 hours of CPU time to transport one chemical species for one model year.

The Lagrangian model, described in detail by Taylor (1989), offers a number of significant advantages over previous modelling approaches. Most importantly, the approach is flexible in that different transport fields, trace gas source functions and initial tracer concentration fields may be readily incorporated into the model, multiple trace species may be advected simultaneously, numerical instabilities associated with Eulerian advection are avoided and conservation of tracer mass is ensured using the Lagrangian approach. The model is computationally efficient so that one year model simulations can be performed in ~140-200 seconds of CPU time on a CRAY X-MP/48. The low computational cost allows a range of possible source functions to be studied and multi-year simulations to be performed for carbon dioxide, methane, CFC-11, radon and methyl chloroform.

The Lagrangian approach is about 400 to 40,000 times faster than the comparable Eulerian models at lower resolutions for three-dimensional transport of one chemical species. If the fact that the Eulerian approach must advect each chemical species independently is also taken into account, then the efficiency, e, defined as the ratio of computational times of the Eulerian methods to Lagrangian methods for n chemical species, n_{chem}, will be $e \sim 4{,}000 n_{chem}$. Further, in the event that Eulerian methods become as computationally fast as Lagrangian methods in advecting one chemical species, the Lagrangian approach would still remain n_{chem} times more efficient. Future studies of atmospheric chemistry will see n_{chem} increasing towards the many thousands of chemical species present within the atmosphere.

4. Lagrangian model results

Examples of the performance of the Lagrangian model (Taylor, 1989) with methane, CFC-11, radon and methyl-chloroform will be presented. Partial model validation is achieved through comparison of the model output obtained for CFC-11 with available observational data and with model results obtained by other workers. Clearly, it is not possible to provide a test which absolutely validates the behaviour of an atmospheric transport model. One reason is that the complexity of atmospheric transport and mixing makes the fully deterministic modelling of this process practically impossible. The complexity of atmospheric transport also means that long time model integrations (> 10 days) represent no more than one possible realisation of the evolving distribution of atmospheric concentrations. This problem is now popularly understood as arising from the chaotic behaviour of the system. The chaotic behaviour of the atmospheric transport of trace gases means that no global atmospheric transport and chemistry model will be able to predict trace gas concentrations to an accuracy within the natural variability produced by atmospheric transport and the sources and the sinks of the trace gas. It is therefore important that models are able to generate an ensemble of concentration estimates.

Apart from fundamental theoretical restrictions to model validation, practical constraints associated with the available observational data also apply. No observational dataset of trace gas concentrations with global tropospheric coverage exists. What data that is available largely consists of point observations, often representing no more than a few litres of air sampled once a week, scattered over the entire globe with the intention of determining long term trends rather than model validation. Unfortunately models cannot easily predict point concentrations as they generate trace gas concentration estimates which represent large volume averages. In order to compare model output with observations it is necessary to compute long term average values, such as monthly means. This tends to smooth but not remove, the effects of the natural variability of the system. Another approach is to compare the outputs of different model studies. Again, problems of differing model resolutions and assumptions contained within the model make this approach qualitative rather than quantitative. Ultimately model validation is achieved through a combination of approaches aimed at invalidation rather than an absolute validation.

In summary, comprehensive quantitative model validation of atmospheric transport and chemistry models will remain problematic until datasets of trace gas concentrations are available with global coverage and high spatial and temporal resolution. Even if these data become available, the large natural variability will make selection between models difficult.

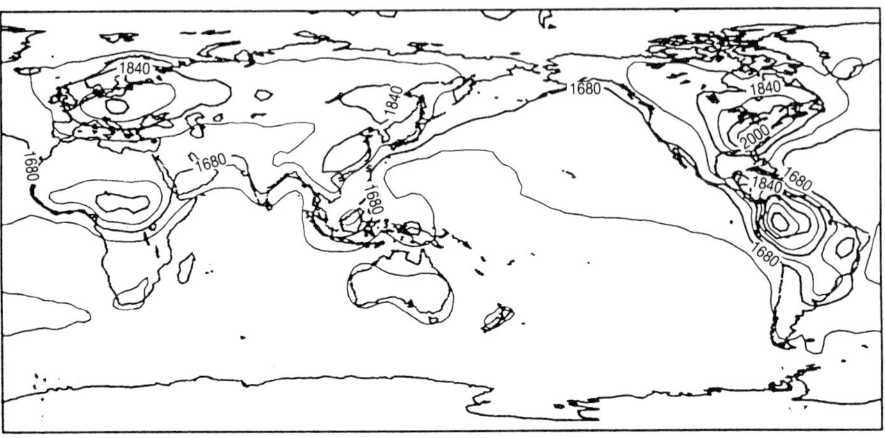

Fig. 1: The model predicted methane concentrations (ppb) at the model surface layer (1000-925hPa) for the month of June.

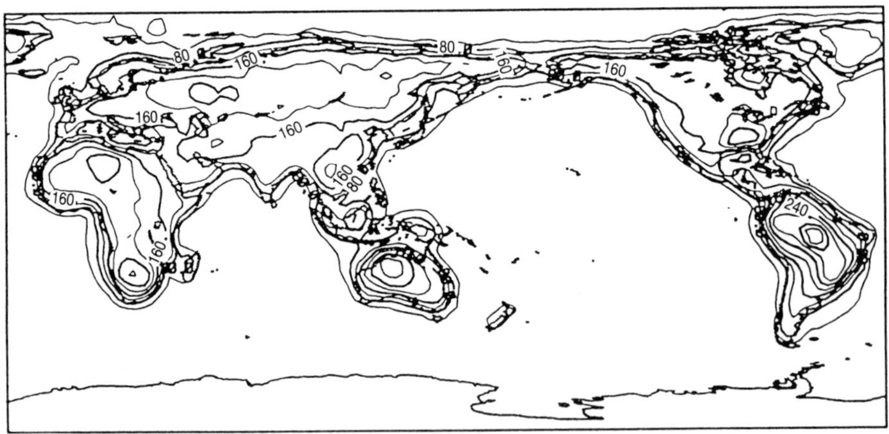

Fig. 2: The model predicted radon concentrations (pCi m^{-3}) at the model surface layer (1000-925hPa) for the month of June.

Other criteria, such as the conservation of mass of tracer, which can be determined directly from model runs, and the ability to generate an ensemble of model predictions representing the natural variability of the system, will be fundamentally important model performance assessment criteria.

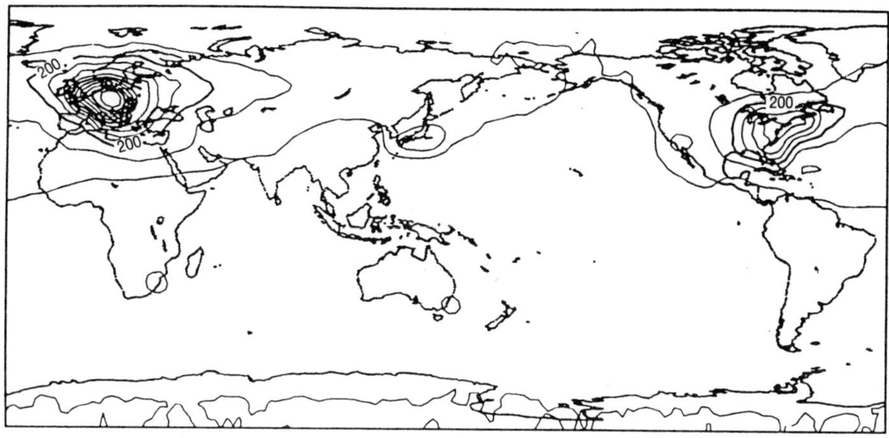

Fig. 3: The model predicted methyl-chloroform concentrations (ppt) at the model surface layer (1000-925hPa) for the month of June.

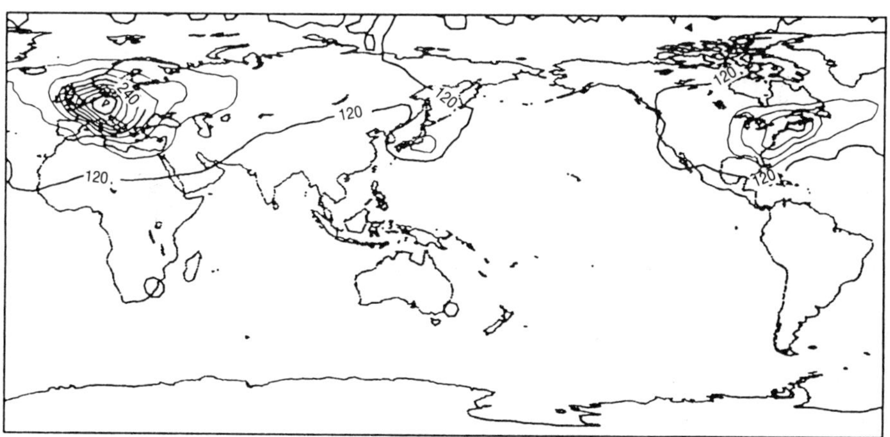

Fig. 4: The model predicted CFC-11 concentrations (ppt) at the model surface layer (1000-925hPa) for the month of June.

Model results are presented for methane, methyl-chloroform, radon, and CFC-11. Results for carbon dioxide are presented by Taylor (1989). In all cases model runs were initialized with a predicted concentration distribution based on observed data and run forward the equivalent of two years. The results reported here are taken from the second year of the model run. Fig. 1 shows the predicted distribution of methane concentration (ppb) based upon a source function described by Taylor et al (1990) for the month of June for the model surface layer which is assumed to be the lowest level of the model corresponding to the pressure range 1000-925 hPa. The methane flux is highest in the northern hemisphere in June.

Fig. 2 similarly illustrates the predicted model surface layer concentration predicted for radon for the month of June. Radon is assumed to be emitted from land surfaces at the constant rate of 1 molecule $m^{-2}.s^{-1}$.

Fig. 3 presents the predicted methyl-chloroform concentration (ppt) for June and Fig. 4 presents the predicted CFC-11 concentrations for June. Both emission source functions are based upon the source function derived for the chlorofluorocarbons developed by Prather et al (1987). The model results obtained using the Lagrangian model are in excellent qualitative agreement with those reported by Prather et al (1987) who employed a Eulerian model with transport parameters derived from a global GCM.

Figs. 1-4 show the importance of employing high resolution models when studying the problem of the global biogeochemical cycles of greenhouse gases. The results clearly illustrate the substantial spatial variation associated with the sources and sinks of trace gases which would be lost in models of lower dimensionality and resolution.

5. Conclusions

A global three-dimensional Lagrangian tropospheric tracer transport model has been developed and successfully applied to study the sources and sinks of trace gases important to climatic change. The model has a number of advantages over existing approaches. The Lagrangian model conserves the mass of tracer by definition and is capable of predicting an ensemble of possible tracer distributions associated with the natural variability of atmospheric transport and mixing.

The model also offers the practical advantages of being at a much higher resolution than other models and of having employed observational wind field transport data to parameterize tracer transport rather than GCM derived wind fields. In addition the Lagrangian formulation is computationally much more efficient than Eulerian models. Finally, the relative efficiency of the Lagrangian approach is likely to improve with the increasing number of trace species included in tropospheric chemistry models.

6. Acknowledgements

The model simulations were undertaken with a grant from the Australian National University Supercomputer Facility and while a Visiting Scientist at the United States National Center for Atmospheric Research. The National Center for Atmospheric Research is sponsored by the United States National Science Foundation. The author would like to thank Shelley Santoso for assistance with word processing.

7. References

Prather, M., McElroy, M., Wofsy, S., Russel, G. and Rind, D. (1987). Chemist of the global troposphere: Fluorocarbons as tracers of air motion, Journal of Geophysical Research, 92:6579-6613.

Rood, R.B. and Kaye, J.A. (1989). Chemistry and transport in a three-dimensional stratospheric model: Chlorine species during a simulated stratospheric warming. Journal of Geophysical Research, 94:1057-1083.

Taylor, J.A. (1989). A stochastic Lagrangian atmospheric transport model to determine global CO_2 sources and sinks - a preliminary discussion. Tellus, 41B:272-285.

Taylor, J.A., Brasseur, G., Zimmerman, P. and Cicerone, R. (1990). A study of the sources and sinks of methane using a global 3D Lagrangian Tropospheric tracer transport model, Journal of Geophysical Research (submitted).

Zannetti, P.(1989). Simulating short-term, short-range air quality dispersion phenomena. In: Encyclopedia of Environmental Control Technology, Vol.2, Air pollution control, P. Cheremisinoff, editor. Gulf Publishing Company.

Modelling the behaviour of radioactive tracers in the environment: three case studies

E.M. Scott

University of Glasgow

M.S. Baxter, P. Bradley, F. Begg

Scottish Universities Research and Reactor Centre

M. McCartney

Ministry of Agriculture, Fisheries and Food

1. Introduction

The impact on the environment, and indirectly on man, of the release of radioactive pollutants is an area of considerable importance and one which is frequently tackled through coupled experimental and modelling programmes. The models used cover a considerable range of complexity and are used as both descriptive and inferential tools. Frequently, their major use is for prediction, in both the short- and long-term, and sensitivity analyses are used to identify particular parameters or pathways, to which the response variable (e.g. dose) is particularly sensitive (Emanuel et al, 1984; Killough, 1980).

A wide variety of models exist for different applications, the majority of which are, however, based on simple linear systems. The character of the models is determined by the chemical nature of the radionuclide and the nature of the release environment. Increasing complexity is often not appropriate due to the lack of the necessary experimental results which would allow all the parameters to be estimated and the model to be validated. Expectations of "fit" of model output are, as a result, correspondingly low: in some applications, agreement within an order of magnitude is regarded as acceptable. The definition of the compartments and their boundaries may be notional rather than physically meaningful, and transfer is determined by first-order kinetics, the coefficients representing mathematical convenience, rather than process driven transfer mechanisms. Important features in the choice of model concern the spatial and temporal resolution required and the availability of data. In this paper, we describe local (within 40km of a source), regional (i.e. Scottish coastal waters) and global impact studies which have made use of simple environmental models. These divisions of resolution reflect the current position of the International Atomic Energy Agency on investigation of the apportionment of dose to population (IAEA, 1989).

Model validation is a critical but very difficult process for a variety of reasons - not least the lack of sufficient appropriate experimental data and the impracticality of including all conceivable processes and mechanisms within the model. The uncertainties on model predictions can be difficult to quantify, due to natural variability within the systems to be modelled, yet are essential in interpretation of the results. The sources of uncertainty in some instances can be quantified from literature values or experimental results, in others they must be subjectively ascribed values.

The processes of validation and sensitivity analysis should be seen as part of the assessment of the reliability of the model, other processes to be considered include parameter uncertainty analysis, taking into account the highly interdependent nature of the parameters and where possible, the comparison with results from other models.

We present the results of three case studies involving the radioactive pollutants radiocarbon, ^{14}C, and radiocaesium ^{137}Cs and ^{134}Cs released to the environment by the nuclear industry; discuss models used to predict their dispersal, and assess their reliability.

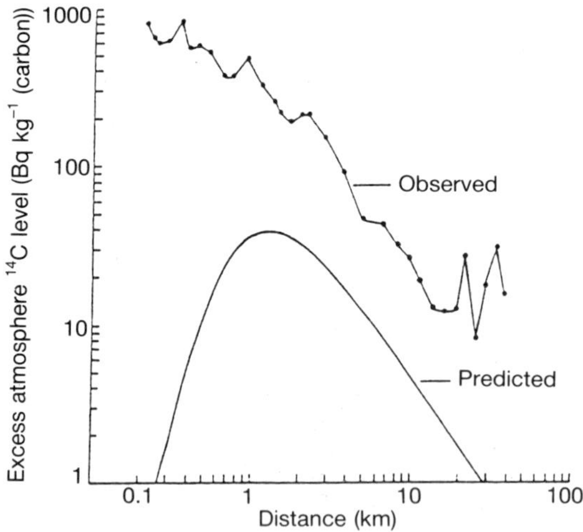

Fig. 1: Atmospheric ^{14}C levels around Sellafield in 1985 as predicted by the Gaussian plume model.

2. Illustrations of simple environmental models

2.1 Radiocarbon discharges from the nuclear fuel cycle

^{14}C is produced and released as part of the nuclear fuel cycle. Its long half-life of 5730 years, mobility in the environment and incorporation into man via the food-chain, make it of some considerable radiological significance. As installed nuclear energy capacity increases, the potential dose detriment resulting from future ^{14}C discharges requires to be assessed and divided into local, regional and global components.

Local effects (McCartney et al, 1988b): The dispersal of atmospheric pollutants from a single point source is widely modelled using a Gaussian plume model, where the basic model equation is given by

$$X(x,y,z,h) = \frac{Q}{2\pi\sigma_y\sigma_z u}\exp\{-\frac{1}{2}[\frac{y}{\sigma_y}]^2 + [\frac{z-h}{\sigma_z}]^2\} + \frac{Q}{2\pi\sigma_y\sigma_z u}\exp\{-\frac{1}{2}[\frac{y}{\sigma_y}]^2 + [\frac{z+h}{\sigma_z}]^2\} \qquad (2.1)$$

where: Q is release rate or total release; u the mean wind speed affecting the plume ($m.s^{-1}$); h effective height of the plume centre-line (m); σ_y, σ_z the standard deviations of the plume spread in the vertical and horizontal directions.

Although very simple, the Gaussian plume model retains some flexibility through careful choice of the parameters σ_y and σ_z, which are in part determined by the value of x (distance down wind), the meteorological conditions during release and by the terrain roughness (Clarke, 1979). This model was deemed appropriate to be used in an assessment of the local effects of ^{14}C discharges from BNF plc's fuel-reprocessing plant in Cumbria. An experimental programme was developed whereby in each of 3 years, grass samples were collected up to 40km from the plant, used to validate the model and hence determine the radiation dose to the population within 40km of Sellafield (McCartney, 1987).

Results: The agreement between observed and predicted ^{14}C levels was found to be poor, not only was the predicted maximum excess 20 times smaller than that observed, but also situated 4 times further from the source. The Gaussian plume model can be seen to be a rather coarse instrument, and likely sources of inaccuracy include: the definition of the effective height, h; the assumption of a single source of ^{14}C; and the assumptions of constancy of both $^{14}CO_2$ release rate from Sellafield and its uptake by plants throughout the growing season.

Fig. 1 demonstrates the discrepancies between observed and predicted atmospheric ^{14}C excess. These discrepancies can be considerably reduced by fine-tuning the model using the effective height parameter h. The requirement for site-specific information in the form of meteorological information is also relevant in determining the direction of maximum dispersal and failure of the model might result from inappropriate meteorological parameters; in this instance, frequency distributions of wind speed and direction can be calculated from wind rose data.

A simpler empirical derivation of the Gaussian plume model was also used, namely, a hyperbolic model which describes the fall-off with distance from source in a single direction. This model, when fit in the form

$$\chi(x) = \alpha + \frac{\beta}{x} \tag{2.2}$$

was validated through comparison of the constant α with the global baseline level. Agreement was found to be extremely good, as shown in Fig. 2 and thus this simpler model was used to empirically describe the atmospheric dispersion pattern for $^{14}CO_2$ at this site.

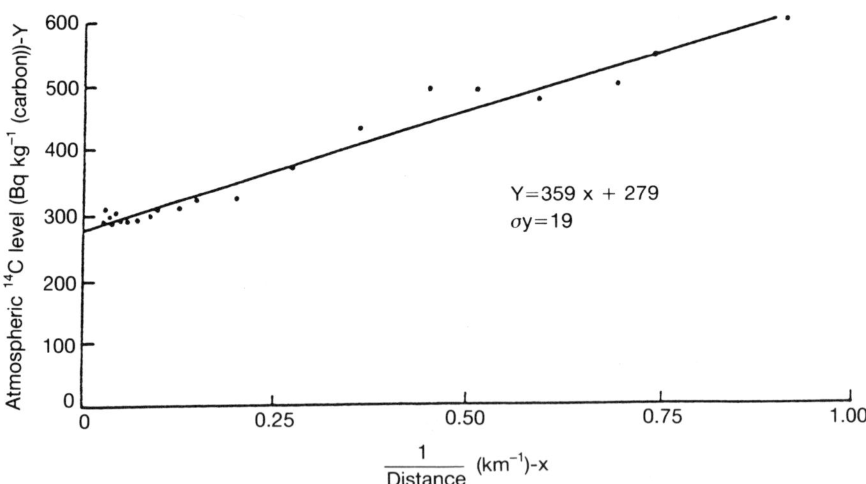

Fig. 2: Atmospheric ^{14}C levels versus distance^{-1} for Sellafield, 1985.

This study illustrates the importance of matching model complexity to existing experimental data and to model function as well as the ability to validate the model either on further experimental data or with extraneously determined targets (e.g. the global baseline level).

2.2 *Regional Effects: Sellafield radiocaesium in UK Coastal Waters (Bradley et al, 1987)*

Radiocaesium has been discharged into the Irish Sea from the Sellafield nuclear fuel re-processing plant in Cumbria for more than three decades and has been used as a tracer of water movement in the western UK coastal system. Extensive radiocaesium and hydrographic data have been collected for the last 12 years from many cruises, and this data has been used to characterise the controlling oceanographic flow parameters.

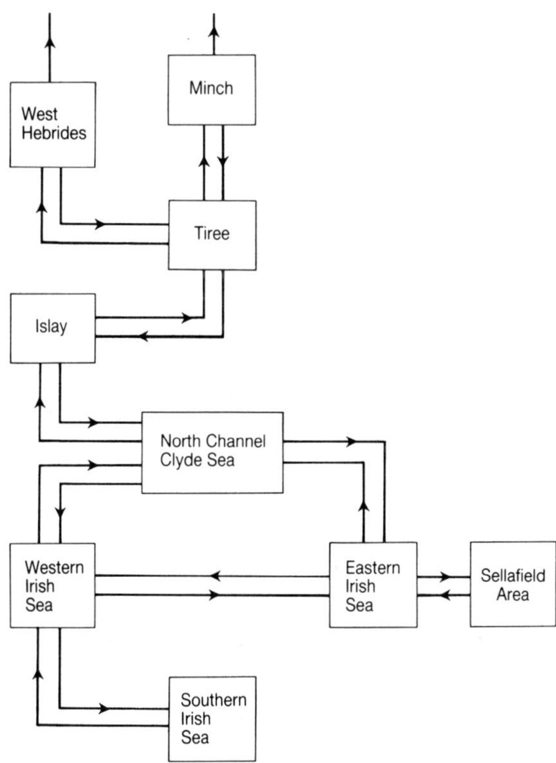

Fig. 3: Compartments of the western coastal UK system.

Modelling: The Western UK coastal system has been compartmentalized to allow the use of linear systems of equations to describe the transfer of ^{137}Cs and ^{134}Cs through the compartments. The definition of the compartments was made on the basis of known hydrographic features, including depths and current information. Fig. 3 shows a schematic diagram of the system.

Radioactive isotopes have, of course, considerable advantages as water tracers as they exhibit decay during transport - the half-life of ^{134}Cs is highly compatible with the rate of coastal current movement. In particular, for this application, the Sellafield discharges have been seasonally pulsed, thus making the matching of short-term temporal trends possible, and allowing current rates to be derived.

Experimental data: A different number of measured activities are available for each compartment in the model (i.e. multiple time series of differing lengths and frequencies) - the best characterised compartments being the North Channel and Clyde Sea Area. Some transfer coefficients are taken as known (measured using flow meters), while others require to be estimated (in particular transfers in the Clyde Sea Area, and northern Scottish waters). The estimation method which we have used is due to Hallstadius (1986) and is known as "inverse compartment analysis". The basis of this method for estimation of the transfer coefficients, involves an iterative procedure defined as follows:-

Step 0: Initial set of transfer coefficients defined.
1: Results for each compartment are simulated.
2: A comparison of the observed and simulated activities are made, as a result of which a new set of transfer coefficients are defined and the procedure repeated.

Convergence is achieved when the agreement between simulated and observed results is acceptable.

The procedure was applied to the radiocaesium data, within the framework of the model described earlier. Examples of the model fit are given in Fig. 4. The method is numerically inefficient and for complex systems, the time taken to convergence is long (of the order of several hours). It handles only one coefficient at a time, where more appropriately the optimisation procedure should be dealt with in a multivariate manner. Model validation is difficult, since the sparsity of data in some compartments argues for using all the existing data to fit the model, however, one approach which has been used previously (Kershaw et al, 1989) and which we currently are investigating is to validate the model on a different radionuclide for which data is available in greater quantities and which has as similar as possible properties to the nuclide of interest.

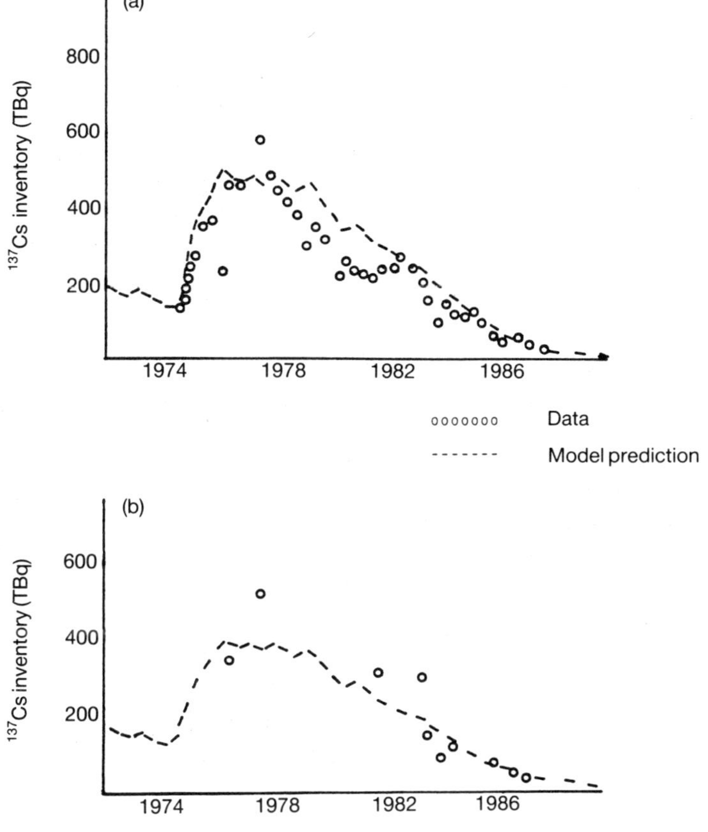

Fig. 4: Comparison of observed (circles) and predicted (dashed lines) ^{137}*Cs levels. Upper (a) Compartment 5: North Channel/Clyde Sea area. Lower (b) Compartment 6: Islay Front area.*

2.3 *Global Effects: The global effect of* ^{14}*C discharges from the nuclear fuel cycle (McCartney et al, 1988a)*

As discussed earlier, the long half-life of ^{14}C and its biological mobility make it of radiological significance to the world population. Its global significance has been assessed using a variety of carbon cycle models; the availability of considerable experimental data, and other appropriate input data has allowed the models to be validated and sources of uncertainty inherent in the dose estimates to be evaluated. The reliability of model predictions has been investigated by comparing the results from a variety of models of differing complexity thus allowing model structure to be assessed as a source of uncertainty.

Carbon cycle models: The carbon cycle models considered include a very simple 3 box model (Baxter and Walton, 1970), an 8 box model with north/south division (Bush et al, 1983) and a 25 box model with detailed biosphere (Emanuel et al, 1984). The models comprise n discrete reservoirs and the requisite number of transfer coefficients. Again linear effects are solely considered, although in the 25 box model, the complexity of the ocean compartments can be interpreted as a discretized version of a box-diffusion model, while the biosphere is described by 5 compartments.

Input data: The various sources of input data include CO_2 records taken at several locations in the southern hemisphere and which have allowed the models to be validated in a preliminary stage. ^{14}C input data have included discharge data from the nuclear fuel cycle and weapons testing production of ^{14}C in the 1960s. CO_2 production as a result of fossil fuel combustion and deforestation has also been incorporated.

Seven different energy use scenarios were considered, and the variation in the prediction of future atmospheric ^{14}C levels (using the 25 box model) is shown in Fig. 5. A range of 205-245 Bq $kg^{-1}C$ is observed.

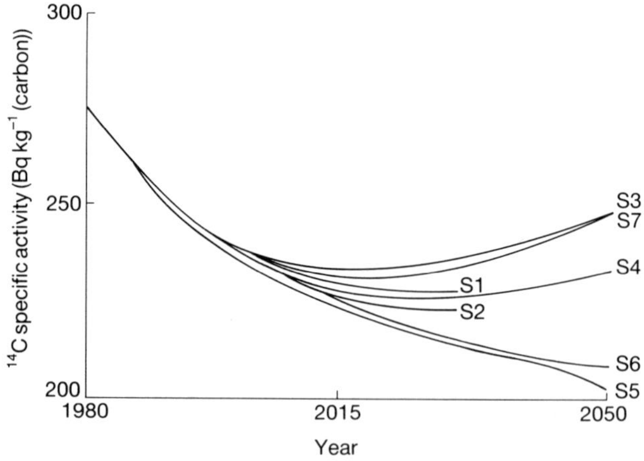

Fig. 5: Future atmospheric ^{14}C levels, 1980-2050, as predicted by the 7 basic scenarios (using 25-box model).

Table 1. The CEDEC (man Sv) to Man from the release of 1 TBq of ^{14}C into different compartments of the carbon cycle, as predicted by various models, assuming the medium fossil fuel scenario.

Model	Discharge to		
	Atmosphere	Surface ocean	Deep ocean
3-box	114	111	97
8-box	118	113	109
25-box	127	125	95

Using a single energy scenario, the values predicted for the atmospheric ^{14}C specific activities by the different box models lie in the range 228-236 Bq $kg^{-1}C$. This consistency is driven largely by the atmosphere-surface ocean exchange rate, which is relatively well defined.

For longer term predictions, the environmental behaviour of ^{14}C discharges must be studied and to this end a nominal input of 1TBq is injected into the different model compartments. Two key features controlling the response are the compartment into which ^{14}C is released and the future rate of combustion of fossil fuels. Model validation by comparison of results obtained by the different models is an appropriate method allowing relative rankings of parameters and processes as contributors to the overall uncertainty. Table 1 indicates the variation in CEDEC (Collective Effective Dose Equivalent Commitment) across 3 models and with differing input compartment.

3. Conclusions

The three studies described here attempt to highlight the different processes required in the assessment of environmental transfer model reliability. The models used are simple in nature, yet even for these models, validation and assessment of reliability require considerable experimental data which are often not available. One approach which would ensure that sufficient data (of high quality) are available is to integrate both the modelling and experimental work. Model building then becomes an iterative process: experimental results are used to validate existing models; large discrepancies and uncertainties define pathways and processes requiring further experimental investigation; and finally the models define the appropriate sampling framework for efficient estimation of parameters. The models developed then become process driven, acquire increasing realism and hence ultimately provide more reliable short and long term predictions.

4. Acknowledgements

One of us (M.McC) thanks the UK Science and Engineering Research Council (SERC), while two (P.B. and F.B.) thank the UK Natural Environment Research Council (NERC) for financial support.

5. References

Baxter, M.S. and Walton, A. (1970). A theoretical approach to the Suess effect. Proc. Royal Society London, A 318:213-230.

Bradley, P.E., Economides, B.E., Baxter, M.S. and Ellet, D.J. (1987). Sellafield radiocaesium as a tracer of water movement in the Scottish coastal zone. In: Radionuclides-a tool for Oceanography, Cherbourg. Elsevier Applied Science.

Bush, R.P., White, I.F. and Smith, G.M. (1983). Carbon-14 Waste Management. AERE-R10543, UKAEA, Harwell.

Clarke, R.H. (1979). A model for short and medium range dispersion of radionuclides released to the atmosphere. NRPB-R91, National Radiological Protection Board, Chilton.

Emanuel, W.R., Killough, G.G., Post, W.M. and Shugart, H.H. (1984). Computer implementation of a globally averaged model of the world carbon cycle. US DOE DOE/NBB-0062. NTIS, Springfield, Virginia.

Hallstadius, L. (1986). Compartment modelling in nuclear medicine: a new program for the determination of transfer coefficients. Nuclear Medicine Comms, 7:405-414.

IAEA. (1989). Principles for the establishment of upper bounds to doses to individuals from global and regional sources. IAEA Safety Series No.92, Vienna.

Killough, G.G. (1980). A dynamic model for estimating radiation dose to the world population from releases of ^{14}C to the atmosphere. Health Physics 38:269-300.

Kershaw, P.J., Pentreath, R.J., Gurbutt, P.A., Woodhead, D.S., Durance, J.A. and Camplin, W.C. (1989). Modelling the behaviour of long-lived radionuclides in the Irish Sea - Comparison of model predictions with field observations. In: Methods for assessing the reliability of environmental transfer models predictions. Elsevier Applied Science.

McCartney, M. (1987). Global and local effects of ^{14}C discharges from the nuclear fuel cycle. PhD Thesis, University of Glasgow.

McCartney, M., Baxter, M.S. and Scott, E.M. (1988a). Carbon-14 discharges from the Nuclear Fuel Cycle 1: Global Effects. J. Environmental Radioactivity 8:143-155.

McCartney, M., Baxter, M.S. and Scott, E.M. (1988b). Carbon-14 discharges from the Nuclear Fuel Cycle 2. Local Effects. J. Environ. Radioactivity 8:157-171.

Lagrangian modelling of an atmospheric convective plume and an analysis of the comparative advantages of Lagrangian and Eulerian methods

A. Gadian

University of Manchester Institute of Science and Technology

1. Introduction

Traditionally, modelling of atmospheric fluid flow has concentrated on solving sets of partial differential equations using analytical or numerical techniques. Eady (1949) is one of the first to apply analytical techniques to the growth of baroclinic eddies. Green (1960) amongst others, extends this theory to produce numerical solutions for an advanced set of these perturbation equations. Phillips (1956) first produces a numerical general circulation model, using a simplified set of the Navier Stokes equations. The numerical models generally require some kind of spatial finite difference discretisation. Derivatives of the leap-frog type timestepping schemes are often used in these early models. Later (e.g. Gordon and Davies, 1977) more sophisticated schemes, such as Adams Bashforth, are employed, although these present other problems associated with a damped flow development. More recently, spectral (e.g. Doron, 1974) and finite element (e.g. Cullen, 1974) spatial discretisations improve the modelling of certain regimes of fluid flow, while semi-implicit techniques are introduced to reduce the time integration errors.

In the first part of the paper solutions to the barotropic vorticity equation are presented using Eulerian and Lagrangian schemes. Sawyer (1963) suggests the distinct advantages of longer timesteps using the Lagrangian framework. More recently Salmon (1985) examines the shallow water equations and Cullen (1984) develops the semi-geostrophic equations of Hoskins (1975), both suggesting novel approaches to fluid flow computation. In this paper Sawyer's (1963) work is extended and the relevant limitations of the two schemes analysed. The advantages and restrictions of the Lagrangian solution provides the necessary discussion of the general principles.

The more specific Lagrangian two-dimensional plume convective model is described in the later sections. Smooth Particle Hydrodynamics is used to describe the fluid as a collection of overlapping elements of fixed radius and a polynomial density distribution within each element. Although the density field is derived from the summation of all neighbouring particles, it is possible to trace the distinct fluid elements. Results can be presented with the vapour and liquid water phases included. The problems associated with the choice of the density kernel, the interaction of the dynamics with the thermodynamic equation of state are discussed. The relevance, or lack of it, of the inter-particle mixing and diffusion processes on the entrainment of elements into the buoyant plume are observed. It is apparent that diffusion timescales, as deduced by Baker and Latham (1979), are broadly inconsistent with realistic type model flows. Results of particle trajectories for different mixing regimes are presented.

2. Barotropic vorticity equation: an Eulerian approach

Numerical solutions of the vorticity advection equation

$$\frac{\partial \eta}{\partial t} + J(\Psi,\eta) = 0 \qquad (2.1)$$

are usually obtained in a Eulerian form. η the absolute vorticity is defined as

$$\eta = \nabla^2\Psi + f \qquad (2.2)$$

where: Ψ is the stream function; f is the coriolis parameter; and J the Jacobian operator. Sawyer (1963) describes a typical procedure but is modified here to include better numerical space and time discretisations. After briefly describing the approach used, the results are compared in the following section with those from the Lagrangian model.

2.1 The numerical technique

A square grid of 51 x 51 elements is used in the computation, with an implicit cyclic continuity at the two side inflow/outflow boundaries. The stream function is defined along the two upper and lower rows of gridpoints. Although, this overspecifies the boundary conditions (Charney et al, 1950), the insignificant distortion of the near boundary flow is compensated by the compatibility with the Lagrangian approach. To complete an iteration cycle, the Jacobian is first calculated, then a forward timestep implemented before finally recalculating a new stream function field from the modified vorticity.

The numerical approximation to the Jacobian is important in trying to conserve vorticity. The Arakawa (1966) formulation, which conserves mean square vorticity is used in this computation. Non-linear instability is also reduced (Haltiner and Williams, 1980) but the results may not necessarily conserve vorticity, depending on the time discretisation. However, the relative change per timestep, in the summation of the vorticity at all gridpoints, is found to be less than 10^{-12} and therefore within reasonable limits.

An advanced embedded Runge-Kutta (Dormand and Prince, 1980) enables the the magnitude of the local truncation errors for each timestep to be closely controlled. Traditionally, a leap-frog or modified Euler (in reality a 2nd order Runge-Kutta) type schemes have been used (Haltiner and Williams, 1980). Smoother time integration schemes which remember previous gridpoint vorticity values, such as Adams Bashforth (e.g. Gordon and Davies, 1977) are common. Unfortunately, these often exhibit inertia to rapid changes in flow pattern. Recently, 5th order accurate Runge-Kutta's (e.g. Dormand and Prince, 1980) have provided fast accurate implicit methods with known adjustable truncation errors. These implicit schemes, first developed for ordinary differential equations, prove to be very computationally efficient. This very stable, fifth order, seven stage Runge-Kutta is used in this paper for the time integrations. The model local truncation error is set as 10^{-6}.

At this stage the derived vorticities are used in the equation

$$\zeta = \nabla^2 \Psi \tag{2.3}$$

where ζ is the relative vorticity. An optimised and speedy Successive Over Relaxation (SOR) scheme (e.g. Smith 1978) is used to derive a new field of stream function values, given the cyclic continuity at the inflow and outflow boundaries. A five point evaluation for the ∇^2 term is used and the iteration tolerance is defined as the same size as the time truncation error given above.

The initial flow field is given by the analytical expression

$$\Psi = 2 - \frac{1}{1+(x^2+y^2)a^{-2}} - \frac{y}{b} \tag{2.4}$$

where a is proportional to the radius of the initial disturbance and $-\partial\Psi/\partial y$ represents the zonal wind. This represents a vortex moving in a uniform flow field and for the sake of comparison is the same as that used by Sawyer (1963). The relative results are given later.

3. Barotropic vorticity equation: a Lagrangian approach

In Lagrangian form the equation may be written

$$\frac{D\eta}{Dt} = 0 \tag{3.1}$$

Consequently, the vorticity allocated initially to each element remains unaltered through the iteration. However, this implies that a further equation is required to describe the elements motion. In this problem we can write

$$\frac{D\underline{r}}{Dt} = \underline{v} \tag{3.2}$$

where $\underline{r}$ are the vector co-ordinates and $\underline{v}$ is the velocity at that point.

3.1 Numerical technique

The initial flow field used is defined in Eq. 2.4. At the start of the integration, an element is positioned at each of the interior co-ordinates of the 51 x 51 grid in the Eulerian framework. The associated vorticity is calculated using the initial stream function field and the (2.2). To complete an iteration cycle, a forward timestep is implemented using the 5th order Runge-Kutta described above, the new gridpoint vorticity values are then interpolated, and finally a new stream function field deduced from (2.3). A velocity field is required to predict a new element position from (3.2). Wiin-Nielsen (1959) suggests a smoothing procedure to derive a mean flow in which the disturbance or vortex is embedded. In this example the uniform stream function field is directly derivable from the initial flow. Fjortoft (1952) shows that the errors produced from an assumption of a constant velocity field were not as severe as at first might appear. The scalar product of the varying velocity field component and the gradient of the absolute vorticity is relatively small. The longer the time integration, the larger the errors become (Fjortoft, 1952) and this disadvantage has to be considered when comparing the methods.

The Runge-Kutta's local truncation error is produced by comparison between 4th and 5th order accurate solutions. Its accuracy is only really specifically applicable to ordinary differential equations. The use for the partial differential equation solutions of the Eulerian model will be preferable to lower order methods, but will be dependent on the accuracy of the Jacobian.

The second stage of the iteration cycle requires an interpolation from the element to the grid point values. A simple 2nd order polynomial (Wiin-Nielsen, 1959) allocating the vorticity to the surrounding four gridpoints (i.e. weighted by area), gives sufficiently accurate results. The two rows of gridpoints with fixed values of stream function, are consistent with Sawyer (1963), the Eulerian scheme, and ensure that no Lagrangian particles cross the model walls.

The final Poisson's equation was solved with the same SOR procedure as described in the Eulerian scheme.

4. Lagrangian and Eulerian barotropic results

Two sets of results are described below: one set uses a narrow vortex where a=1 (Fig. 1), corresponding to a vortex of less than two elements wide, and the other with a value of a=3 (Fig. 2), a vortex of less than six gridpoints in diameter.

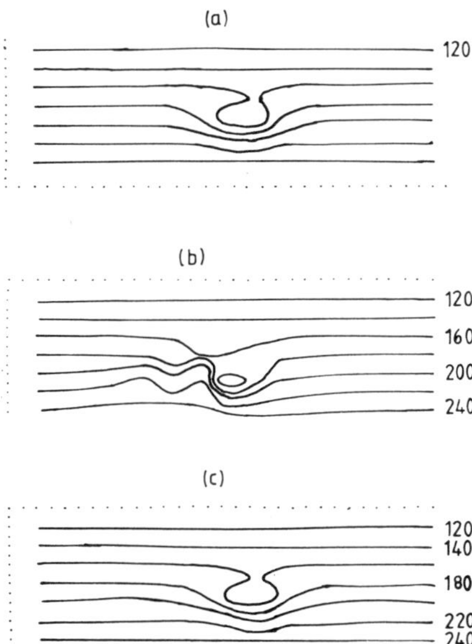

Fig. 1: In diagram (c) the original vortex stream function with a=1 is linearly transposed to its analytically calculated value. Diagram (b) gives the stream function using the Eulerian approach, and diagram (a) the Lagrangian discretisation. The grid spacing, of one length unit, is shown by the dots, and all three indicate the stream functions x100 after 42 model seconds.

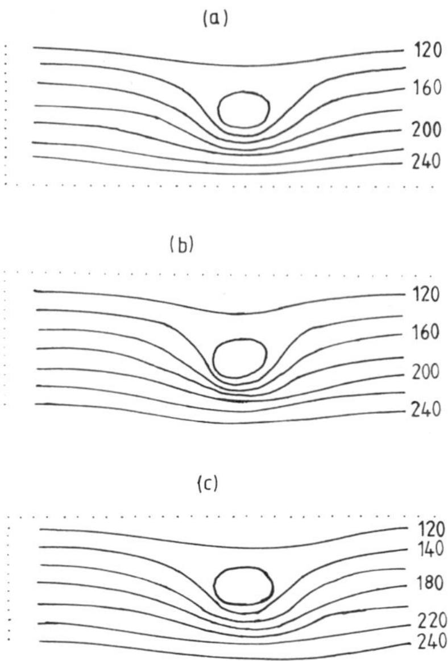

Fig. 2: As in Fig. 1, but with a=3 and at a time of 42 model seconds.

The x and y grid-spacing are defined as 1 unit length and b=10. These values were chosen to illustrate the differences in numerical techniques. The results (Fig. 1) for the small perturbation, give the largest differences. Within an error of 0.1%, there are no differences between the horizontal advection of the original flow (Fig. 1c) and the Lagrangian solution, Fig. 1a. This is as expected, from the known truncation errors of the Runge-Kutta. The Eulerian solution (Fig. 1b) shows the bunching of the stream function contours behind the vortex, thus sharply increasing the fluid speed in this region. Also indicated is the development of several wave-like perturbations in this area. This is connected to the reduction in phase speed of the vortex, of between 5-10%. This is significantly smaller than the 20% observed by Sawyer, using his simpler model. The implicit Runge-Kutta and the use of an improved Jacobian have significantly reduced the errors.

For the second case study (Fig. 2), with a=3, the two schemes show smaller differences. The speed of the Eulerian vortex is less than 5% slower than the horizontal translation of the initial field. The shape of the Eulerian vortex is now similar to the intial shape and the gridpoint values are within 3%, except for the 4 points in the centre of the vortex. Again the Lagrangian field is almost identical (within 0.1%) to the transposed initial field. When the initial value of the vortex was set at a=4 or larger, there was less significant difference (≤2%) between the Lagrangian and the Eulerian schemes.

In this simple case, the Eulerian integration for the smallest vortex relatively rapidly diverges from the analytical and Lagrangian solutions. The use of a 5th order timestepping scheme does reduce the errors by approximately a factor of more than two, when compared with a second order Runge-Kutta. The finite difference errors generated by a small vortex (i.e. having a large high order derivative component of the stream function) are not mitigated by having a Jacobian which theoretically conserves the mean square vorticity, and actually conserves the spatial sum of the gridpoint vorticities (Section 2.1). The Eulerian approach thus does not seem suitable for predicting the movement of perturbations on the same scale as gridpoint scheme.

In the Lagrangian technique, Eq. 3.2 determines the trajectory of the elements. A constant space-smoothed velocity-stream function field is required for the integrations. In this problem the mean fluid flow, the y gradient of the stream function, is used. The assumption and determination of such a field are limitations on the use of this procedure. However, Wiin-Nielsen (1959) does strongly suggest that this approximation is not as severe as might appear when applied to the atmosphere. The method requires a suitable choice of the space smoothing (his a_n constant) giving the size of the area over which the smoothing is averaged and which depends on the physics of the perturbation.

Computationally, the Lagrangian technique is very efficient. A small amount of extra memory is required to store the element coordinates as compared with the Eulerian approach. For a grid of 51 x 51, the computations can easily be carried out on a fast microcomputer. In the experiment with the small vortex (Fig. 1 with a=1) the Lagrangian scheme is 20 times faster than the Eulerian approach. For the larger vortex with a=3 (Fig. 2) the Lagrangian Scheme is only 10 times more rapid. A vortex size of a=5, approximately halves the speed difference again. The Lagrangian scheme, with its smoothed velocity field is controlled by the minimal truncation error of the Runge-Kutta, and therefore the timesteps can be large. The Eulerian technique, as well as having to calculate the Jacobian at each timestep, has grid-points which rapidly change their vorticity values etc. as the perturbation passes. Thus the timesteps are much shorter, and the computation longer for the same Runge-Kutta tolerance. However, the Runge-Kutta methods are designed specifically for ordinary, not partial differential equations, and it is likely that such a strict tolerance should not really be applied to the time integration, given the spatial discretisation errors. For the Lagrangian flow, the implicit physical assumption of a spaced smoothed flow field is often a simplification of reality, and one which becomes more dubious as the number and size of perturbations increases.

Accepting these Lagrangian limitations, in the subsequent sections, a simple Lagrangian plume model, using the same Runge- Kutta, is used to look at diffusion and entrainment. The entrainment scale is of the same order as the element size suggesting that a Lagrangian approach might be useful and tractable with limited computer resources.

5. Lagrangian model of plume convection

Smooth Particle Hydrodynamics has been used to model astrophysical fluid dynamics (Ginghold and Monoghan, 1977) and in plasma physics simulations (Birdsall and Fuss, 1969). The fluid is represented by a finite set of particles, each with its own density field described by a symmetrical function centred on the elements' coordinates (Fig. 3).

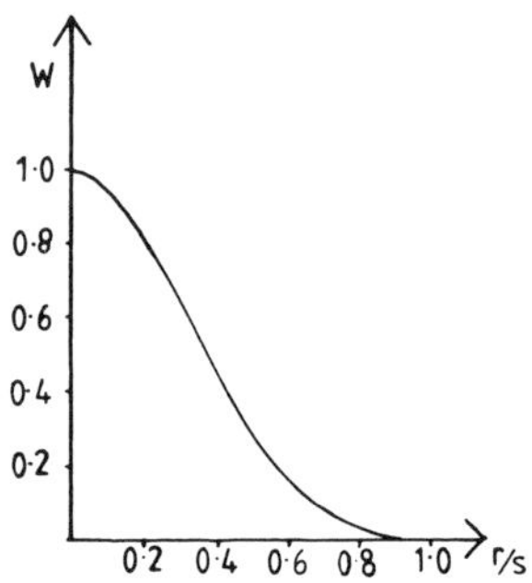

Fig. 3: The smoothed particle kernel shape, using the polynomial $W = W_m(1+mr/s)(1-r/s)^m$ *where:* $m=3$*;* r *is the distance from the centre; and* s *is the smoothing length. 73% of the area lies under the curve from* $r/s=0$ *to* $r/s=0.5$*.*

A third order differentiable polynomial is used to define the kernel, or density field, although alternative smoothing functions can be used. A more detailed description of the approach may be found in Gadian et al (1990).

This model describes fluid motion in a two-dimensional box, 4km across and with undetermined vertical extent. Using 5000 elements, the properties of a heated plume 300 metres across are described below. The condensation and evaporation of water is incorporated in the thermodynamics, although the microphysics of droplet size distributions and ice phase calculations could easily be added. The vertical and horizontal velocities are determined by the buoyancy and horizontal pressure gradients respectively.

5.1 Numerical model

The density at a given point is determined by a summation of all the contributions within the given smoothing distance. In this model it is assumed that all the particles have the same radius. A variable smoothing length makes the diffusive mixing more complicated, but it is planned to develop this aspect in later models. At a given point, the density is defined as

$$\rho = \int_{Domain} W(r,s)dr \tag{5.1}$$

where the radially symmetric polynomial kernel function W is

$$W = W_m(1+\frac{mr}{s})(1-\frac{r}{s})^m \tag{5.2}$$

for $0 \leq r/s \leq 1$ and for m=3. s is the smoothing distance or the fixed radius of the particle. Larger values of m produce steeper function gradients and local pressure gradients but made no significant flow differences. W_m is a constant to ensure each element has a unit mass. Gaussian kernels of the type $W \propto exp(-r^2/s^2)$ have been widely used and might merit further study. However, density discontinuities at the edges of the particles presents a problem in this type of discretisation.

The pressure is defined from the equation of state

$$p = \rho RT \tag{5.3}$$

where R is the gas constant and T is the element temperature. The horizontal pressure gradient is obtained by differentiation of (5.3) and hence the density kernel, (5.2).

The simplistic horizontal vector equation of motion is

$$\frac{D\underline{v}_h}{Dt} = -\frac{1}{\rho}\nabla_h p + \underline{F} \tag{5.4}$$

with the left hand side the Lagrangian substantial derivative. The frictional forces $\underline{F}$ were defined such that $\underline{F} \sim k|\underline{v}|\underline{v}$.

The aim was to remove any unwanted oscillations and to increase the computational efficiency. The value chosen for $k \sim 5x10^{-3}$ or a scale length of 200 metres, produces a largest difference of 18% in the maximum velocities obtained when compared with the undamped flow, and almost no difference to the averaged flow velocities over 50 model seconds.

The vertical acceleration is replaced by the buoyancy term

$$\frac{Dw}{Dt} = -g\frac{\delta T}{T} \qquad (5.5)$$

where T is the mean temperature at a height z and δT is the deviation from this mean. The vertical pressure gradient, assumed to be balanced by the gravitational term, is ignored in this buoyancy approximation. The potential temperature conservation due to the vertical motion of a particle is satisfied explicitly by incorporating a height dependency in the temperature. This simplification is used to allow for a fixed smoothing length to allow for conservation of elemental properties during diffusive mixing. Within 200 metres of the side wall boundaries, the vertical pressure gradient and the gravitational term are included to allow a return circulation without affecting the plume.

A smoothing length of 200 metres is chosen for this case study. Hill and Choularton (1985) observed a "blobby" dynamical structure to clouds. In reality, this produces the computationally feasible requirement of 5000 particles, with a spatial resolution of the order of 100 metres (Fig. 3).

The representation of the thermodynamics presents difficulties in the particle system. The process of heating a parcel will not produce a vertical acceleration, thus the need for the buoyancy equation. It is clearly feasible to vary the smoothing length to include density variations. However, although planned for further studies, this would require omission or severe modification of the mixing processes described later.

At each timestep the temperature change due to any liquid to vapour phase change is calculated, using a standard formula (Pruppacher and Klett, 1978)

$$\frac{\rho v}{\rho_A} = 0.622\frac{E_s}{p} \qquad (5.6)$$

where: ρ_v/ρ_A is the ratio of vapour to dry air; E_s is the saturation vapour pressure; and p is the atmospheric pressure. A polynomial formulation, from Pruppacher and Klett (1978) is used to calculate the saturated vapour pressure over water, E_s.

A simple iterative scheme (Newton Raphson) calculates the temperature change due to the latent heat release, given that the saturated vapour is itself a function of temperature

$$\delta T = L.H. \times \delta r_s | \rho | C_v \tag{5.7}$$

L.H. is the latent heat of the phase change, δr_s the change in liquid water, and C_v the specific heat at constant volume.

The inter-element diffusion processes are modelled by a redistribution of water vapour, liquid water and temperature on time scales commensurate with empirical values. Since the smoothing lengths are fixed, the elemental properties (e.g. temperature) can be interchanged with other neighbouring elements in a proportion related to their intersecting areas or volumes. This is not generally true for elements of varying smoothing length. Since this first model formulation is directed at studying the role of diffusive processes, the model required the use of the buoyancy equation to determine the vertical velocity, (see above) so that the particle characteristics are conserved (Gadian et al, 1990).

Baker and Latham (1979) suggested a mixing scale time constant τ of approximately 100 seconds, where $\tau \sim (X^2/e)^{1/3}$. X is the relevant space scale and e the kinetic energy dissipation via turbulent mixing. This leads to unrealistically rapid mixing, and an alternative scheme is tried where the mixing time constant $\tau \sim 1000/(|\delta u| + |\delta v|)$, δu and δv are the velocity differences with neighbouring elements.

The boundary conditions are chosen to be as simple as possible. At the side walls, cyclic continuity is ensured (i.e. particles are introduced to represent the corresponding elements at the other side boundary). At the lower boundary, each particle is mirrored by one below the ground. The gravitational term $g\delta T/T$ traps the particles in the vertical. Within two smoothing lengths of the boundary, the vertical pressure gradient is included in the vertical momentum equation

$$\frac{DW}{Dt} = -\frac{1}{\rho}\frac{\partial p}{\partial z} - g + F \tag{5.8}$$

This allows a return downward flow, while not interfering with the convection in the centre of the grid. Also in this region the resistive scale length is reduced to 10 metres to suppress any reflected waves at the side boundaries.

In a complete iteration cycle, (5.4) and (5.5) are used to calculate new coordinates using the Runge-Kutta scheme. The known truncation properties and the good stability of the Runge-Kutta enables a local time relative truncation error of 5×10^{-6} to be used. Vertical movements of the elements results in an adiabatic change in temperature, and in addition any thermodynamic phase changes are included at this stage. Finally the diffusive mixing processes are introduced each timestep. For a speed improvement, the elements are allocated on a specified area grid at the beginning of each timestep, so that the list of all the neighbouring elements can be readily obtained, without searching through the complete particle set. To set up an initial atmosphere, the full momentum equations, with the vertical and horizontal pressure gradients, are integrated until there is a stable and almost stationary arrangement of particles.

6. Results and discussion of the convective plume model

The model environmental atmosphere used is illustrated in Fig. 4.

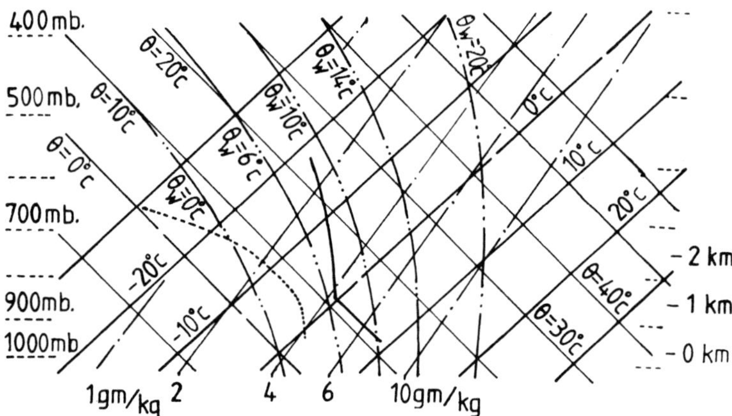

Fig. 4: A Tephigram for the environmental temperature profile used in the model. The dry bulb temperature (full-line) and the dew point temperature (dashed line) profiles are given to a height of 3km.

An adiabatic boundary layer, about 100mb thick lies underneath a stable troposphere, and the data is based on measurements made by the Meteorological Office aeroplane (Jonas, 1987). To drive the plume, for Figs. 6 the temperature was uniformly increased by 1°C and the mixing ratio given a constant value of 4.3 $g.kg^{-1}$, for a height, z, of ≤ 1200 metres and $|x|$

≤150 metres. In Fig. 5, this height was reduced to 1000 metres to reduce the flow modification due to water (Gadian et al, 1990).

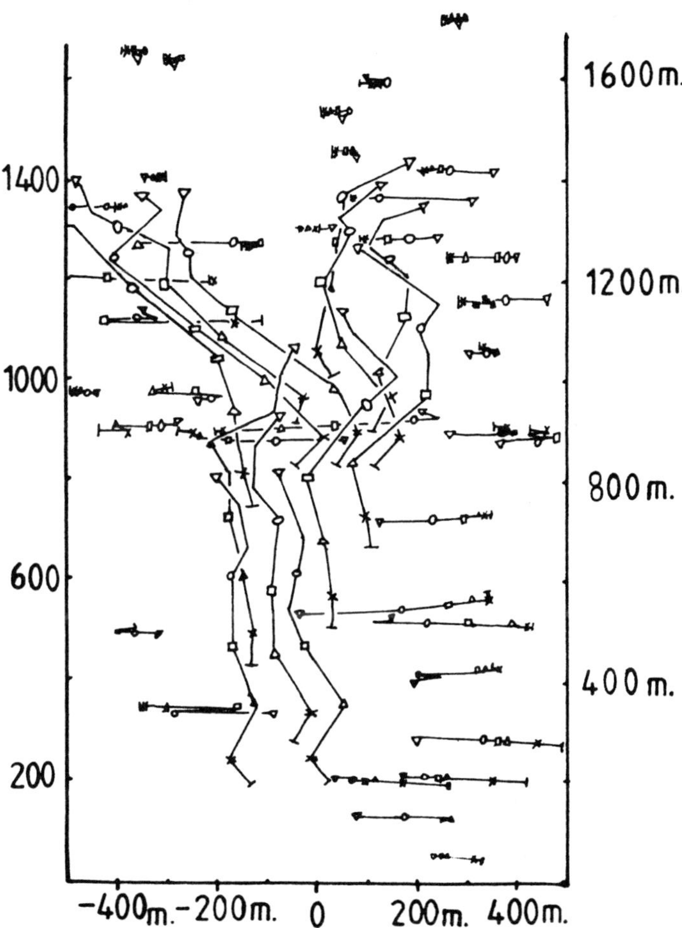

Fig. 5: Trajectories of a few particles calculated with the central area of the model, 1000 metres by 1600 metres displayed. Elements within the area of z ≤ 1000 metres and |x| ≤ 150 metres are heated by 1°C. Time samples occur at 1 second (−), 48 seconds (×), 94 seconds (△), 142 seconds (□), 200 seconds (○), and 280 seconds (▽). A diffusive e-folding time of 10^3 seconds is used.

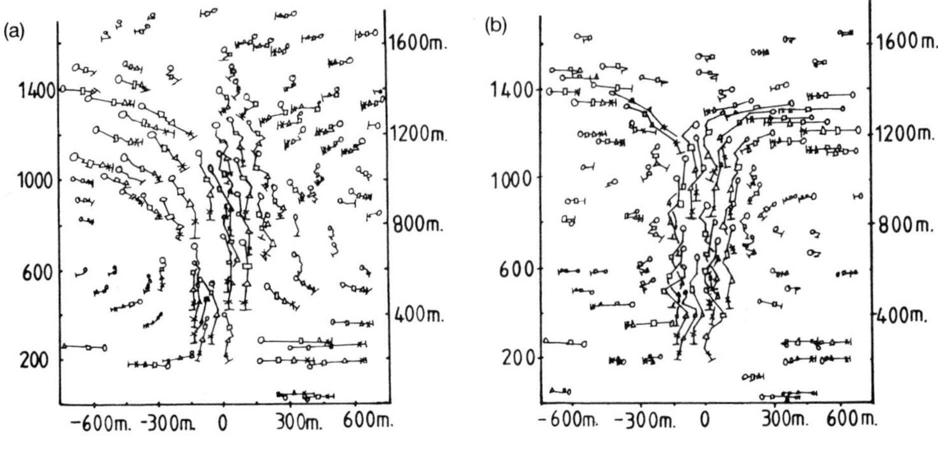

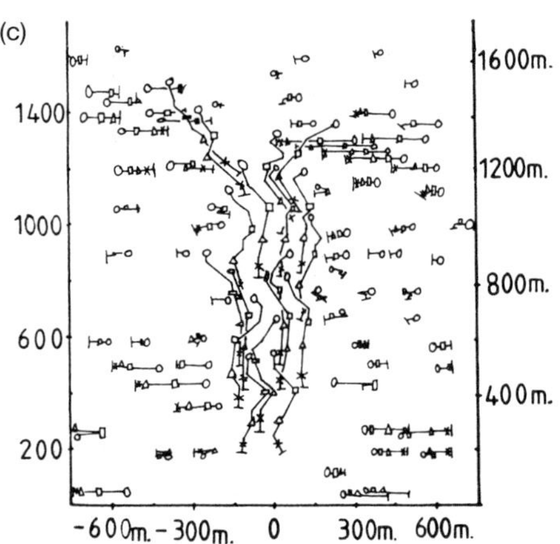

Fig. 6: Trajectories of a few particles calculated with a central area 1500 metres by 1600 metres. Elements within the area of z ≤ 1200 metres and |x| ≤ 150 metres are heated by 1°C. Time samples occur at 1 second (−), 48 seconds (×), 99 seconds (△), 147 seconds (□), and 203 seconds (○). Diagram: (a) a diffusive length scale of 10^3 seconds; (b) of 10^4 seconds; and (c) of 10^5 seconds.

The Figs. 5-6 show particle trajectories, marked at approximately 50 second intervals, displaying a plume type structure in the centres. The parcels rise until the potential temperature is equivalent to the environmental value, with a corresponding convergence at the base. In all the model computations, the flow is not symmetrical due to minor initial perturbations (of the order of centimetres/sec) in the initial flow field.

The effects of the model parameterisations of diffusion are clearly illustrated in the computational results. Baker and Latham (1979) suggested a diffusive e-folding time constant of approximately 100 seconds would be appropriate. Fig. 5 gives trajectories for e-folding times of 10^3 seconds. Figs. 6 provide an alternative mixing scheme such that the diffusive time scale is $\tau \sim K/(|\delta u|+|\delta v|)$ where K has the values of 10^3, 10^4, 10^5, respectively. Calculations for a constant e-folding time of 10^2 seconds are not significantly different from the results in Fig. 6a. In this case the vertical velocities are noticeably reduced, and the rapid diffusion of temperature quickly broadens the plume. It is clear that the mixing processes of time scales many times larger are required for the plume-like behaviour to be simulated. Fig. 5, with a constant time scale of 10^3 seconds, exhibits more expected behaviour. However, in this case certain elements are entrained and even pass through the buoyant plume.

Using the velocity dependent diffusion, these horizontal plume traversing trajectories are not observed, even in Fig. 6c with the smallest length scale. Although most horizontal (and all vertical) velocities are less than 5 $m.s^{-1}$, instantaneous values of up to 8 times are observed in Fig. 5 for entrained or convergent elements. The large inter-particle velocity gradient produces much larger diffusive effects for these elements. In addition the limitation of the two-dimensional approach will increase the convergence speeds. Hill and Choularton (1985) produces evidence that entrained elements of 100m in size are often observed in the middle of clouds and convective plumes; exactly the motion inhibited by this diffusion. In Figs. 6a and 6b the diffusion properties can be seen to dominate. Figs. 5 and 6c also indicate larger vertical velocities of approximately 2 $m.s^{-1}$. The liquid water and vapour fields are discussed in detail in Gadian et al (1990), but the more active diffusion (Figs. 6a,b) has correspondingly similar effects, delaying and reducing the cloud formation.

7. Summary and general comments

If this model visualisation of entrainment in plume convection is at all realistic, then the diffusion processes are of minor importance and of relatively long time scales. In future work, variable smoothing length descriptions could produce useful results and produce a more useful model.

The clear delineation of the magnitude of the time truncation error in this Lagrangian description provides a novel approach.

The comparison of the Lagrangian and Eulerian solutions for the barotropic vorticity equation, indicates that the Lagrangian schemes seem to be more representative for perturbations of the same size as the grid. In the barotropic Lagrangian solution, there had to be an assumption of a smooth mean flow. The interacting particle pressure field overcomes some of the limitations. The inclusion of chemical, as well as water, properties can be readily incorporated.

Several of the advantages of both Lagrangian approaches are the computational ease and accuracy. The time integration schemes define a known local truncation error for each timestep. The models, although implicit in nature, are accommodated on fast microcomputers which have over 4 megabytes of memory. For the convective plume model, advantage can be made of the parallel nature of the computation problem. From a theoretical viewpoint, by judicious manipulation of the smoothing lengths, it may well be possible in future to use a convex analysis (Cullen and Purser, 1984) to further improve the speed and accuracy of computation.

8. References

Arakawa A. (1966). Computational Design for Long-term Numerical Integrations of the Equations of Atmospheric Motions. Jour. Comp. Phys. 1:119-143.

Baker M. and Latham J. (1979). The Evolution of Droplet Spectra in Small Cumulus Clouds. Jour. Atmos. Sci. 36:1612-1615.

Birdsall D. and Fuss, D. (1969). Clouds in Clouds, Clouds in Cells, Physics for many Body Plasma Simulations. Jour. Comp. Phys. 3:494-511.

Charney et al (1950). Numerical Integration of the Barotropic Vorticity Equation. Tellus, 2:237-254.

Cullen, M. (1974). Integrations of the Primitive Equations on a Sphere using a Finite Element Method. Quart. Jour. Roy. Met. Soc. 100:555-562.

Cullen M. et al (1984). An Extended Lagrangian Theory of Semi-geostrophic Frontogenesis. Jour. Atmos. Sci. 41:1477-1497.

Dormand J. et al (1980). A Family of Embedded Runge-Kutta Formulae. Jour. of Comp. and Appl. Maths, 6:19-26.

Doron E. et al (1974). A Comparison of Grid Point and Spectral Methods in a Meteorological Problem. Quart. Jour. Roy. Met. Soc., 100:371-383.

Eady E. (1949). Long Waves and Cyclone Waves. Tellus, 1:159-172.

Fjortoft B. (1952). Integration of the Barotropic Vorticity Equation. Tellus, 4:179-194.

Gadian A., Dormand J., and Green J. (1990). Smooth Particle Hydrodynamics as Applied to 2-D Plume Convection. Atmos. Res, Vol 24 (in press).

Ginghold R. and Monoghan, J. (1977). Smooth Particle Hydrodynamics. Monthly Not. of Roy. Astr. Soc. 181:375-389.

Gordon H. and Davies D.R. (1977). The Sensitivity of Response of Response of Climate in a Two Level General Circulation Model. Tellus, 29:484-501.

Green J.S.A.G. (1960). A Problem in Baroclinic Instability. Quart. Jour. Roy. Met. Soc., 86:237-251.

Haltiner G. and Williams R. (1980). Numerical prediction and Dynamic Meteorology. Ch.5, John Wiley.

Hill T. and Choularton T. (1985). An Airborne Study of the Microphysical Structure of Cumulus Clouds. Quart. Jour. Roy. Met. Soc. 111:517-544.

Hoskins B. (1975). The Geostrophic Momentum Approximation and the Semi-geostrophic Equations. Jour. Atmos. Sci. 32:233-242.

Jonas P. (1987). An observed Maritime Temperature Profile. Private Communication.

Phillips N. (1956). The General Circulation of the Atmosphere: a Numerical Experiment. Quart. Jour. Roy. Met. Soc. 82:123-164.

Prupppacher H. and Klett J. (1978). Microphysics of Clouds and Precipitation. D. Rheidol Publ. Co.

Salmon R. (1985). New Equations for Nearly Geostrophic Flow. Jour. Fluid Mech. 153:461-477.

Sawyer J.S. (1963). A Semi-Lagrangian Method of Solving the Vorticity Advection Equation. Tellus, 15:336-342.

Smith G. (1978). Numerical Solution of Partial Differential Equations. Ch. 5, Oxford University Press.

Wiin-Nielsen A. (1959). On the Application of Trajectory Methods in Numerical Forecasting. Tellus, 11:180-196.

Modelling global atmospheric chemistry with the FACSIMILE/CHEKMAT package

A.R. Curtis

ARC Scientific Limited

R.G. Derwent, A.M. Hough, C.E. Johnson

AEA Environment and Energy

1. Introduction

The increasing emission into the atmosphere of substances such as NO and hydrocarbons, combined with the observed increases in the concentrations of trace gases such as ozone, NO_x and methane necessitate the study of the chemistry of the global atmosphere. Models of the chemistry of the atmosphere such as the two-dimensional global tropospheric model described here link equations of atmospheric transport and diffusion with those describing chemical reactions. The problem of the routine solution of the resulting sets of differential equations from this type of problem has stimulated the production of a program which solves the discretised equations accurately in a routine manner.

The program developed at Harwell (FACSIMILE/CHEKMAT) for the automatic solution of ordinary differential equations is described here, together with its application to a model of global tropospheric chemistry. This model is briefly described, and used to show the changes in atmospheric chemistry which occur with global temperature rise.

2. The CHEKMAT program

2.1 Historical background

The computer program used in our modelling, CHEKMAT, has been developed at Harwell since 1968. This was when Gear (1969) first popularised a method, now known as the Backward Difference Formula (BDF) method, for solving sets of ordinary differential equations (ODEs) which have a property, known as stiffness, common in ODEs arising from chemical kinetics. CHEKMAT was derived as an acronym from CHEmical Kinetics with MATching; the last word refers to the parameter-fitting feature - the ability to adjust parameters (e.g. rate coefficients) for best fit between computed and experimental values of solution components (e.g. species concentrations). Although the numerical method used to solve the differential equations is an implementation of BDF, it has been refined and developed at Harwell, and incorporates many functional improvements, designed to make it more automatic and reliable, over published BDF programs. For example: all time steps, even the first one, are chosen automatically; fewer matrix recalculations are needed; convergence testing is much more rigorous; and the wide range of numerical values of solution components, which occurs in this kind of problem, is properly handled. The method used follows in general that described by Curtis (1980), with some further improvements.

Until 1975, CHEKMAT was limited to straightforward reaction kinetics problems, except when modified by experts. However, experience with it suggested many desirable features, and a much more powerful piece of software, called FACSIMILE, intended for easy use by non-experts, was specified and written. This has been used and developed since 1975. It now forms the basis for two different application programs: CHEKMAT (Curtis and Sweetenham, 1987), which is a far more powerful and flexible ODE solver with special features for mass action kinetics and for diffusive and advective transport; and HOWGOOD (Curtis, 1988), which is a tool for studying how the results from any continuous modelling program are affected by uncertainties in the data supplied to it.

2.2 *General features*

CHEKMAT has many features which make it suitable for use by people who are not experts in computing or mathematics, but who wish to solve modelling problems, especially in reaction kinetics. It provides a powerful, special-purpose high-level programming language in which the user describes his/her problem. One of its main advantages for the user is single-shot running - it reads this description as data, compiles it into an internal pseudo-code (Curtis, 1985) for interpretive execution, and solves the problem, all in a single computer run. Although interpretative execution of the user's code carries a penalty in computer time, the saving in user's time which it makes possible greatly outweighs this, except in the largest problems (such as the current one). Work is in hand to offset this for large problems, using the fact that they do not need the full generality offered by the FACSIMILE language.

Another advantage of the FACSIMILE/CHEKMAT approach is that it enables the program to run, with unchanged user image, on an extremely wide range of computers. This is especially useful now that networks are common, since the model can be developed on a workstation and production runs on a more powerful server or mainframe. It is in fact available for most computers, from IBM-compatible PCs to CRAY supercomputers, although naturally the larger problems require more powerful computers.

2.3 The user language

The CHEKMAT user language enables many kinds of problem to be specified, while containing features special to mass action kinetics and to transport by diffusion and advection. There are features for element-by-element numerical operations on scalars and on arrays of up to five dimensions; matrix operations; powerful built-in output, numerical with predefined or user-programmed formats, and graphical for plotting at coarse resolution on a character-oriented printer. There are command features for initiating and controlling the solution, describing output structures, separating the user's code into subroutines, and reading experimental data for the parameter-fitting option. For example, the solution process is controlled by statements beginning WHEN, which specify what actions are to be taken at various stages; other run control statements, such as BEGIN, RESTART and STOP, have obvious meanings.

CHEKMAT user-written calculation is organised into a series of program routines. Some of these have CHEKMAT system names, and these are executed automatically as required. For example, routine FTIME is called once per time step to calculate parameters which depend only on time, not on the solution components; routine EQUATIONS is called to compute time rates of change of solution components; routine GENERAL is called before EQUATIONS to compute parameters which do depend on the solution. Other routines are given non-system names by the user, and (for example) can be called at stages through the solution specified in a WHEN statement. In the present model, routine ANNBGT calculates annual budgets and is called at the end of each year.

All identifiers (names) have global scope; that is, once a scalar or array has been defined it may be referred to from any routine. The following main types of quantity may be declared: VARIABLE, for solution components of the ODE system; PARAMETER, other real-valued quantities to be manipulated by the user's program; CONSTANT, real-valued quantities which the program may not change; and INTEGER, integer-valued quantities for use in dimensions, indexing and logical decision-making. A declaration may contain array dimensions, which apply to all subsequent names in it until superseded; each name declared may be followed by initial values; the last array in a declaration may have a list of sub-arrays or scalars equivalenced to it.

As an example of the power of the language, a reaction which a chemist might write in the form

$$O(^1D) + H_2O \rightarrow 2OH \tag{2.1}$$

with a rate coefficient K, could be coded for CHEKMAT as

% K : O1D + H2O = OH + OH; (2.2)

Here the percent sign introduces the rate coefficient (which can be an arithmetic expression if desired), the colon introduces the reaction, and the semicolon is a standard statement terminator. This equation would be the same, whether the quantities appearing in it were scalars or arrays.

The user has to code the discretisation of transport (diffusion and/or advection) partial differential operators on his chosen grid, since at present CHEKMAT does not do this. However, once this is done and coefficient and index arrays are set up, the action is specified by a simple statement, in routine EQUATIONS, of the form

TPORT <dim.> {coeff. array} {index array} : {species list}; (2.3)

This causes an internal CHEKMAT subroutine to be called to add the transport terms to the rates of change of solution components.

To illustrate the value of the array features, the chemical reaction coding for a problem is often developed and tested in single-box form, each species concentration being represented by a single scalar VARIABLE. It can be converted to reaction code for a spatially non-homogeneous problem simply by declaring these VARIABLEs (and possibly some PARAMETERs used as rate coefficients) to be arrays of the grid dimensions, and enclosing sections of code in suitable ARRAY statements. Adding the transport processes consists of defining and initialising index and coefficient arrays for the partial differential operators, discretised on the chosen grid, and inserting a TPORT instruction to compute and add the transport terms contributions.

Thus the whole user code for a large problem can be reasonably compact. For the present model, approximately 4,600 lines of code (including comments) were written, broken down roughly as follows:

A. Declarations and comments, 1100 (including initial values, 390).
B. Preliminary calculations (including interpolation of temperature, wind and eddy diffusion data to grid points, and computation of transport coefficients), 900.
C. Calculation of time-dependent parameters such as temperatures, rate coefficients and photolysis rates, 800.
D. Computation of time rates of change, 540.

E. Output and budget calculation and editing (with user-defined formats), 1270.

3. Description of the Harwell Global Tropospheric Model

The model used in this study is a two-dimensional latitudinally averaged Eulerian grid model, with a domain which extends from pole to pole and from the earth's surface to a height of 24km. Within this domain, the model is divided into 288 grid cells, 24 in the horizontal direction and 12 in the vertical.

3.1 Treatment of atmospheric diffusion and advection

Two-dimensional transport data, derived by Plumb and Mahlmann (1987) from numerical experiments with a three-dimensional global circulation model, consist of a stream function for advection and eddy diffusion coefficients K_{xx}, K_{yy} and $K_{xy} = K_{yx}$. A skew-symmetrical part of the eddy diffusion tensor had been absorbed into the stream function. Curtis (1987a,b) shows how to convert this information into transport coefficients of the kind needed by CHEKMAT [all the terms arising from the stream function in Curtis (1987b) have the wrong sign]. The method achieves numerical stability by artificially increasing the diagonal terms of the diffusion tensor by just enough to ensure ellipticity of the modified total transport operator, while retaining second order accuracy because the changes are of second order in the grid spacings.

The presence of mixed derivatives, as evidenced by the K_{xy} terms, requires a nine-point finite-difference operator, instead of the more familiar five-point operator in two space dimensions. Earlier studies (Derwent and Curtis, 1977) had shown the value of using equal-volume grid cells, achieved by taking a grid equally spaced in the sine of latitude. The vertical coordinate was taken as the logarithm of pressure normalised to the surface value, and scaled by a scale height H_z, of 7.2km.

Monthly transport and temperature data (Barnett and Corney, 1985, 1986) were interpolated onto the present model grid and are also interpolated with time within the model so as to provide smooth and continuous variations with time. Studies of inert tracers such as CCl_3F (CFC-11) and ^{85}Kr have been used to evaluate the operation of the model (Hough, 1989a), and the model results were in good agreement with surface observations of these species over a wide latitude range.

With the exception of the transport and temperature data referred to above, the model is driven completely by the emissions, chemistry, deposition processes and upper boundary conditions. No fixed concentration fields are maintained in the model other than those of oxygen and the total molecular number density. The concentrations of all chemical species are determined as solutions of the differential equations which define the model transport and chemistry.

All species occurring in the model are subjected individually to the transport operator. The alternative of transporting only "family groups" (e.g. the NO_x species), as is done in some other models, has been considered. In the latter method, after the transport contribution to the rate of change of the total group concentration in each grid cell has been computed, it must be shared out among the species making up the group. To do this properly requires the solution at each integration step of a large system of equations, which is generally non-linear because (a) species (e.g. nitric acid) may belong to more than one family group, and (b) species in a group may react with each other or themselves. The most efficient way of solving these equations to reasonable accuracy seems to be our choice, to include both transport and chemistry for each species, taking special action to avoid excessive computation due to interaction between them. This could arise because of excessive "fill-in" of initially zero elements in the (very sparse) iteration matrix used when it is decomposed for equation solving. We avoid this at present by omitting the transport terms from the iteration matrix, and accepting a consequent limit on the size of time step (typically to a few hours, in the present model). In the longer term, we plan to partly use transport terms in the iteration, while avoiding the resulting fill-in.

3.2 The chemical mechanism

In the model it is necessary to restrict the number of species and the complexity of the mechanism in order to produce a tractable problem. However, sufficient detail must be retained so as to represent the chemical processes in a realistic manner. Residual uncertainties in our knowledge of tropospheric chemistry complicate the selection of the chemical scheme.

The mechanism used in the present work contains 56 chemical species, including water vapour, 12 hydrocarbons, 8 carbonyl compounds and 5 organic peroxides. There are 91 thermal reactions and 27 photolytic processes. Diurnal average calculations are made throughout, but night-time NO_3 chemistry is modelled with a further 7 reactions. The rate coefficients for the thermal reactions were taken from recent reviews such as that of Atkinson et al (1989). The photolysis rates are calculated for each grid cell as diurnal averages for each month, and are interpolated to intermediate times (Hough, 1988).

The model emissions inventory contains 17 species thought to be important in controlling tropospheric chemistry on a global scale. These include N_2O, NO, H_2, CO and CH_4, together with 12 hydrocarbons including isoprene and terpene. The distribution of the emissions in the model is described by ten separate source categories, each having a separate dependence on latitude and time of year. The derivation of these functions of latitude and time of year is discussed by Hough (1989b), who also describes the full chemical mechanism.

3.3 Water vapour and clouds

The model treats evaporation from the surface using the local values of the vertical diffusion coefficient. In each grid cell, the local precipitation rate into the grid cell immediately below is parameterised as a smooth function of the relative humidity. The water vapour pattern produced is similar to that observed, including the observed minimum in the region of the tropical tropopause (Hough, 1989a). The predicted rainfall pattern is similar to a latitude average of the observed annual rainfall. The cloud cover in each grid cell in the lowest few kilometres of the model is calculated using the algorithm of Buriez et al (1988) based on relative humidity and atmospheric pressure.

3.4 Removal and boundary processes

The model includes both wet and dry deposition. Dry deposition velocities for each species removed are used, taking into account the fraction of land, sea, snow and ice in each latitude range. The species removed by dry deposition are: O_3, NO_2, PAN, H_2O_2, H_2, CO and organic peroxides. Wet deposition is parameterised as a function of the local liquid water content, the rainfall rate, and the solubility of the species concerned. Species transported and removed by wet processes are: HNO_3, H_2O_2, HCHO, organic peroxides, and α-dicarbonyl compounds.

Transport across the upper boundary of species with very low mixing ratios in this region can be ignored. However, this is not the case for ozone, NO_x and methane, and the exchange of these species across this boundary needs to be modelled. This exchange is represented as diffusion, using the local value of the vertical diffusion coefficient. The required mixing ratios of the species above the model domain were taken from observations, including satellite observations for ozone (Hough and Woods, 1988).

4. Results

A complete description of the model results for a simulation of present day conditions is given by Hough (1989b). The model has also been used to simulate future atmospheric concentrations of trace gases from estimates of emission changes. The assumption made in this approach is that the other model input parameters, including atmospheric temperatures, remain constant during each simulation. However, the change in tropospheric chemistry with temperature is one of the feedbacks between climate and the chemistry of the atmosphere (Ramanathan et al, 1987), and the resultant change in the radiatively active gases in the model (O_3, CH_4 and N_2O) is clearly of interest, and the effects of altering the temperature globally are investigated here.

Several changes to the model conditions are caused by changes in temperatures. The increase in the saturated water vapour pressure with temperature results in increased water vapour concentrations throughout the domain, as the rainfall amount is controlled by an algorithm which tends to maintain the relative humidity at the same values. The increase in the global inventory of water vapour is 6.7 $\%.K^{-1}$, slightly higher than the increase in water vapour content per unit surface area of 5.5 $\%.K^{-1}$ found by Ravel and Ramanathan (1989) from a regression of water content with sea surface temperature. This increase has a direct radiative feedback to global climate, but also influences tropospheric chemistry, mainly through the reaction (2.1).

The global inventory of the hydroxyl radical (OH) was found to increase by 3.7 $\%.K^{-1}$, which has important implications for many other species due to the reactivity of OH. Many species which react with this radical such as methane and the non-methane hydrocarbons decrease by 1-3 $\%.K^{-1}$. Other radicals interdependent on OH have also increased: HO_2 by 2.2 $\%.K^{-1}$ and CH_3O_2 by 4 $\%.K^{-1}$.

The increased evaporation rate results in higher rainfall amounting to about 5 $\%.K^{-1}$. This has the effect of increasing the washout of species which are soluble, but this is compensated by the reduced solubility at higher temperatures and by the change in concentrations, as well as the fact that highly soluble species are already efficiently removed in lower levels of the atmosphere. The net effect is that the washout remains the same for HNO_3, decreases slightly for HCHO, CH_3COO_2H, HCOCHO and CH_3COCHO and increases slightly for H_2O_2 and CH_3OOH.

Many of the thermal rate coefficients used in these calculations are temperature dependant (Atkinson et al, 1989), and four of these have much greater temperature coefficients than any others, in the range 16-22 $\%.K^{-1}$. These refer to the thermal decomposition reactions

$$HO_2NO_2 + M \rightarrow NO_2 + HO_2 + M \quad (4.1)$$
$$N_2O_5 + M \rightarrow NO_2 + NO_3 + M \quad (4.2)$$
$$CH_3O_2NO_2 + M \rightarrow CH_3O_2 + NO_2 + M \quad (4.3)$$

$$PAN + M \rightarrow CH_3CO_3 + NO_2 + M \quad (4.4)$$

The high decomposition rate of PAN at high temperatures is well known, and limits its distribution in the atmosphere, and also creates large seasonal changes. The rate of change in the global inventory for PAN is -8 $\%.K^{-1}$ in January and -13 $\%.K^{-1}$ in July. This change reduces the proportion of the total reactive nitrogen in the form of PAN, though the seasonal change is greater. The dry deposition of PAN to the surface is also reduced, but this represents only a small term in the NO_x budget. In addition to changes in the thermal rate coefficients, small increases were observed in the photolysis rates due to temperature sensitive cross sections and to reductions in the concentrations of the absorbing gases.

The amount of ozone in the model decreased with increasing temperature at a rate of about 1.5 $\%.K^{-1}$. This is largely due to the reaction $O(^1D) + H_2O$ proceeding more rapidly due to the increased water vapour concentrations, but changes due to increased radical concentrations also increase the rate of loss of ozone. The loss is partially compensated by increases in photochemical production caused by increases in the radicals HO_2 and CH_3O_2. The increased reactivity of the model reduces the turnover time for ozone in the model from 57 days to 54 days when an increase of 2 K is applied. Of the other radiatively active gases in the model, nitrous oxide has a long lifetime in the atmosphere and is not studied here. The final change in the methane concentrations due to the temperature rise has not been determined because its lifetime is several times longer than the model run time. However the loss of methane proceeds mainly through its reaction with OH, and an increase of around 1.2 K is sufficient to change the model from having an increasing inventory of around 0.6 % per year to one with a stable inventory. Further increases in temperature will result in a net loss of tropospheric methane assuming that emissions remain constant.

The treatment adopted here has been idealistic, as global temperature rise has been applied uniformly throughout the model domain. Only the changes in atmospheric chemistry have been modelled here, ignoring changes in emissions with increasing temperature and also treating the atmospheric transport as invariant. The maintenance of relative humidity in the model applies two constraints to its behaviour; the increase in water vapour concentration is restricted to that indicated by the saturated vapour pressure curve, and the cloud amounts remain constant because they are also linked to relative humidity. Changes in cloud amounts may also need to be considered, as these influence the photolysis rates used in the model.

5. Conclusions

The FACSIMILE/CHEKMAT program for solving sets of ordinary differential equations is briefly described. Many features of this program make it easy to use by scientists who are not expert mathematicians, including a user language with special features for reaction kinetics and diffusive and advective transport. As an example of its use, a large Eulerian grid model of the global troposphere is briefly described, together with an example of its use to show the sensitivity of tropospheric chemistry to temperature.

The increase in atmospheric water vapour with increasing temperature is well known, and forms a positive feedback to the greenhouse effect. The increased water vapour concentrations also influence the tropospheric chemistry, notably on the radical concentrations in the troposphere. This has the effect of decreasing the concentrations of two radiatively active gases in the model, ozone and methane, and thus forms an additional negative feedback to global temperature change.

6. Acknowledgements

This work was funded by the United Kingdom Department of the Environment as part of its Air Pollution Research Programme, and by the United Kingdom Department of Energy. The authors are grateful to Dr A. Plumb for supplying the global circulation coefficients and to Dr J. Barnett for supplying printouts of the global temperature fields.

7. References

Atkinson, R., Baulch, D.A., Cox, R.A., Hampson, R.F., Kerr, J.A., and Troe, J. (1989). Evaluated kinetic and photochemical data for atmospheric chemistry: supplement III. J. Phys. Chem. Ref. Data. 18:881-1097.

Barnett, J.J. and Corney, M. (1985). A middle atmosphere temperature model from satellite observations. Adv. Space Res. 5:125-134.

Barnett, J.J. and Corney, M. (1986). Middle atmosphere reference model derived from satellite data. In: Middle Atmosphere Programme Handbook 16, K. Labitzke, J.J. Barnett and D. Edwards, editors. SCOSTEP Secretariate, University of Illinois, Urbana, USA, pp.47-85.

Buriez, J.-C., Bonnel, B., Fouquart, Y., Geleyn, J.-F., and Morcrette, J.-J. (1988). Comparison of model-generated and satellite derived cloud cover and radiation budget. J. Geophys. Res. 93:3705-3719.

Curtis, A.R. (1980). The FACSIMILE numerical integrator for stiff initial value problems. In: Computational Techniques for Ordinary Differential Equations, I. Gladwell and D.K. Sayers, editors. Academic Press, London.

Curtis, A.R. (1985). FACSIMILE Release H internal instruction code. AERE, Harwell Report M 3482, HMSO, London.

Curtis, A.R. (1987a). Accurate conservative discretisation of transport equations retaining diagonal dominance. AERE, Harwell Report R-12523 HMSO, London.

Curtis, A.R. (1987b). Discretisation of the zonally-averaged transport equation for use in global atmospheric pollution studies. AERE, Harwell Report R-12524, HMSO, London.

Curtis, A.R. (1988). HOWGOOD User's Manual. AERE, Harwell Report R-12864, HMSO, London.

Curtis, A.R. and Sweetenham, W.P. (1987). FACSIMILE/CHEKMAT User's Manual. AERE, Harwell Report R-12805, HMSO, London.

Derwent, R.G. and Curtis, A.R. (1977). Two-dimensional model studies of some trace gases and free radicals in the troposphere. AERE, Harwell Report R-8853, HMSO, London.

Gear, C.W. (1969). The Automatic Integration of Stiff Ordinary Differential Equations. In: Information Processing 68, A.J.H. Morrell, editor. North Holland, Amsterdam, pp.187-193.

Hough, A.M. (1988). The calculation of photolysis rates for use in global tropospheric modelling studies. AERE, Harwell Report R-13259, HMSO, London.

Hough, A.M. (1989a). The development of a two-dimensional global tropospheric model - 1. The model transport. Atmospheric Environment, 23:1235-1261.

Hough, A.M. (1989b). The development of a two-dimensional global tropospheric model: The model chemistry. (submitted).

Hough, A.M. and Woods, K.J. (1988). Ozone concentrations in the global atmosphere. An analysis of data from the SBUV instrument on the Nimbus-7 satellite. AERE, Harwell Report R-13271, HMSO, London.

Plumb, R.A. and Mahlman, J.D. (1987). The zonally-averaged transport characteristics of the GFDL General circulation/transport model. J. Atmos. Sci. 44:298-327.

Raval, A. and Ramanathan, V. (1989). Observational determination of the greenhouse effect. Nature, 342:758-761.

Ramanathan, V., Callis, L., Cess, R., Hansen, J., Isaksen, I., Kuhn, W., Lacis, A., Luther, F., Mahlmann, J., Reck, R. and Schlesinger, M. (1987). Climate-chemical interactions and effects of changing atmospheric trace gases. Rev. of Geophys., 25:1441-1482.

Note in proof: *The program FACSIMILE/CHEKMAT is now being referred to as FACSIMILE alone to avoid conflict of name with another program.*

Modelling dry deposition of SO_2 in Britain

J.N. Cape, R.I. Smith and D. Fowler

Institute of Terrestrial Ecology

1. Introduction

Most models in the environmental sciences fall between the two extremes of a purely empirical model, where past observations are parameterized in order to predict the future, and a purely process-based model where fundamental laws of physics and chemistry are integrated to predict an outcome a priori. Empirical models can be constructed so that the parameters chosen are related to physical or chemical processes, and therefore have an apparent relationship to actual mechanisms. Process-based models in general have insufficient input data to permit a priori calculations, with the result that processes have to be approximated. The end result in both cases is a mixture of measured data, fundamental theory, empiricism and temporal or spatial averaging. As understanding of fundamental processes improves, and as computing power increases, compartments of a model which have had to rely on very simple parameterisation may be developed to include explicitly the underlying physical and/or chemical processes involved.

In air pollution modelling, the complexities of meteorology, atmospheric chemistry and uncertainties in emissions have led to the adoption of largely empirical statistical models, where underlying processes are simplified to single variable parameters. Most models use a Lagrangian framework, i.e. parcels of air are followed in time as they move through space. Trajectories for such air parcels are based on actual or average weather patterns, and are tied either to sources (emission points) or, more usually, to receptors (measurement points). The availability of data on appropriate scales of time and space dictates the spatial resolution of the model.

For Britain there are three models which are used to estimate the transport and removal of air pollutants in the atmosphere. The Harwell Statistical Long-Range Transport Model (Metcalfe et al, 1989) is tied to receptor sites, and includes explicitly the chemical reactions which convert pollutant gases such as SO_2 to the acids which ultimately are removed from the atmosphere as "acid rain". The Warren Spring Laboratory Statistical Long-Range Transport Model (Perrin, 1986) and the CERL Statistical Long-Range Transport Model (Fisher, 1988) are applied on a grid over Britain incorporating emission sources and a statistical climatological treatment of windspeed, direction and weather type. In all three current models the removal of a gas such as SO_2 by direct absorption at the earth's surface (dry deposition) is included as a single parameter which may be "fitted" by comparison of model results and measured data.

The model described below is an attempt to apply current knowledge of the mechanisms controlling the dry deposition process to produce the "best estimate" of the rate of SO_2 removal at the ground. Comparison with the simple parameters used in long-range transport models should show whether the two approaches are compatible.

2. Dry deposition

The continuous measurement of the flux of a gas from the atmosphere to the ground is difficult and expensive. Estimation of fluxes over long times and large spatial scales requires an understanding of the factors controlling the flux. One of the simplest representations is to consider the process as analogous to the movement of an electric current, with individual transfer resistances in series or in parallel contributing to an overall transfer resistance from the atmosphere to the surface. For the case of a gas like SO_2 the driving force is the concentration in air at some reference height above the surface. The transfer resistance may be split into three: aerodynamic resistance (r_a), viscous sub-layer resistance (r_b) and surface resistance (r_c). The values of r_a and r_b are dictated by atmospheric turbulence and surface structure, and may be estimated from knowledge of windspeed and the aerodynamic "roughness" of the surface being studied. These two resistances in series will dictate the maximum rate at which a gas can be delivered to a leaf surface, for example. The gas may then be absorbed, and behaviour at the surface may be described by a number of parallel resistances which represent the affinity of different elements of the surface for the gas in question. For a water-soluble gas, such as SO_2, one can envisage at least 3 routes of deposition to a leaf:

(i) sorption at the outer leaf surface (cuticle)
(ii) solution in water on the leaf surface
(iii) entry into the leaf and subsequent solution

When stomata are open, sorption of SO_2 at a leaf's outer surface is not very effective in comparison with entry through a stomatal pore into the leaf's interior. The net uptake, described by the parallel resistances of the three possible routes, is therefore controlled by a surface resistance (r_c) which is strongly dependent on the extent to which SO_2 molecules can enter a leaf. If the atmospheric resistances are not limiting (i.e. $(r_a + r_b) < r_c$)

then the overall removal (dry deposition) of SO_2 is controlled by the physiological state of the plant, which determines r_c.

In principle, the physiological state of every plant species and the factors controlling atmospheric turbulence would need to be known in order to model dry deposition over Britain. In practice, adequate information on the interactions between SO_2 and plant surfaces is only known for a small number of plant species, and assumptions and approximations must be made, based on the best available data. The resultant sum of series and parallel resistances is then usually represented by a "deposition velocity" (v_d) which is the reciprocal of the overall resistance. The product of v_d ($m.s^{-1}$) and air concentrations ($g.m^{-3}$) then gives the flux directly ($g.m^{-2}.s^{-1}$), provided that v_d is independent of air concentration, as is usually the case at ambient concentrations of SO_2. If SO_2 gas concentrations are available, then knowledge of the deposition velocity will allow the estimation of flux (i.e. dry deposition) for comparison with the dry deposition removal rate parameters of the long-range transport models.

3. Requirements of the model

In order to estimate the components of the transfer resistance, and thence the effective deposition velocity for SO_2, information is needed both spatially over Britain and temporally on an hourly, daily and monthly basis within the year. Such information may not be available, and then estimates must be made based on what is available. The model described here uses a 20 x 20 km grid as its unit of spatial variation.

3.1 Atmospheric resistance (r_a and r_b)

The calculation of the atmospheric components of the transfer resistance requires knowledge of the windspeed and characteristic roughness length of the surface for each grid square. In principle, this information should be available at a time resolution of hours, but is only available as a seasonal mean. Seasonal (3-month) mean wind speeds for each grid square have been corrected for the mean altitude of each grid square, and combined with roughness lengths typical of 5 different land-use categories within each grid square: forest, arable, permanent grass, moorland and urban. Characteristic lengths are allowed to vary seasonally (for example, to account for the growth of cereal crops) but are otherwise assumed to be constant for each land use. Estimates for each type of land use are aggregated within each grid square based on the proportion of each land use as derived from the ITE database for Britain (Bunce and Heal, 1984). Values for each grid square are therefore available as 3-month means throughout the year. This is not thought to be a major limitation for SO_2, as the largest component of the transfer resistance is that at the surface (r_c), which is calculated more precisely. This part of the model would need to be improved for other gases which have relatively small surface resistances, such as nitric acid vapour.

3.2 Surface resistance (r_c)

This part of the model is a modified version of the "Big Leaf" model of Hicks et al (1987). The surface resistance is made up of two parallel paths, transfer to the outer (inert) leaf surface and transfer through the stomata to the inside of the leaf. As the latter is more important, it is necessary to model stomatal opening in response to changes in temperature and light throughout the year. The model uses diurnal temperature data for each hour of the day, averaged over each month of the year, to assess whether or not stomata will open. A maximum, minimum and optimum temperature is set for each vegetation type. Stomatal closure occurs below T_{min} and above T_{max}. The timing and extent of stomatal opening is also dependent on solar radiation, which is calculated for each hour of the day, averaged over each month and assumes clear skies. The dependence of stomatal opening on light includes a factor to allow for shaded leaves within a plant canopy, based on the leaf area index (the ratio of total leaf area to ground area) and the zenith angle of the sun. In practice, the assumption of clear skies is not a major source of error as stomatal opening in summer is triggered at low light levels relative to mid-day maxima, and stomatal opening in winter is controlled more by temperature than light.

The overall surface resistance (r_c) is therefore defined in terms of light and temperature, and has been "tuned" for each vegetation type by comparison with literature values of minimum canopy resistance. The rules governing stomatal resistance are taken to be uniform across Britain, but in principle could be calculated for each grid square. At night, stomata are assumed to be closed, so that the surface resistance is given by the transfer resistance to (inert) leaf surfaces. Although the model could use data on surface wetness to modify the surface resistance term, it appears from measurements that dry deposition of SO_2 to rain wetted surfaces is not significantly greater than to dry surfaces, at least for some vegetation types (Fowler and Cape, 1983), and no dependence on surface wetness is included in the current model.

3.3 Flux estimates

The flux of SO_2 gas is calculated from the product of the deposition velocity, aggregated over vegetation type for each hour within each month, and the annual average concentration of SO_2 for each grid square. The calculations here use measured data for SO_2 concentrations in rural and urban areas (G.W. Campbell, Warren Spring Laboratory, pers.comm), but in principle, estimates from long-range transport models could be used as input.

4. Description of algorithm

The program is written in strict Fortran 77 code and consists of one main program and six subroutines. There are 339 lines of code and 422 lines of comment. Arrays and variables are usually passed into and out of the subroutines using common blocks. The basic unit for calculation is a grid square.

The main program requests from the user the land use type, the time period for the calculation and the name of an output file. The main program then calls the main subroutine DDEPOS.

The user is given a choice of land use types: arable, forest, grassland, moorland, urban or all combined. The time period is specified as a day, a month or a year. If it is specified as a day then the month in which the day occurs is also requested. The representative day used is always in the middle of the month. For a period of one month the calculations are done for the representative day in the month and then multiplied up by the number of days in the month. The yearly values are the summation of all twelve monthly values.

The subroutine DDEPOS sets up the arrays required for the calculations and then calls two subroutines INSQAR and INYEAR to input data. The subroutine then cycles over each grid square for the calculations. For each representative day in each selected month the subroutine DDEPOS cycles over each hour of the day calling the subroutine ZENGEN to calculate the sunlit and shaded leaf area and then calling the subroutine LNDUSE to calculate the hourly deposition velocities for each required land use type. The hourly deposition velocities are then multiplied by the SO_2 concentrations to give daily fluxes of SO_2 for each required land use type. If the request was for all land use types, the percentage of the grid square in each land use type is used to give a weighted average daily flux. Monthly and yearly fluxes are calculated by appropriate cumulations of daily fluxes. The required flux for each square is written to a file and shown on the terminal. A cumulated flux over all squares, weighted by the percentage of land in each square, is output as the total UK SO_2 flux.

The subroutine INSQAR inputs the data relevant to each grid square. This comprises the easting and northing, latitude and longitude, percentage cover of each of the five land use types, percentage of the square which is land (as opposed to sea), the monthly mean wind speed and the monthly mean SO_2 concentration.

The subroutine INYEAR inputs a representative diurnal cycle for each of solar radiation and air temperature for each month.

The subroutine ZENGEN calculates the position of the sun at any time from the latitude and longitude supplied. It then splits the photosynthetically active fraction of the received global radiation into direct beam and diffuse components.

The subroutine LNDUSE has values for leaf area index, minimum stomatal resistance (single leaf, both sides) and canopy height for each land use type in each month. It determines the atmospheric resistance from the wind speed and canopy height, calls the subroutine STMRES to produce the stomatal canopy resistance for each land use type and then combines these along with the soil resistance to calculate a deposition velocity for the requested land use type.

The subroutine STMRES produces leaf surface resistances for sunlit and shaded leaves using the information on received solar radiation and air temperature to calculate stomatal opening. Using the leaf area indices for exposure to direct beam and diffuse radiation the subroutine determines an effective stomatal canopy resistance from combining the parallel stomatal resistances for the sunlit and shaded leaves.

5. Computing requirements

The program currently calculates results for 768, 20 x 20 km grid squares covering the United Kingdom mainland and offshore islands but excluding all parts of Ireland. For a run covering all five landclasses and one day in each of the twelve months in a year, the CPU time taken on a VAXstation 3100 (a machine rated at 3mips) is fractionally over 8 minutes. The input and output routines have been reduced to a minimum so an approximate calculation indicates that to run the program for one landclass for one day requires 24 million instructions and takes about 8 seconds CPU on a VAXstation 3100 and 27 seconds CPU on a microVAX II.

6. Model results and discussion

Dry deposition of SO_2 for each 20 x 20 km grid square of Britain is shown in Fig. 1 for a winter month and a summer month. There is a clear similarity with the pattern of SO_2 gas concentration (Fig. 2) as would be expected. The much greater deposition rates in summer reflect the greater physiological activity of vegetation, with longer periods of stomatal opening during the day. A seasonal cycle of SO_2 gas concentration (winter 1.2 x mean, summer 0.8 x mean) has been included, although the effect on deposition is small relative to the differences in deposition velocities through the year. Deposition in winter is small, and mostly to leaf surfaces, whereas in summer there is considerable internal uptake of SO_2. To set the results in context, Fig. 3, which shows annual modelled dry deposition of SO_2, should be compared with Fig. 4, which shows the pattern of wet deposition of sulphur over Britain, derived from the UK national network of rainfall chemistry (Campbell et al, 1988). Close to areas of pollutant emission, dry deposition is the more important source of deposited sulphur. In remote areas to the north and west, and in areas with high rainfall, wet deposition predominates.

Integration over the model results for the whole year over all of mainland Britain gives an estimate of total dry deposition of 3.1×10^5 tonnes SO_2. This is considerably smaller than earlier estimates (Barrett et al, 1987), and only a small fraction of annual emissions of SO_2 from Britain, which are about 3.6×10^6 tonnes (Anon, 1988). The discrepancy with earlier estimates arises primarily from the relatively small annual average deposition velocities which are calculated by the model. Averaged over Britain, the deposition velocity for SO_2 is estimated as 3.6 $mm.s^{-1}$. The effective deposition velocity usually used in long-range transport models to account for the removal of SO_2 from the atmosphere, is 8 $mm.s^{-1}$ as a time-average over the year.

The value of 8 $mm.s^{-1}$ is not based on a physical model of dry deposition, but results from "fitting" transport models to "calibration" measurements, and is supported by early measurements of SO_2 dry deposition to grass and cereals (e.g. Garland et al, 1973; Fowler, 1978). There are several reasons why the current process-based dry deposition model may have underestimated the true deposition velocity for SO_2, related to lack of data and the exclusion of important processes.

Data on the stomatal resistance to SO_2 uptake are only available for a few agricultural plants and one coniferous tree species in Britain. Although available data from elsewhere in the world are consistent with British data, it is possible that surface resistances could have been systematically overestimated. There are no data for vegetation types covering large areas of Britain, i.e. moorland and urban land use categories.

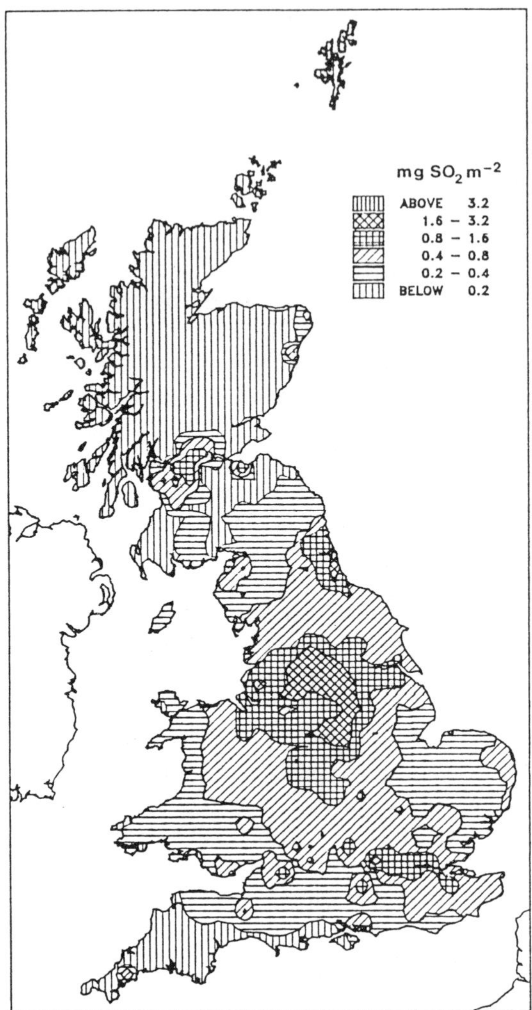

Fig. 1a: Dry deposition of SO_2 for a day in December (mg.SO_2.m^{-2}).

No chemistry is included in the model, but it is likely that deposition rates to surfaces, including leaf surfaces, are enhanced in the presence of materials which would neutralize the acidity formed when SO_2 dissolves in water. Alkaline dusts may be important in some areas, but more important is the potential for reaction and neutralisation with ammonia gas, which is present, albeit at low concentrations, throughout Britain, particularly in summer (Metcalfe et al, 1989). If surfaces were wet it is conceivable that the

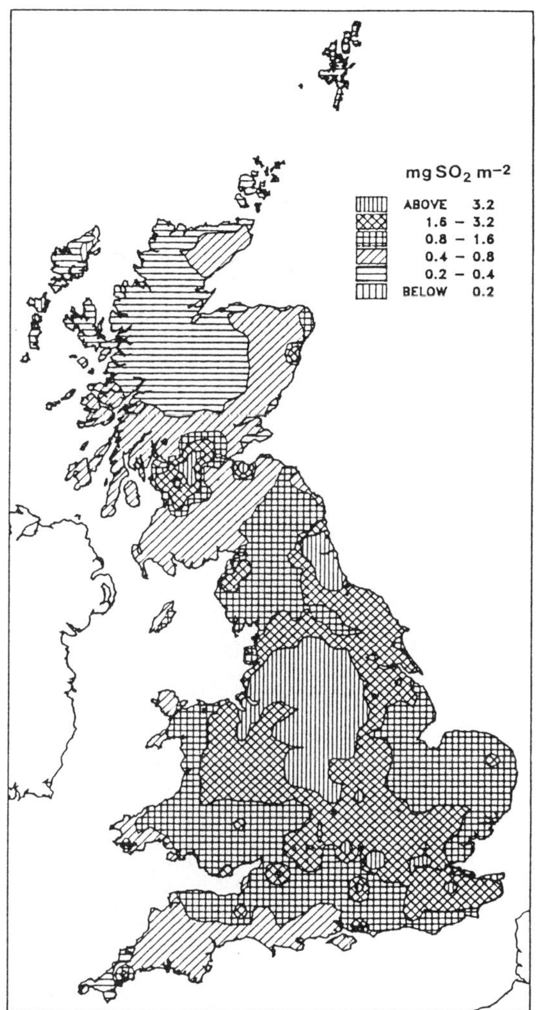

Fig. 1b: Dry deposition of SO_2 for a day in June ($mg.SO_2.m^{-2}$).

surface resistance term (r_c) would become close to zero in the presence of ammonia gas, and that deposition of SO_2 would be limited by atmospheric resistance and the availability of ammonia. In principle, especially for forests where the atmospheric resistance terms ($r_a + r_b$) are small, this could lead to very large deposition velocities (several $cm.s^{-1}$) for some of the time. Incorporation of ammonia data into the model is the priority for the future.

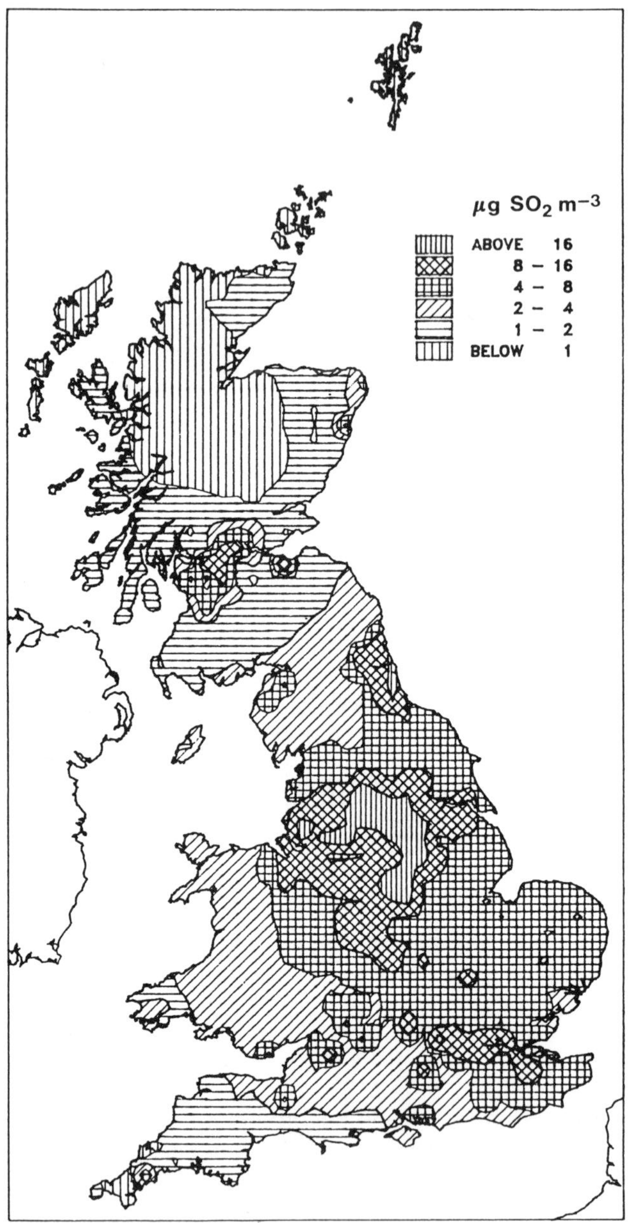

Fig. 2: Annual average concentration of SO_2 over Great Britain ($\mu g.m^{-3}$). Data from Warren Spring Laboratory.

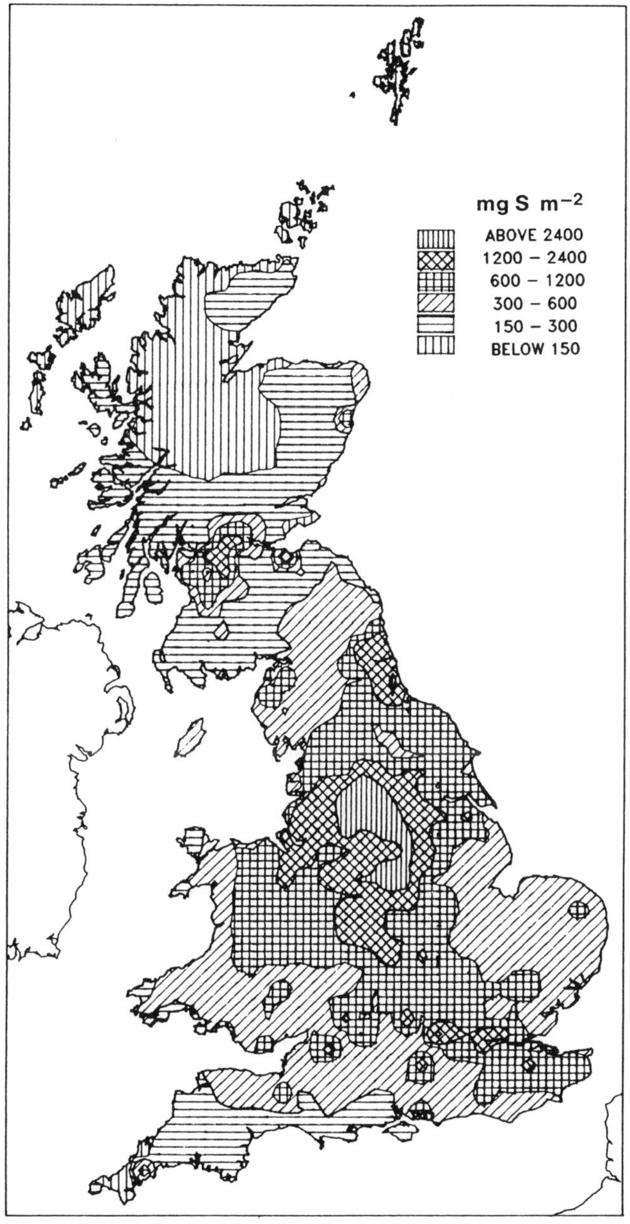

Fig. 3: Annual modelled dry deposition of SO_2 (mg.S.m^{-2}). Note logarithmic scale of class intervals.

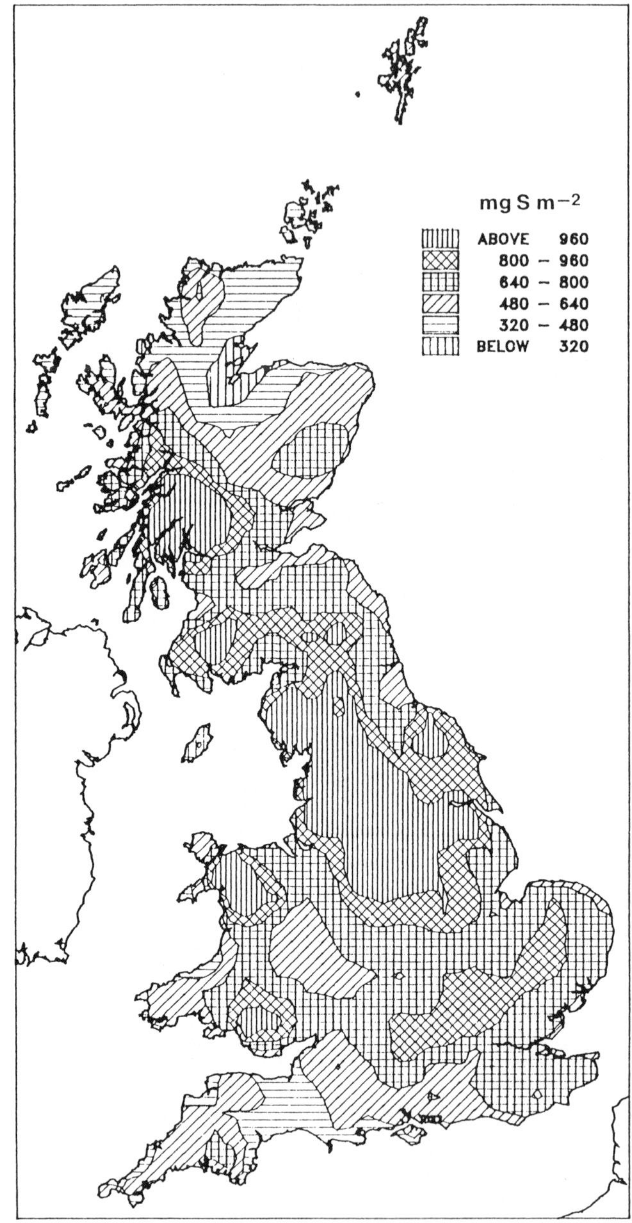

Fig. 4: Annual measured wet deposition of non-marine sulphate ($mg.S.m^{-2}$). Note linear scale of class intervals. Data from UK secondary rainfall chemistry network, Warren Spring Laboratory.

7. References

Anon. (1988). Digest of Environmental Protection and Water Statistics, No.11. HMSO, London.

Barrett, C.F., Atkins, D.H.F., Cape, J.N., Crabtree, J., Davies, T.D., Derwent, R.G., Fisher, B.E.A., Fowler, D., Kallend, A.S., Martin, A., Scriven, R.A. and Irwin, J. (1987). Acid Deposition in the United Kingdom 1981-1985. Warren Spring Laboratory, Stevenage.

Bunce, R.G.H. and Heal, O.W. (1984). Landscape evaluation and the impact of changing land use on the rural environment: the problem and an approach. In: Planning and Ecology, R.D.Roberts and T.M.Roberts, editors. Chapman and Hall, London, pp.164-188.

Campbell, G.W., Cocksedge, J.L., Coster, S.M., Dennis, A.L., Devenish, M., Heyes, C.J., Perrin, D.A., Stone, B.H. and Irwin, J.G. (1988). Acid rain in the United Kingdom: spatial distributions and seasonal variations in 1986. Report LR 691 (AP)M, Warren Spring laboratory, Stevenage.

Fisher, B.E.A. (1988). A case study of sulphur transport modelling over Great Britain. In: Air Pollution Modelling and Its Application VI, H. van Dop, editor. Plenum Press, New York, pp.293-303.

Fowler, D. (1978). Dry deposition of SO_2 on agricultural crops. Atmospheric Environment, 12:369-373.

Fowler, D. and Cape, J.N. (1983). Dry deposition of SO_2 onto a Scots pine forest. In: Precipitation Scavenging, Dry Deposition and Resuspension, H.R. Pruppacher, R.G.Semonin and W.G.N.Slinn, editors. Elsevier, New York, pp.763-773.

Garland, J.A., Clough, W.S. and Fowler, D. (1973). Deposition of sulphur dioxide on grass. Nature(London), 242:256-257.

Hicks, B.B., Baldocchi, D.D., Meyers, T.P., Hosker, R.P.Jr and Matt, D.R. (1987). A preliminary multiple resistance routine for deriving dry deposition velocities from measured quantities. Water, Air and Soil Pollution, 36:311-330.

Metcalfe, S.E., Atkins, D.H.F. and Derwent, R.G. (1989). Acid deposition modelling and the interpretation of the United Kingdom secondary precipitation network data. Atmospheric Environment, 23:2033-2052.

Perrin, D.A. (1986). Modelling the transport and removal of sulphur dioxide emissions in the United Kingdom. Warren Spring Laboratory Report LR560(AP), Stevenage.

From free-hand curves to chaos: computer modelling in ecology

J.N.R. Jeffers

University of Kent at Canterbury

1. Introduction

Modelling and simulation have a long history in some of the more applied aspects of ecology. As early as the mid-1880s, the Central European foresters were modelling the growth and production of forests by drawing free-hand curves through points plotted on graphs. The resulting models of plant growth were used as basis for forest management and in the preparation of forest working plans. These graphical methods continued to be used until the mechanical calculators, at first hand-cranked and then electrical, in combination with the new ideas of statistical analysis, enabled yield and volume tables to be calculated more efficiently. In agriculture, too, graphical methods for relating yields of crops to management, weather, and site only slowly gave way to the complex calculations that were required for the application of the statistical methods that had been developed from the 1920s onwards. The extensive application of modelling and simulation to forestry and agriculture began in earnest as the electronic digital computers were released from the secrecy which surrounded their original use in defence and space research, and became generally available for use by "peaceful" scientists.

The important developments of multivariate and regression analysis had taken place before the early 1950s when this release occurred, but the necessary computations were too difficult and took too long for them to be contemplated in practice, even with electrically-driven mechanical calculators. The early computers, using difference equation approximations to the analytical solutions of differential equations and efficient algorithms for matrix inversion, were sufficient to provide a huge impetus to the application of modelling to practical problems in applied ecology. The development of modelling and simulation was thus closely parallel to, and made possible by, the development of the new technology of computing and the associated developments in mathematical and statistical theory.

Ecosystem modelling has a shorter history, if only because it was first necessary to discover the paradigm of the ecosystem, originally defined by Tansley (1935). Perhaps the greatest stimulus to ecosystem modelling came from the International Biological Programme (IBP) which conveniently coincided with the emergence into civilian use of the first computers and the development of high level programming languages. Although few of the whole-ecosystem models that were envisaged by IBP ever became operational, the international ecological community became aware, as never before, of the potential of simulation and modelling for the integration of widely disparate research topics (Worthington, 1975).

That awareness has continued to the present day, stimulated by the parallel developments that have continued to take place in computing and in the theory of mathematics and statistics. Today, much of applied ecology would be impossible without a direct appeal to methods of modelling and simulation because of the need to retain the complexity of the interactions of organisms with themselves and with their environment if we are to remain within the paradigm of the ecosystem. Nevertheless, problems remain in the ways in which modelling is related to the wider issues of the scientific method, and these problems are, if anything, exacerbated by the dependence of modelling on the use of computers and computer algorithms. In this paper, I want to identify some of these problems and to look, particularly, at the possible effects of some of the developments which are likely to take place in the near future.

2. Models and systems analysis

In the sense in which I use the word in this paper, a *model* is a formal expression of the relationship between defined entities in physical or mathematical terms. The inclusion of the word "formal" in this definition indicates that the expression must ultimately be capable of being tested against reality, as a validation of the implied relationships. The expression must be capable of making predictions against which such a test can be made, and modelling is not, therefore, a purely theoretical exercise, but part of the formal logic of the scientific method.

The use of mathematics to define the relationships between organisms, or between organisms and their environment, in ecology is necessary because of the complexity of those relationships. Mathematics is a symbolic language which has been specially developed to describe and manipulate complexity, so it is the appropriate way of coping with the non-linearity and feedback that is characteristic of ecological systems. There is then the added advantage that these mathematical expressions abstract the essential elements of a system from a much larger set of possible factors, and do so in such a way that there is no ambiguity about either the relationships or the elements. Mathematical formulae and their representations in computer algorithms are precise and verifiable statements of the interactions between entities in a defined system, and should, if our explanation is to be scientific, be capable of verification.

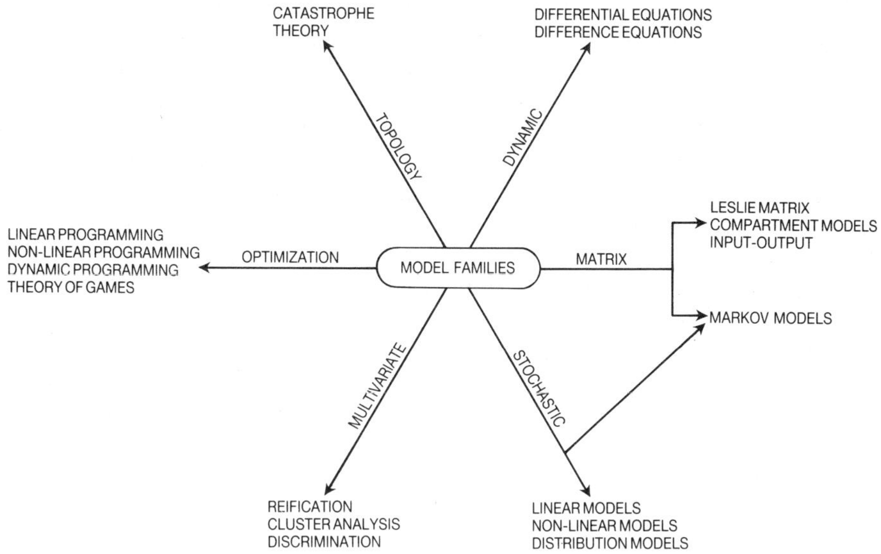

Fig. 1: Model classification.

Maynard Smith (1974) distinguishes between models and simulations by the criterion that a model should contain as little detail as possible, whereas a simulation should include as much detail as possible. He suggests that a model cannot be used to predict the future behaviour of whole ecosystems, and that its utility lies mainly in analysing a particular case. In contrast, simulations have a practical purpose, and the better a simulation is for its own purpose, by the inclusion of all relevant details, the more difficult it is to generalise its conclusions to other species and systems. Not least because the dividing line between these two types of mathematical descriptions is indeterminate, I will not follow this distinction. In this paper, a model is any mathematical description of the relationship between defined entities, and the process by which such models are created and used I will call simulation.

Elsewhere (Jeffers, 1982), I have suggested that models can be grouped within families, and some of these families are indicated in Fig. 1. Dynamic models are based on equations expressing the levels of state variables in terms of rates of change controlled by decision functions. The use of such models depends on the ability of modern computers to solve large numbers of equations or to trace the changes in the state variables by small-step iterations. Their popularity arises from the great flexibility of the methods used to describe ecological systems, including non-linear responses of components to controlling variables, and both positive and negative feedback - the carrying back of some of the effects of a process to their source so as to strengthen or modify those effects. However, this flexibility can also be a disadvantage, making the outcome of models difficult to predict for quite small departures from the parameter values. Dynamic models are heavily orientated towards deterministic solutions and do not usually reflect the inherent variability of ecological systems. Nevertheless, they may well be helpful in the early stages of modelling by concentrating attention on the critical variables and subsystems.

Matrix models represent an important family of models in which the "realism" of the model is sacrificed in order to obtain the benefits of a convenient mathematical formulation and the computational advantages of matrix algebra. Mathematical analysis of a matrix model is sufficient to reveal all the properties of the model, without the need for time-consuming "experimentation". Markov models have as their basic format a matrix of entries expressing the probabilities of the transition of one state of a system to another in a specified time step. The value of such models is confined to situations in which the future development of the ecological system is determined by the present state of the system, and is independent of the way in which the present state has developed.

Models that incorporate probabilities fall within the family of stochastic models and are particularly valuable in simulating the variability of the response of organisms to change in ecosystems. Such models include the vast array of statistical methods on which almost all modern techniques of data analysis now depend. The mapping of the spatial and temporal patterns of communities, and the analysis of variance and regression are all examples of such methods. Multivariate models are often useful to deal with the many situations in ecology where it is necessary to capture the behaviour of several, and perhaps many, variates.

Optimisation models are used to find the maximum or minimum of some mathematical expression or function by setting values to variables that we are free to alter within defined limits. Linear, nonlinear, and dynamic programming all fall within this category. Closely related are the game theory models that represent an interesting and surprisingly little-explored approach to the solution of strategic problems. The metagame theory of Fraser and Hipel (1984) is a particularly valuable extension of such models for the assessment of environmental impact, and for environmental planning. Finally, topological models embrace a wide range of interesting and relatively unexplored approaches to the modelling of complexity, and include such esoteric concepts as catastrophe theory, fractal geometry, and chaos mathematics (Gleich, 1987)

The choice of the appropriate family of model can only be made efficiently if the modelling is itself embedded in some larger and more strategic framework. Increasingly, it has come to be recognised that systems analysis, defined as the orderly and logical organisation of data and information into models, followed by the rigorous testing and exploration of these models necessary for their validation and improvement, provides such a framework (Jeffers, 1978). Systems analysis is designed to help decision makers to choose a desirable course of action, or to predict the outcome of one or more courses of action that seem desirable to those who have to make decisions. In particularly favourable cases, the choice of action that is indicated by systems analysis will be the "best" choice in some specified or defined way. Alternatively, we can think of systems analysis quite simply as the application of the scientific method to the study of complex systems.

3. The objectives of modelling

There are three principal objectives of modelling and simulation:

Descriptive: We use models as a parsimonious description of the process or processes that underlie the problem with which we are concerned. By abstracting just the essential elements of those processes, we hope to concentrate attention on those elements, freed from the distractions of the many other factors that surround the problem. Our choice of essential elements represents an hypothesis of relevance, and this hypothesis will ultimately have to be tested as part of the scientific process in which modelling and simulation are embedded by systems analysis. Until such time as the model is rejected by explicit tests, it represents our understanding of the ecological system.

Predictive: We also use models to make predictions about the future behaviour of the system or systems, possibly as a response to changes that are induced in one or more of the driving variables, or possibly as an extension of existing conditions. Such predictions may form part of a deliberate programme of determining "What if?", but they may also represent an exploration of the dynamics of the processes designed to extend our understanding of the system being modelled.

Decision making: In ideal situations, we may also use models for the making of practical decisions when confronted by difficult choices about the management or future direction of complex systems. The effects of the decision can be tested on the model rather than on the real system, usually very much more quickly than by waiting to see what will happen over a period of months or years. If the predicted effect is to produce an unwanted side effect, or even the total destruction of the system, the manager has the opportunity to rethink the decision.

More simply, models are used as a way of explaining how complex systems function. By definition, the models are simpler than the actual system, and can be used to explain the underlying mechanisms that govern the functioning of the system. Combined with verbal and diagrammatic descriptions, they form part of the communication between the scientist and the resource manager, administrator or politician.

4. Limitations and potential

There is now a huge literature on the modelling of ecological systems, and I will make no attempt to review that literature in this paper. Broadly, most of the published papers emphasise the mathematical basis of the models, and concentrate on the details of the model formulation, usually at the expense of the logical framework within which the model is intended to fit. Recently, scientists who have been closely involved in the application of modelling to the problems of managing ecological systems have become increasingly critical of the use of formal models (e.g. Clark and Munn, 1986). In part, this criticism arises because of the universal failure of modelling to predict ecological surprise. The role of acid rain in forest decline, the depletion of the ozone layer by chlorofluorocarbons, the accumulation of pesticides at the ends of food chains, and the seal deaths in the North Sea are all examples of ecological surprise that should, with hindsight, have been predicted by the use of mathematical models.

An essential element in ensuring that we are not overcome by ecological surprise in the future must be an increased emphasis on systems analysis in our modelling and simulation of environmental systems. By linking our models to the broader context of major environmental problems, and by carefully defining and bounding the extent of the models we construct, we can retain a holistic view while working on the details of the subsystem models. In this way, the wider concepts of the analysis help to direct and inform the use of the models, while the models themselves add incrementally improved structure to the concepts. We are less likely to be overtaken by global or local surprise if we have retained an overview of the whole system, and updated that view with the results of our simulations. During the last 10 years, both the use and teaching of system analysis have declined in the face of greatly increased concentration on computational techniques - a trend that needs to be reversed if we are to retain control of the methodology that we are developing.

Some of the current excitement generated by chaos mathematics arises from the recognition that widely different phenomena can be represented with the same basic equations by changing the parameters of those equations. Even the discontinuities shown by catastrophe theory can be shown to belong to apparently stable systems at particular intersections of critical parameters

(Gleich, 1987). Similarly, the full dynamic behaviour of ecosystems at an aggregate level can be represented by the sequential interaction of four ecosystem functions, namely exploitation, conservation, creative destruction, and renewal (Holling, 1986). The progression of events is such that these functions dominate at different times, the two properties of degree of organisation and the amount of capital accumulation and retention being determined by the speed and amplitude of the cycle.

Concern has also been expressed about our current ability to cope with what has come to be called "the scaling problem", the difficulty of evaluating interactions of biological, chemical, and physical processes in the face of small-scale spatial and temporal heterogeneity in terrestrial, freshwater, marine and atmospheric environments. I have argued elsewhere (Jeffers, 1988) that much of our difficulty with spatial and temporal changes within the geosphere/biosphere arises from a failure to apply the necessarily strict rules of the logic of the scientific method. That method demands the clear definition of the population about which we will seek to make inferences in any investigation, and the adoption of valid methods of sampling or experimental design in order to estimate the parameters - including the heterogeneity - of the target population. Computational methods alone will not bypass the intellectually challenging and methodologically strict formulation of hypotheses, and the subsequent collection of data in order to:

(a) integrate scales, conceptually and practically, between and within disciplines;
(b) aggregate local scales meaningfully into larger spatial scales; and,
(c) evaluate how global scale processes influence local processes.

Each of these goals, separately or in combination, requires the formulation of a priori hypotheses, and the collection of data in carefully prescribed ways so as to test these hypotheses explicitly. The search for a method, computational or conceptual, which will convert a haphazard collection of data into a useful working hypothesis is a delusion!

Inevitably, much of the attention of this symposium has been focused on computing. However, important as good computing tools are, they do not in any way substitute for good statistical knowledge about how to analyse structured data, or, indeed, how to synthesise those data into valid models. Good tools will allow bad methods to be encoded just as easily as good methods, and the difficulty lies in identifying what has come to be called *metadata* - systematic descriptive information about data content and organisation that can be retrieved, manipulated and displayed in various ways. From the modeller's point of view, the principal problems are in knowing what methods can be regarded as valid in processing a particular data set, or, in its simplest form, what data can be combined or compared legitimately. The need for metadata in statistical data processing is self-evident, and experience shows that the only solution is to capture the metadata at source with the data, and to use integrated tools to handle both together. There is, however, a fundamental lack of integration between database management systems (DBMS) and classical statistical analysis - the conventions of almost all DBMS systems available today result in the loss of the metadata necessary to ensure valid subsequent analysis of the data.

The two most promising research areas for finding generic solutions to the metadata problems of modelling and statistical analysis are object-orientated databases and expert systems. There is some indication that these two areas of research are coming closer together, especially if rule-based systems are used to process messages as the objects in object-orientated database systems become more complex, and database management systems become essential if expert systems are to encompass major environmental problems.

We have been less than successful in using formal mathematical models as a means of communication with decision makers. Resource managers, administrators, and politicians, in particular, are inevitably strongly committed to policies and practices that they have advocated, and that, in some cases, has been the reason for their attaining their present position in their organisation or political party. Where model predictions conflict with their preconceptions, or even with their intuitive judgements, formal models are likely to be rejected, especially where their mathematical basis is beyond the experience of non-mathematicians. There is a gap to be filled between the results of our research based on modelling and simulation and the implementation of that research by those who have to make practical decisions in the management of ecological systems and the environment.

One possible solution lies in the wider use of expert systems, defined as knowledge-based computer systems capable of giving intelligent advice, and whose basis can be readily justified. Expert system shells now exist for the development of knowledge-based systems, and these can be integrated with models so as to provide an intelligent interface between the model and the user. Such an interface can, for example, be used to determine why the model behaves in a way that is unexpected or counter-intuitive to the user. It may also help the user to explore a wide range of possible scenarios or strategies in order to obtain a clear view of the sensitivity of the underlying ecological systems to proposed change. However, a great deal still needs to be done to develop expert systems adequate for the capture and communication of metadata, perhaps by creating generalised semantic rules and frameworks within which rule-based systems can be activated.

The current trends in statistical data collection, dissemination and analysis highlight the inadequacies of today's data processing systems. The evolution of such systems is not providing the necessary facilities to handle the growing demands of complexity, accessibility, timeliness, and completeness. Because modelling and simulation depend on valid analysis of statistical data, a major advance is needed in the ability of computer systems to provide semantic capability for the handling of metadata. Expert system techniques may provide the necessary mechanisms and models for the needs of the future, and may, coupled with advances in database technology, enable us to both model and communicate the complex relationships which are the characteristic of ecological systems.

As the Chinese proverb says, "Prediction is difficult, particularly about the future". We must all surely be impressed, and probably bemused, by the speed of development of computer technology. Today, we have machines on our desks, or even in our briefcases, that were wild dreams only a few short years ago. It seems likely that some of the technologies that are currently becoming available will help us to overcome the inadequacies that I have highlighted above. The use of videodiscs as a medium for data storage and retrieval in Steve Job's NeXt computer opens up possibilities of integrating maps, diagrams, photographs, text, tabulated information and computer programs in new ways. One of my colleagues at the Virginia Polytechnic Institute and State University is already exploiting videodisc technology in his Cyberquest innovative support system. Several major encyclopaedias already publish their information in this medium, providing very rapid access through the use of keywords.

Viewdata and packet-switching systems are already available for the transmission of information between multiple points of access through ordinary telephone lines, and perhaps we should be giving more attention to the exploiting of such systems to provide decision makers and scientists with access to ecosystem models. The introduction of satellite television and communication as a commercial system offers further possibilities for enabling resource managers, administrators, and politicians to remain in contact with scientists who have information relevant to the decisions that they have to make, increasingly derived from models of complex ecological systems and the reaction of such systems to proposed changes in policy or management.

As responsible scientists, it is our task to embrace these new technologies and to use them constructively, whilst remembering the need to maintain the philosophical basis and logic of the science that we practice. The effects of "technology push" through commercial software have already created acute difficulties for professional statisticians, who have to cope with the results of wholly invalid analyses obtained with the aid of statistical packages. We are beginning to see much the same result in modelling, and it is our task, in the coming years, to ensure that this powerful tool, aided by the rapid developments in computer hardware and software, is not misused.

5. References

Clark, W.C. and Munn, R.E. (1986). Sustainable development of the biosphere. Cambridge University Press, Cambridge.

Fraser, N.M. and Hipel, K.W. (1984). Conflict analysis: models and resolutions. North Holland, New York, Amsterdam, Oxford.

Gleich, J. (1987). Chaos: Making a new science. Heinemann, London.

Holling, C.S. (1986). The resilience of terrestrial ecosystems: Local surprise and global change. In: Sustainable development of the biosphere, W.C. Clark and R.E. Munn, editors. Cambridge University Press, Cambridge.

Jeffers, J.N.R. (1978). An introduction to systems analysis: with ecological applications. Edward Arnold, London.

Jeffers, J.N.R. (1982). Modelling. Chapman and Hall, London and New York.

Jeffers, J.N.R. (1988). Statistical and mathematical approaches to issues of scales in ecology. In: Scales and global change, T. Rosswall, R.G. Woodmansee and P.G. Risser, editors. John Wiley and Son, Chichester.

Maynard Smith, J. (1974). Models in ecology. Cambridge University Press, Cambridge.

Tansley, A.G. (1935). The use and abuse of vegetational concepts and terms. Ecology 16:284-307.

Worthington, E.B. (1975). The evolution of IBP. Cambridge University Press, Cambridge.

Ecological modelling on personal computers

P.J. Radford

Plymouth Marine Laboratory

1. Introduction

Personal computers are now to be found on the desks of nearly all scientists, even those from disciplines not traditionally associated with mathematics. In many cases their primary application is word processing, although increasingly spreadsheets are being used for basic data handling and graphical interpretation. New and imaginative problem-orientated computer languages are now available which open up the possibility of introducing computer simulation modelling to this wide group of specialists without the need for extensive training in mathematics and computer science. They are of particular value to ecologists concerned with highly interactive parallel systems best represented by differential equations and solved as open ended recursive systems using numerical simulation methods. There is nothing new in the mathematical basis for these languages, the principles having been clearly presented by Forrester (1961) and implemented in DYNAMO (Pugh, 1963) and CSMP (IBM, 1967) but the provision of user friendly front ends and improved graphical output greatly facilitates the novice modeller in defining and solving his problem.

2. Malaysian mangrove model

Whilst on a short study leave in Penang, financed by the Malaysian National Science Council, I worked with Dr W.K. Gong and Dr J.E. Ong to apply simulation modelling to problems of mangrove management. Our first elementary model (Fig. 1), formulated in the space of a few hours, serves as an illustration of the power and usefulness of STELLA (Richmond et al, 1987). This language, based on the principles of industrial dynamics, requires that the system to be modelled is first drawn as a flow diagram composed of only four fundamental building blocks. Rectangles represent state variables (integrators), valved pipes represent rates of material flow, circles represent auxiliary variables and parameters and arrowed lines represent relationships (information flow) as in DYNAMO. The diagram is built up by "dragging" "icons" to their desired positions on the screen and by assigning them names. A stick-of-dynamite icon provides a very satisfying

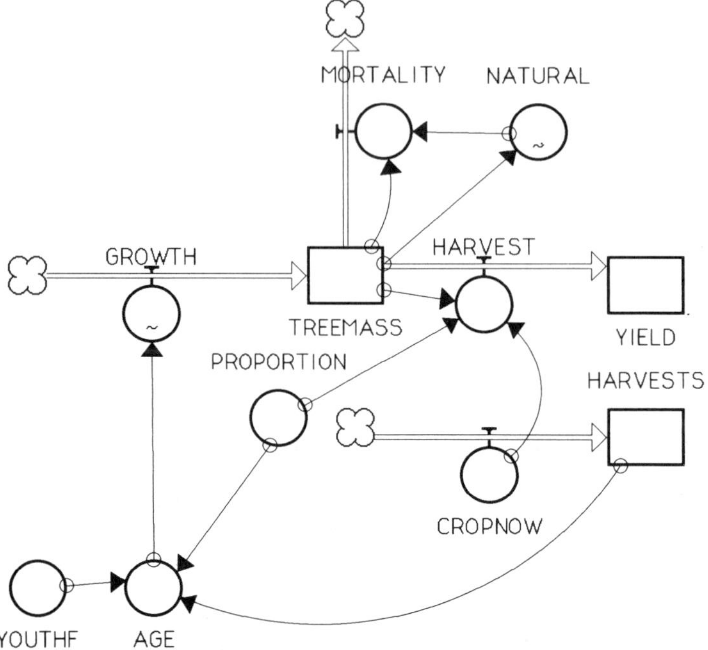

Fig. 1: STELLA flow diagram from the "DIAGRAM" window, representing the growth and harvesting of mangrove trees in Malaysia.

means of demolishing unwanted components and clouds form a neat way of dealing with undefined sources and sinks!

The mangrove model has been designed to demonstrate the impact of alternative harvesting regimes open to farmers who wish to maximise yield and minimise impact on the mangrove ecosystem. The key state variable is the standing stock of timber (TREEMASS) which increases with natural growth (GROWTH) and is diminished by natural mortality (MORTALITY) and harvesting (HARVEST). The implicit thesis of the model is that natural growth of a mangrove stand depends upon its relative age (AGE) since its last harvest. Complete destruction of a stand reduces its age to zero but cropping by a thinning out process of a certain proportion (PROPORTION) of the crop rejuvenates the stand by a factor (YOUTH) which is empirically related to that proportion. The remainder of the model simulates the signal to harvest (CROPNOW), which when integrated determines the number of harvests (HARVEST) taken in a given period of time. This signal triggers the harvest which transfers a given proportion of the standing stock (HARVEST) to produce the yield (YIELD).

Table 1. *A complete listing of the mangrove model represented in Fig. 1 from the STELLA "EQUATIONS" window. The icons before each equation refer to the component types shown in the flow diagram.*

```
□ HARVESTS = HARVESTS + dt * ( CROPNOW )
  INIT(HARVESTS) = 0
□ TREEMASS = TREEMASS + dt * ( -HARVEST - MORTALITY + GROWTH )
  INIT(TREEMASS) = 5
□ YIELD = YIELD + dt * ( HARVEST )
  INIT(YIELD) = 0
○ AGE = TIME-YOUTHF*HARVESTS*PROPORTION*PROPORTION
○ CROPNOW = PULSE(1.,15,5)-PULSE(1.,25,5)
○ HARVEST =CROPNOW*TREEMASS*PROPORTION
○ MORTALITY = TREEMASS*NATURAL
○ PROPORTION = .5
○ YOUTHF = 20
⊘ GROWTH = graph(AGE)
   0.0 -> 0.0
  5.000 -> 2.800
  10.000 -> 18.000
  15.000 -> 12.500
  20.000 -> 12.200
  25.000 -> 12.000
  30.000 -> 12.000
  35.000 -> 12.000
  40.000 -> 12.000
  45.000 -> 12.000
  50.000 -> 12.000
⊘ NATURAL = graph(TREEMASS)
   0.0 -> 7.771e-318
  25.000 -> 0.0105
  50.000 -> 0.0235
  75.000 -> 0.0390
  100.000 -> 0.0545
  125.000 -> 0.0705
  150.000 -> 0.0700
  175.000 -> 0.0705
  200.000 -> 0.0705
  225.000 -> 0.0695
  250.000 -> 0.0700
```

The computer language STELLA translates the flow diagram into a working simulation model. First it writes the equations for the state variables as first order difference equations. Next, by inserting a question mark into each component of the flow diagram it prompts the user to provide information regarding: (i) the initial conditions for the state variables; (ii) the relationships which govern the rate variables; (iii) the values of parameters. Relationships can take the form of equations or be entirely empirical; in the extreme case they can be represented as free-hand curves drawn with the "mouse" on an X-Y graph. Rate equations may only include variables which are shown to be imputed in the flow diagram by lines of information flow. These are listed and may be selected by "mouse" and combined with functional operators (+, -, etc) and numerical constants available from a simulated keyboard. In addition STELLA provides some 40 built-in functions including time (TIME) dependent analogue components such as RAMP, STEP and PULSE and logical operators (ELSE, IF, AND). The mathematical rigour introduced by the insistence upon a consistent flow diagram makes it almost impossible to make silly mistakes in model formulation (such as spelling mistakes etc). A full listing of the equations is displayed in the EQUATIONS window as shown in Table 1. The state variables are shown as first order difference equations although they may be solved by selecting second or fourth order Runge-Kutta integration formulae. The optimal integration interval (dt) may be selected by trial and error.

As soon as a simulation run is initiated, graphical output starts to be plotted in the GRAPH PAD window so ensuring that grossly erroneous results can be noticed and the simulation run cancelled at the earliest possible time. An example of such graphical output is given in Fig. 2 which also illustrates how harvesting of a proportion of the mangrove stand helps to maintain maximal growth for a sustained period. The lines, labelled 1, 2 and 3 for identification purposes may also be printed in colour if the Apple Imagewriter (dot matrix printer) is fitted with a multi-coloured ribbon.

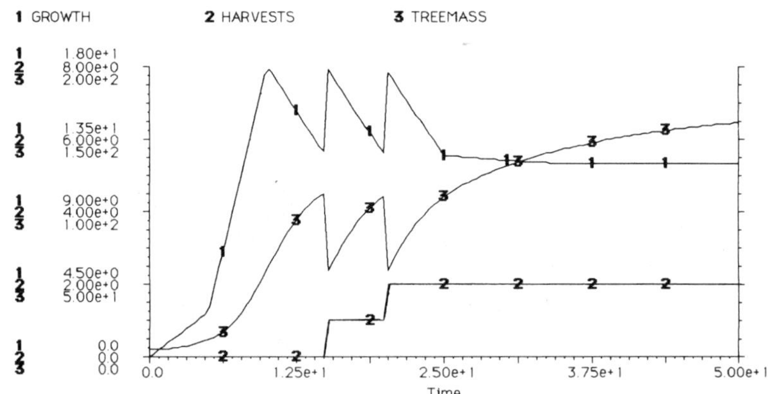

Fig. 2: Graphical output from the "GRAPH PAD" window of the model listed in Table 1. Time is measured in years; growth and biomass terms are in arbitrary units.

The very rapid feedback provided by STELLA at all stages of formulation and simulation makes the language particularly rewarding to use. Very rarely is the modeller left in any doubt on how to proceed with the modelling methodology since prompts are continually given of the possibilities open for use via the menu keywords and the more detailed instructions to which they provide access.

3. Model of copepod defaecation

Herbiverous copepods represent a section of the oceanic zooplankton which graze on phytoplankton. Copepod defaecation might seem to some as a rather esoteric subject but it constitutes one of the important processes by which carbon dioxide from the atmosphere, having been fixed by photosynthesis, is sedimented out and need not be returned to exacerbate the greenhouse effect. A crucial factor which determines the direction of this process is the density of faecal pellets relative to that of sea water. I have worked with Dr Nan Duncan of the Royal Holloway and Bedford College, using copepod defaecation as a test problem to introduce her to computer modelling using the STELLA (Richmond et al, 1987) simulation language.

The first model, developed in a few hours, was very simple indeed consisting of only three state variables (food, zooplankton biomass and faeces weight) and five processes (primary production, feeding rate, respiration, defaecation rate and faecal pellet sinking rate). At this level of abstraction the equations which govern the processes are necessarily empirical and the results not sufficiently responsive to changes which occur in the field. Within a few days a much more sophisticated model was produced which reached the heart of the problem by considering the fate of phytoplankton cell numbers, cell biomass and cell volume which enabled densities of faecal pellets to be simulated. Finally, after a week or two, a more realistic model was attained which included representations of many of the relevant processes which have been researched and pointed to other processes which need to be studied if realistic simulations are to be obtained. The flow diagram given in Fig. 3 shows the degree of complexity that the model embodied after about three weeks of development. It is doubtful that such progress would have been made without the user friendly facilities provided by STELLA. The drafting package, MacDraw, has been used to emphasise the different symbols, all of which are illustrated in the key.

The logic of the model is, that for a given copepod, the characteristics of its faecal pellets will be a function of its food intake exemplified by the phytoplankton parameters, viz. cell concentration (cellconc), cell weight (cellwt) and cell volume (cellvol). These parameters of the food source determine the copepod ingestion rate (ingestrt) in terms of number

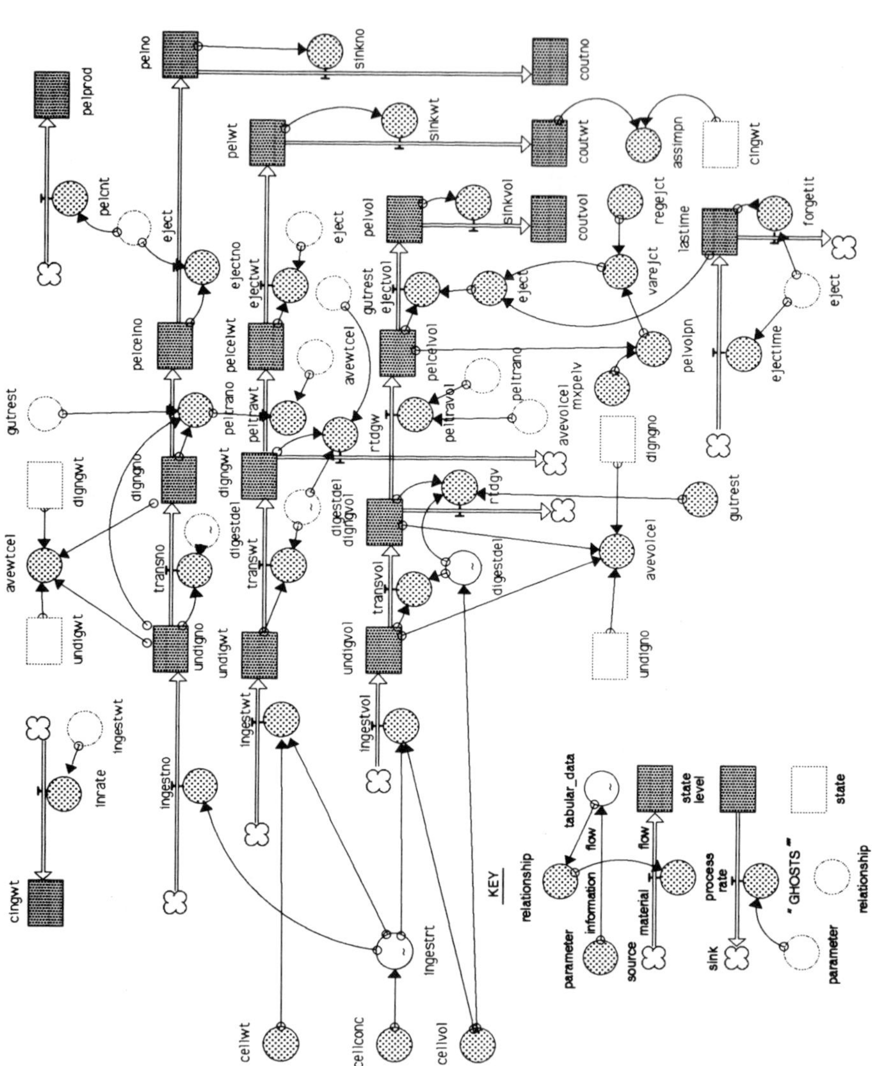

Fig. 3: STELLA flow diagram of copepod defaecation. The icons have been shaded to emphasise the three main material flows of phytoplankton viz. (i) numbers (no); (ii) weight (wt); (iii) volume (vol).

(ingestno), weight (ingestwt) and volume (ingestvol). The undigested cells (undigno) are broken down after a size dependent delay time and then are digested in the gut of the copepod so losing weight and volume but not number. Partially digested cells are then passed into the hind gut where they start to form a faecal pellet with a conserved number of cells (pelcelno) with reduced weight and volume (pelcelwt, pelcelvol). When the hind gut is full or after a maximum defaecation interval, whichever comes first, the faecal pellet is ejected (eject) and sinks. For the present use of this model the cumulative volume and weight (coutvol, coutwt) are computed for accounting purposes. The other variables of the model control the frequency of defaecation and calculate useful supplementary variables such as the average weight and volume of a cell within the gut (avewtcel, avevolcel) and the cumulative number of faecal pellets produced by the copepod (pelprod). It is one of the rules of STELLA that every variable used in the model must be represented on the flow diagram. One novel feature of the language is the ability to simplify the overlapping network of lines in the flow diagram by copying any variable to enable it to be drawn close to where it is needed. This is termed "ghosting" and it may be detected on the flow diagram because the "ghost" has a lighter outline than the original symbol.

A full listing of the equations generated by STELLA for the copepod model is given in Table 2. These are listed in groups according to type, state variables, parameters and process equations, and tabular data. Within each group they are arranged in alphabetical order for easy identification. STELLA organises the correct computational sequence required for execution and this may be accessed, if it is so desired.

Typical examples of model output are given in Fig. 4. In both cases the simulations represented four hours (240 mins) of real time and they show the response of an initially starved copepod to two different diets; the first (Fig. 4a) composed of a low concentration (100 cells.l^{-1}) of large phytoplankton cells (0.9 g.$cell^{-1}$) and the second (Fig. 4b) the same cell concentration of much smaller cells (0.015 g.$cell^{-1}$). The four variables plotted as time series represent, in each case: 1. Volume of undigested cells in the gut (undigvol); 2. Volume of digesting cells (digngvol); 3. The cumulative volume of faecal pellets produced (coutvol); 4. The assimilation efficiency (assimpn) defined as the ratio of the cumulative input weight minus output wt, to the cumulative input wt. The first line (labelled "1") illustrates the way in which larger cells are immediately crushed by the copepod whereas small cells offer a certain resistance to the digestive process so gradually accumulating in their undigested state until their cell walls are penetrated. This factor has an impact upon the time taken for the digesting cells (line 2) to reach their equilibrium values. For the large cells

Table 2. A complete listing of the copepod model represented in Fig. 3. The state variables are here shown in alphabetical order in the right hand column, together with their initial values. The parameters and rate relationships are shown in alphabetical order in the left hand column. Tabular data is given at the end of the model.

```
○ assimpn = (cingwt-coutwt)/cingwt
○ avevolcel = (undigvol+digngvol)/(digngno+undigno)
○ avewtcel = (undigwt+digngwt)/(undigno+digngno)
○ cellconc = 100
○ cellvol = 1.0
○ cellwt = .09
○ eject = PULSE(1.,lastime,varejct)
○ ejectime = eject*TIME
○ ejectno = pelcelno*eject
○ ejectvol = pelcelvol*eject
○ ejectwt = pelcelwt*eject
○ forgetit = lastime*eject
○ gutrest =30.0
○ ingestno = ingestrt
○ ingestvol = ingestrt*cellvol
○ ingestwt = ingestrt*cellwt
○ inrate = ingestwt
○ mxpelv = 2000
○ pelcnt = eject
○ peltrano = (undigno+digngno)/gutrest
○ peltravol = peltrano*avevolcel
○ peltrawt = peltrano*avewtcel
○ pelvolpn = pelcelvol/mxpelv
○ regejct =30.
○ rtdgv = digngvol/(gutrest-digestdel)
○ rtdgw = digngwt/(gutrest-digestdel)
○ sinkno = pelno/DT
○ sinkvol = pelvol/DT
○ sinkwt = pelwt/DT
○ transno = undigno/digestdel
○ transvol = undigvol/digestdel
○ transwt = undigwt/digestdel
○ varejct = IF(pelvolpn<1.) THEN regejct ELSE 1.
⊘ digestdel = graph(cellvol)
  ( 0.0, 1.30e+1),( 1.00e-1, 1.30e+1),( 2.00e-1, 1.30e+1),( 3.00e-1, 1.30e+1),( 4.00e-1,
  1.30e+1),( 5.00e-1, 1.30e+1),( 6.00e-1, 1.00e+0),( 7.00e-1, 1.00e+0),( 8.00e-1, 1.00e+
  0),( 9.00e-1, 1.00e+0),( 1.00e+0, 1.00e+0)
⊘ ingestrt = graph(cellconc)
  ( 0.0, 0.0),( 1.00e+2, 3.20e+1),( 2.00e+2, 7.00e+1),( 3.00e+2, 1.00e+2),( 4.00e+2, 1.00e+
  2),( 5.00e+2, 1.00e+2),( 6.00e+2, 1.00e+2),( 7.00e+2, 1.00e+2),( 8.00e+2, 1.00e+2),(
  9.00e+2, 1.00e+2),( 1.00e+3, 1.00e+2)
```

```
□ cingwt = cingwt + dt * ( inrate )
  INIT(cingwt) = 1.
□ coutno = coutno + dt * ( sinkno )
  INIT(coutno) = 0.
□ coutvol = coutvol + dt * ( sinkvol )
  INIT(coutvol) = 0.
□ coutwt = coutwt + dt * ( sinkwt )
  INIT(coutwt) = 0.
□ digngno = digngno + dt * ( transno - peltrano )
  INIT(digngno) =1.
□ digngvol = digngvol + dt * ( transvol - peltravol - rtdgv )
  INIT(digngvol) = 0.
□ digngwt = digngwt + dt * ( transwt - peltrawt - rtdgw )
  INIT(digngwt) = 0.
□ lastime = lastime + dt * ( ejectime - forgetit )
  INIT(lastime) = 0.
□ pelcelno = pelcelno + dt * ( peltrano - ejectno )
  INIT(pelcelno) =0.
□ pelcelvol = pelcelvol + dt * ( peltravol - ejectvol )
  INIT(pelcelvol) = 0.
□ pelcelwt = pelcelwt + dt * ( peltrawt - ejectwt )
  INIT(pelcelwt) = 0.
□ pelno = pelno + dt * ( ejectno - sinkno )
  INIT(pelno) =0.
□ pelprod = pelprod + dt * ( pelcnt )
  INIT(pelprod) = 0.
□ pelvol = pelvol + dt * ( ejectvol - sinkvol )
  INIT(pelvol) = 0.
□ pelwt = pelwt + dt * ( ejectwt - sinkwt )
  INIT(pelwt) = 0.
□ undigno = undigno + dt * ( -transno + ingestno )
  INIT(undigno) =1.
□ undigvol = undigvol + dt * ( ingestvol - transvol )
  INIT(undigvol) = 0.
□ undigwt = undigwt + dt * ( ingestwt - transwt )
  INIT(undigwt) = 0.
```

this takes about 90 mins compared to 120 mins for the small cells. The much greater throughput of food made possible by the consumption of large cells causes a much higher frequency of faecal pellet production driven by gut fill, whereas small cells are defaecated at less frequent but regular intervals according to copepod bio-rhythms (line 3). A longer simulation run confirms (line 4) the slight tendency for large cells to be more efficiently digested than small cells (50% vs. 40%).

It is not within the scope of this paper to discuss the scientific assessment of these results but the process of model building has given stimulus to the

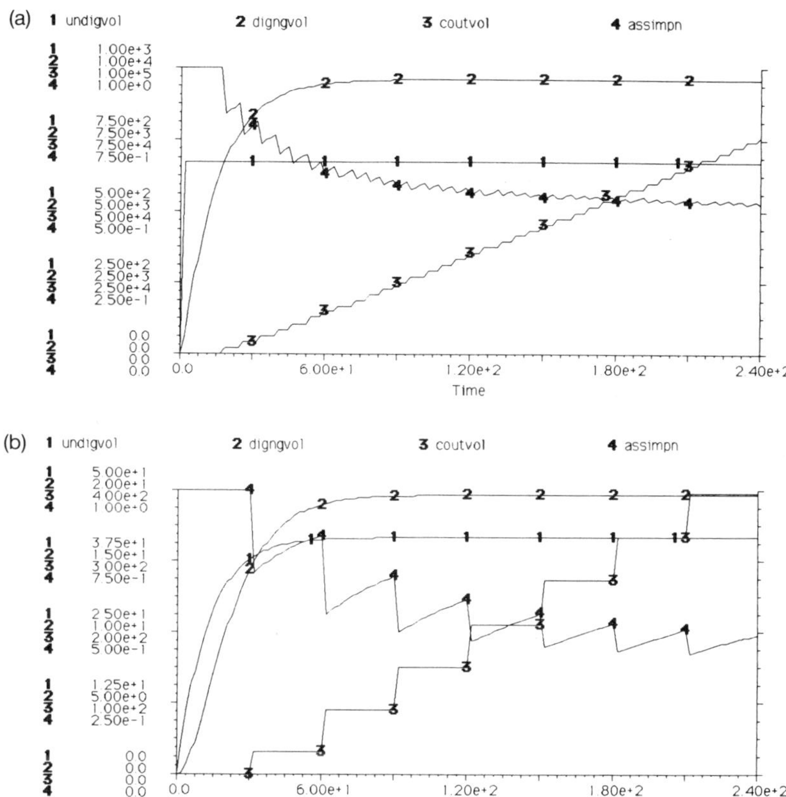

Fig. 4: The graphs trace the volume of phytoplankton cells, both undigested (undigvol) and digesting (digngvol) in the gut of a copepod. Each step in the cumulative output volume line (coutvol) indicates the production of a faecal pellet. The cumulative assimilation efficiency (assimpn) settles down to an asymptotic value after about 240 days. The upper graph (a) represents the case where the copepod diet is of very large phytoplankton cells (.9 g.cell^{-1}) and the lower (b), the case of a very small phytoplankton cell (.015 g.cell^{-1}).

experimental programme. The "hands on" experience of the ecologist has provided insight into the dynamics of copepod defaecation pointing to areas where further research is needed and where collaborations with other experimentalists would be beneficial.

4. Limitations of STELLA

In many ways STELLA could be said to become a victim of its own success in that it rapidly educates its users, who soon need facilities which its very philosophy prevents it from providing. In particular, more complex models demand very large and complicated flow diagrams which can become confusing to interpret and rather tedious to draw, being pedantic by their very nature. Also many real-world problems contain repetitious elements such as a series of interacting regions each of which may be represented by the same equations but at any point in time hold different absolute numerical values. Examples of this are problems such as gaseous diffusion through soil substrates (Radford and Greenwood, 1970), nitrification/denitrification processes down a water column (Blackburn, in press), and salt or pollutant transport through estuarine systems (Radford et al, 1981). Unfortunately the only way to represent these systems using STELLA is by the tedious repetition of the relevant flow diagram components.

These problems are circumvented by other simulation languages available on mainframe computers, such as CSMP (IBM, 1967), which are equation, rather than flow diagram oriented. For the IBM-compatible range of personal computers there is a language named SYSL (System Simulation Language; Estrine, 1985) which claims to be at least 95% equivalent to CSMP but with many extra features which further simplify the activity of simulation modelling. These are illustrated in the examples that follow.

5. Atlantic salmon population dynamics

The life cycle of an Atlantic salmon population has been simulated by Gee and Radford (1982). A system flow diagram was prepared to represent the life cycle from eggs (EGGS) to fry (FRY), to pre-smolt parr (PSP), and to adults of either one (ADS1), two (ADS2) or three (ADS3) years duration before spawning. The total life span of the longest living fish is seven years during which time they potentially interact and compete with members of seven contemporaneous cohorts (COHORT). Thus having modelled one cohort it is necessary to arrange for seven other cohorts to coexist in the model. Using STELLA this could only be achieved by the laborious sevenfold repetition of the whole flow diagram and its associated coding. In CSMP and SYSL this is achieved much more simply by enclosing the computer code for one typical cohort by a MACRO command and then by evoking this seven times in the DYNAMIC section of the program as follows:-

```
MACRO ADS1,ADS2,ADS3,EGGS,FRY,PSP=DEVELP(ASD14Y,ADS25Y,ADS36Y,COHORT)

    << Equations which define one life cycle >>

ENDMAC

DYNAMIC
    ADS10,ADS20,ADS30,EGGS0,FRY0,SPS0=DEVELP(ADS13,ADS22,ADS31,0.)
    ADS11,ADS21,ADS31,EGGS1,FRY1,SPS1=DEVELP(ADS14,ADS23,ADS32,1.)
    ADS12,ADS22,ADS32,EGGS2,FRY2,SPS2=DEVELP(ADS15,ADS24,ADS33,2.)
    ADS13,ADS23,ADS33,EGGS3,FRY3,SPS3=DEVELP(ADS16,ADS25,ADS34,3.)
    ADS14,ADS24,ADS34,EGGS4,FRY4,SPS4=DEVELP(ADS10,ADS26,ADS35,4.)
    ADS15,ADS25,ADS35,EGGS5,FRY5,SPS5=DEVELP(ADS11,ADS20,ADS36,5.)
    ADS16,ADS26,ADS36,EGGS6,FRY6,SPS6=DEVELP(ADS12,ADS21,ADS30,6.)
```

This formulation allows, for example, the number of fry summed over all seven relevant cohorts to determine the density dependent mortality which will occur in the river.

6. General ecosystem models

Even the simplest of ecosystem models are composed of a very large number of components. For example the General Ecosystem Model of the Bristol Channel and Severn Estuary (Radford et al, 1988), known as GEMBASE, contains about 20 state variables and 150 rate variables. A STELLA flow diagram would become a convoluted tangle of lines and would be very difficult to design. An abbreviated diagram (Fig. 5) was developed which grouped similar organisms, so greatly economising on the number of rectangles and material flow lines (carbon and nutrient) required. No valves were drawn to represent rates since the existence of a line automatically implies the existence of a group of rate variables. All information flow lines are omitted so preventing a tangled mass of some one thousand or so intersecting curves.

Even with all of these graphical economies it is still possible to refer to each individual rate such as the grazing of phytoplankton (P1) by omnivorous zooplankton (Z2) viz P1Z2, with the added advantage of also being able to refer to a group of processes such as the carnivores (CI) intake of zoobenthos (BJ) viz BJCI. Each process must still be defined individually as a function of state variables, parameters and exogenous variables (such as temperature). These can be coded in SYSL in a systematic way, for example by numbering the groups, to prevent omissions.

A further complication of the GEMBASE model is that the ecological relationships implied by the flow diagram need to be applied to seven interacting geographical sub-regions of the estuary. Although the number of equations required were too many to include in a MACRO sub-program

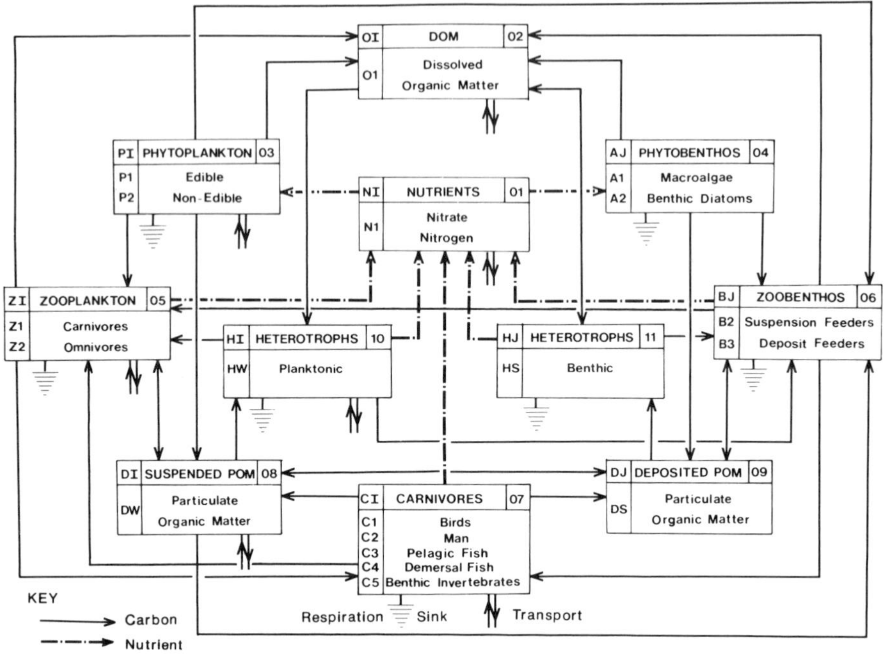

Fig. 5: A flow diagram of the GEMBASE model. Rectangles represent groups of associated state variable and arrowed lines represent groups of processes by which carbon and nutrients are circulated through the ecosystem. Carbon sinks are represented by "earth" symbols and advection and dispersion between geographical regions by a double arrow symbol.

within SYSL it is possible to define them as a FORTRAN subroutine named REGION. This can be achieved because SYSL is a superset of FORTRAN thus allowing expansion of the simulation program to include these features. Care must be taken to write the equations in the correct computational sequence because SYSL will not sort FORTRAN sub-programs for the user. Subroutines may be called from SYSL programs either by using the standard FORTRAN method:-

CALL REGION (input and output parameter list)

or by using a preferred SYSL method:-

output parameter list = REGION (input parameter list).

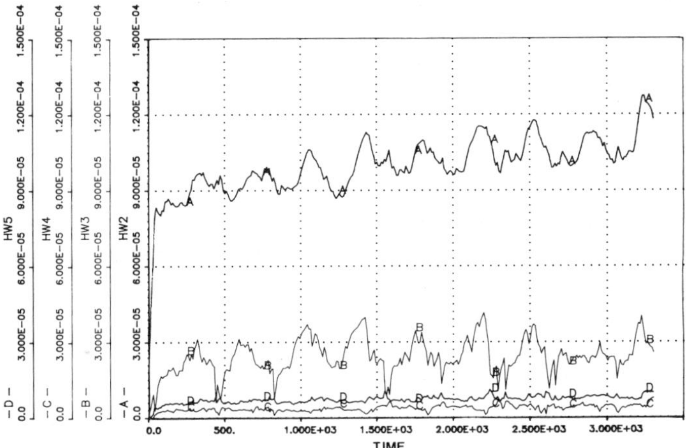

Fig. 6: Graphical output from the GEMBASE model generated from the SYSL package. The graphs show the simulated development of planktonic heterotrophs for four geographical regions of the Severn Estuary over a period of about nine years (3300 days). The original plots were produced on a flat bed plotter in colour.

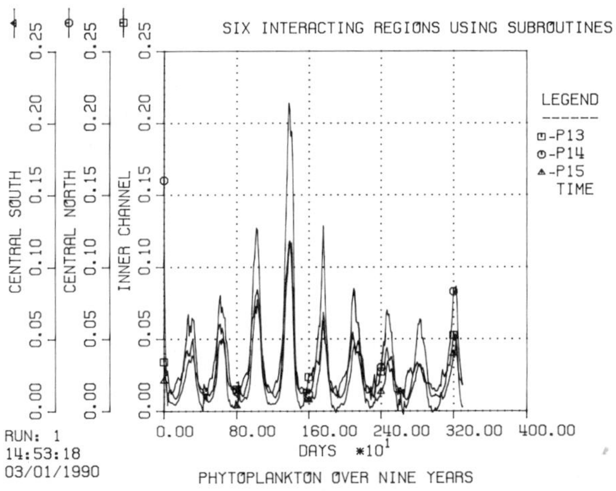

Fig. 7: Graphical output from the GEMBASE model generated from the SYSL post execution plotting package. The graphs show the simulated development of phytoplankton for three geographical regions of the Severn Estuary over a period of about nine years (3300 days). The original plots, produced on a flat bed plotter, are in colour so making the lines easily differentiable.

The SYSL method emphasises the differences between input and output variables and enables subroutines to be sorted into their correct computational sequence within the remainder of the program.

The possibility to include FORTRAN subroutines within SYSL programs was also used to set the estuarine ecology within its correct hydrodynamical context. The transport of all dissolved and particulate matter between geographical regions, shown in Fig. 5 as a double arrow symbol, was computed in a FORTRAN subroutine, HYDROBASE. For input the subroutine took a matrix of transportable state variables by region, and as output it returned the computed transport terms. This facility of SYSL enables, for example, the same ecological relationships to be tested under different hydrodynamic regimes.

Like STELLA the language SYSL provides automatic graphical outputs. These graphs are not produced in real time as the simulation proceeds but are given at the end of any simulation run. They may be produced in various forms including as line printer plots, as dot-matrix printer curves or as flat-bed plotter graphs. An example of the latter, is given in Fig. 6 where graphs are shown of the simulated concentration of planktonic heterotrophs (HW) for four regions of the Bristol Channel (2,-5) over a nine year period (3300 days). The originals of these graphs are in colour so giving a clearer distinction of the correspondence between the scales and the lines. An alternative form of plot which is provided by SYSL is shown in Fig. 7. Here another possibility has been exploited by adding regional names to the vertical axes and a sub-title to identify the state variable plotted (Phytoplankton over nine years) under the horizontal axis. Again colour greatly enhances the separation of the interwoven lines.

7. Conclusions

The simulation language, STELLA, provides the uninitiated with a stimulating introduction into the techniques of simulation modelling. Its insistence upon a consistent and fully defined flow diagram mitigates against users failing to produce logical representation of their systems. It educates users in simple steps providing rewards in terms of instantaneous feedback indicating success or, if not, reason for failure. The language is powerful enough to produce simulation models which can be most useful in understanding ecological systems and in designing critical experiments for future field or laboratory studies.

Certain common environmental problems require, for their solution, features not available in STELLA but the educative value of having used this language prepares novice modellers to graduate to more general purpose packages such as SYSL.

The great value of SYSL is that it adheres to the same basic modelling philosophy embodied in STELLA but allows for expansion into all the computing generalities provided by FORTRAN. Its provision of standard user friendly graphical output greatly facilitates the interpretation of simulation results. This in turn lends to more relevant experimentation, better modelling and greater understanding of one's sphere of interest.

The exercise of simulation modelling in the environmental sciences leads to a more sober estimate of how well we comprehend our sub-system and a clearer idea as to what needs to be done to improve our understanding still further. As a bonus, simulation models may be used to predict the probable outcome of environmental change (Radford, 1990).

Very large ecosystem models which once could only be run on large mainframe computers may now be run on desk top personal computers using a range of new simulation software such as STELLA and SYSL.

8. Equipment for simulation

The simulation language STELLA (Richmond et al, 1987) was implemented on the following Apple desk-top computer system: Macintosh SE HD40 (2mb RAM) with 10" screen; 800kb 3.5" floppy disk drive; SCSI 40mb hard (fixed) disk; Image Writer LQ dot matrix printer (colour ribbon).

The simulation language SYSL (Estrine 1985) was implemented on the following IBM computer system: PS/2 model 70-121 (80386, 20MHz, 2mb RAM); 1.44mb 3.5" floppy disk drive; 120mb hard (fixed) disk; VGA graphics; 80387 20MHz maths co-processor (essential); Epson FX1050 dot matrix printer; Hewlett Packard Color-Pro plotter HP7440A.

9. References

Blackburn, T.H. 1990. Denitrification model for marine sediment. In: Denitrification in Soil and Sediment, J. Sorensen and N.P. Revsbech, editors. Plenum Press, London. (in press).

Estrine, E. 1985. System simulation Language: Users Guide, SYSL/M, Version 1.0. Publ. E^2 Consulting Poway, California.

Forrester, J.W. 1961. Industrial Dynamics. Cambridge Massachusetts, USA, MIT Press.

Gee, A.S. and Radford, P.J. 1982. The regulation of stock characteristics in a simulated Atlantic Salmon population. Fisheries Research, 1:105-116.

IBM, Application Program 1967. System/360. Continuous System Modelling Program (360A-CX-16X). Users Manual IBM.Pugh, A.L. 1963. Dynamo Users manual. Cambridge Massachusetts, USA, MIT Press.

Radford, P.J. and Greenwood, D.J. 1970. The Simulation of Gaseous Diffusion in Soils, J. Soil Science, 21(2):304-313.

Radford, P.J., Uncles, R.J. and Morris, A.W. 1981. Simulating the impact of technological changes on dissolved cadmium distribution in the Severn Estuary. Water Research, 15:1045-1052.

Radford, P.J., Burkill, P.H., Collins, N.R., and Williams, R. 1988. the validation and scientific assessment of an ecosystem model of the Bristol Channel. In: State-of-the-Art in Ecological Modelling, A. Marani, editor. Proceedings of the ISEM conference on ecological modelling, Italy, 1987. Elsevier, p.427-442.

Radford, P.J. 1990. Pre and Post barrage scenarios of the relative productivity of benthic and pelagic sub-systems. In: Evolution and change in the Bristol Channel and Severn Estuary, J. Crothers, editor. The Biological Journal of the Linnean Society. (in press).

Richmond, B., Peterson, S and Vesuso, P. 1987. An Academic Users Guide to STELLA, Structured Thinking Experimental Learning Laboratory with Animation, N. Naville, editor. High performance Systems Inc., Lyme, New Hampshire.

Use of an inductive modelling procedure based on Bayes Theorem for analysis of pattern in spatial data

R. Aspinall

Macaulay Land Use Research Institute

1. Introduction

This chapter describes an inductive modelling procedure integrated with a Geographic Information System for analysis of pattern within spatial data. The aim of the modelling is to predict the distribution or pattern within one dataset by combining a number of other datasets. Dataset combination is carried out using Bayes Theorem. Inputs to the theorem are obtained through an inductive learning process in which attributes of the dataset to be modelled are compared with attributes of a variety of predictor datasets. The statistical significance of these inputs is calculated as part of the inductive learning process.

Use of the modelling procedure is illustrated through analysis of the habitat relationships of Red deer in Grampian Region, North East Scotland to predict winter habitat suitability. The distribution of Red deer in Deer Management Group areas in Gordon and in Kincardine and Deeside Districts is used to develop the model; this is tested against Red deer distribution in Moray District. The habitat datasets used for constructing the model are accumulated frost and altitude, obtained from maps, and land cover, derived from satellite imagery.

An indication of errors associated with model derivation and output is also provided. This is an integral part of the analysis and provides useful support to interpretation of the results of modelling. Potential applications of the model results both as a prediction of distribution and as a map describing habitat suitability are discussed.

2. Background

Growing concern over conservation interests outside specially designated areas such as National Nature Reserves and SSSIs means that information on the distribution and abundance of wildlife is increasingly necessary for integration with other land uses through planning and management.

Frequently such information is not available, partly because the capacity for data collection is restricted by existing field survey methods and partly because of problems associated with objectively assessing quality of habitat for wildlife. This paper presents a method for predicting distribution of wildlife from analysis of habitat information.

The problem addressed is to predict the likely distribution of a species through mapping habitat suitability from an assessment of environmental conditions in areas where the species is known to be present. Part of the solution to this problem lies in application of the method described within the framework provided by Geographic Information System (GIS) methodologies. GIS has been used to monitor and assess wildlife habitat (see for example, Stenback et al, 1987; Scepan et al, 1987) although typically this has involved application of habitat quality criteria ("rules") through GIS overlay facilities to identify co-occurrence of the specified conditions (Agee et al, 1989). Although overlay provides this ability, there is a more general need for incorporation of improved methods of spatial analysis within GIS (e.g. Abler, 1987; Department of the Environment, 1987; Openshaw, 1987; National Centre for Geographic Information and Analysis, 1989). Improved analytical functionality would derive from integration of standard statistical analyses for both data analysis and analysis of error trains consequent on overlaying different data sets. This would also create the potential for inductive learning procedures to be integrated with GIS. Such procedures can be used to generate testable hypotheses concerning relationships between spatial datasets (Openshaw, 1987). Recently Walker and Moore (1988) have presented an inductive method for analysis and modelling of the distribution of kangaroo populations in Australia. The inductive learning procedure they employed used classification and regression of spatial data describing climatic conditions and distribution of kangaroos, to develop decision trees.

The method presented here has been implemented specifically as an inductive spatial modelling procedure operating within GIS and introduces a significant learning capacity to GIS function. The datasets for analysis are habitat information and distribution of Red deer (*Cervus elaphus*) in Grampian Region, North-east Scotland. All data are integrated within a Geographic Information System based on a PRIME 250 series minicomputer and a GEMS Image Processing System. Modelling software has been developed in house.

3. Data analysis: inductive spatial modelling

The inductive spatial modelling procedure is based on application of Bayes' theorem; this approach has been applied on a site specific basis as a method for wildlife habitat evaluation (Grubb, 1988). The procedure has been integrated with a Geographic Information System (Aspinall, 1990). Part of the appeal of the method is that it emulates the way in which a wildlife expert might assess habitat suitability for a species (Grubb, 1988).

Distribution is predicted through a process which derives a probability of occurrence. Bayes' theorem is applied to information on the frequencies with which specific attributes of habitat datasets occur in each of presence and absence categories of the wildlife distribution. The modelling process can be described as a series of steps, each of which matches a procedure in its execution within the implementation in the GIS. Description of the programming procedure also highlights additional facilities not used in this analysis of Red deer and their habitat but which are available for analysis of other species.

3.1 Procedure

(1) Two classes are identified in a distribution to be modelled. These may be either (i) presence and absence of a characteristic or (ii) presence and a large random sample.

(2) The frequency of association between attributes of habitat datasets are calculated for each of the two classes identified in (1). These frequencies form conditional probabilities for discriminating between presence (P_o) and absence (A_o) of the distribution to be modelled. The frequencies are calculated against cutoff values for the habitat attributes, optimum values being found by iteration with the criterion of maximizing the statistical significance of difference between P_o and A_o (see 3). An alternative approach maximizes the absolute difference between conditional probability scores (P_o-A_o) (see Grubb, 1988).

(3) The significance of each habitat attribute for discriminating between classes is tested through analysis of frequencies using chi-square. The chi-square score can also be used in conjunction with the difference between the conditional probability for presence (P_o) and absence (A_o) to decide which attributes should be included in the Bayes' model. Attributes which do not have a significant chi-square score are either excluded or allocated a conditional probability equal to 1.0 for both presence and absence. Appropriate predictor datasets and attributes are then selected. In the standard analysis characteristics of sites with known presence are compared with characteristics of sites with known absence to derive conditional probabilities. For wildlife distribution, and other geographic data, it is often the case that while presence is recorded, absence is not. This programme

option allows the significance of conditional probabilities to be tested for the presence/random case (e.g. Openshaw et al, 1987) and permits the model to be applied even if only presence is known.

(4) Prior probabilities for each category of the model dataset are selected. These are estimates of the probabilities of presence (P_a below) and absence (P_c below) which are to be modified by the calculated conditional probabilities using Bayes' theorem. Two values are required, it being possible to set both to 0.5 if an initial assumption of equal probability for presence and absence is chosen. Other values may be selected; for example, prior probabilities may be set according to the proportion of all sites in which the species is present or in order to model a selected population size.

(5) Conditional probabilities are combined using Bayes' theorem to provide a single probability value; this is the probability that the site will be occupied given the particular combination of habitat attributes at the site; it also provides a measure of the quality of the site for the species being considered. The equation for these calculations is

$$P_p = \frac{P_a * P_b}{(P_a * P_b) + (P_c * P_d)} \tag{3.1}$$

where P_a = prior probability for presence
P_c = prior probability for absence
P_b = calculated by multiplying conditional probabilities for presence according to whether the habitat attribute is greater (P_o) or less ($1-P_o$) than the critical value
P_d = calculated by multiplying conditional probabilities for absence according to whether the habitat attribute is greater (A_o) or less ($1-A_o$) than the critical value
and P_p = probability of presence

4. Distribution and habitat suitability for the winter range of Red deer in Grampian Region

4.1 Data

The distribution of Red deer in Deer Management Group areas within Grampian Region is censused by the Red Deer Commission. Counts are made during the late winter/early spring when the ground is snow covered, number being recorded for 1km squares of the National Grid (Fig. 1). These data give an indication of the winter range of Red deer although a number of management factors influence distribution, most notably the presence of deer fences which restrict movement to lower altitudes (Red

Deer Commission, 1981) and feeding stations where densities tend to be artificially high. Locally, the distribution of deer is a response to social behaviour (Clutton-Brock et al, 1982), the species composition and nutritional quality of herbage, the need for shelter, and disturbance impact (Mitchell et al, 1977). At a regional scale the major factors controlling distribution are climatic and general habitat features (Mitchell et al, 1977). This case study concentrates on a regional scale of analysis which is most appropriate to available environmental data.

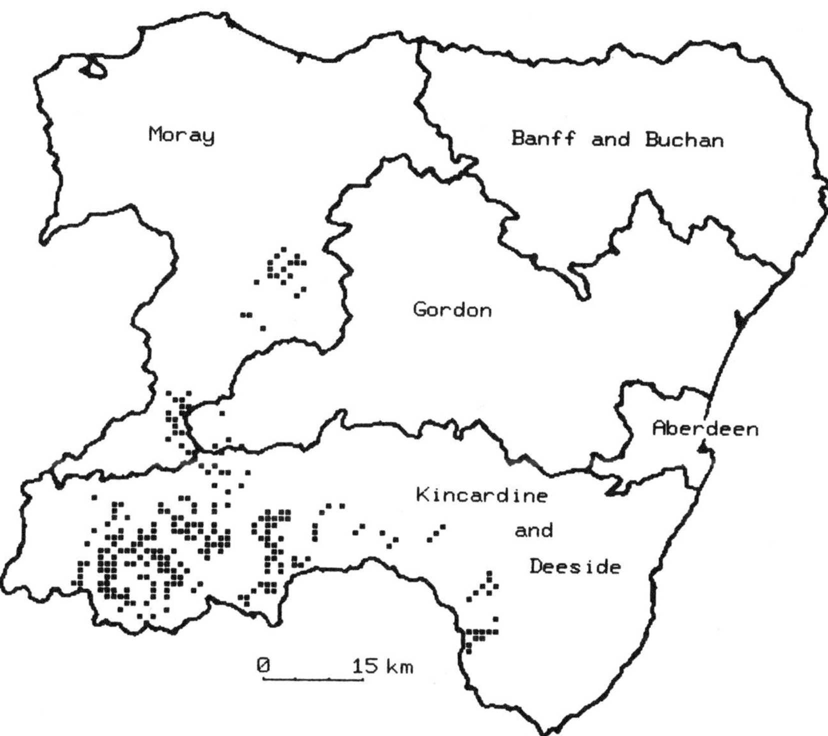

Fig. 1 One kilometre squares in Deer Management Group Areas of Grampian Region in which Red deer were recorded as present during winter 1987. The District Council Areas are also shown.

The presence of deer in 1 km National Grid squares was incorporated into the GIS database. This contains map information for Grampian Region, maps being vector digitised and converted to raster format with 50m pixel minimum resolution (Wright and Morrice, 1988). Data refer to a variety of climatic, edaphic, topographic, biotic, conservation, landscape and administrative information in addition to multi-temporal satellite imagery from both LANDSAT Multi-spectral Scanner (MSS) and Thematic Mapper. Some 50 different datasets are currently available; much of the basic resource information was derived from detailed and long-term field survey carried out since 1939 by the Macaulay Institute for Soil Research.

Three geographic datasets are used in combination with Red deer distribution, namely altitude, land cover, and accumulated frost. Altitude is recorded in nine classes, the minimum contour interval being 60 metres with a larger interval (120m) at higher altitudes (above 300m). Since the pattern of distribution of soils and climatic factors are highly correlated with altitude in Grampian Region (Walker et al, 1982), altitude is used rather than other more direct measures of environmental information. Land cover, which partially reflects land use, is recorded in 9 classes, these being obtained from classification of LANDSAT MSS data (Hubbard and Wright, 1982). Red deer feed on heather and grass but have a preference for *Calluna* vegetation containing grasses rather than for grass-dominated *Agrostis-Festuca* swards (Mitchell et al, 1977). Broad heath, upland scrub and grass-dominated cover types are readily distinguished using LANDSAT MSS data and overall accuracy of the image classification used is high (87.5%). The third dataset, accumulated frost, defined as the annual integrated deficiency of temperature with reference to $0^{\circ}C$ (Shellard, 1959), is mapped in five classes at a scale of 1:625000 (Birse and Robertson, 1970). This dataset gives an indication of winter severity, Red deer distribution being influenced by snow lie and exposure.

4.2 Model coefficients

Conditional probabilities are calculated for squares where Red deer are reported as present compared with a large randomly distributed set of squares from within Kincardine and Deeside and Gordon Districts (Table 1). Conditional probabilities are tested with chi-square to assess their significance for discriminating between habitat conditions in areas where deer are present compared with the random sample of habitat conditions in the same Districts. In this case, the test identifies values for Red deer presence (P_o) which differ significantly from random (A_o) (Table 1). Conditional probabilities are used as calculated for habitat categories in each dataset where chi-square shows significant difference between the conditional probability for presence and random; where the difference is not significant conditional probabilities for both presence and random are set equal to one.

Table 1. Conditional probabilities for Red deer distribution and habitat Suitability Model - calculated from Gordon and Kincardine and Deeside Districts.

(i) Accumulated Frost (day °C below 0°C)	P_0	A_0	Chi-square	p
0- 20	0.00	0.00	n.a.	
21- 50	0.00	0.12	18.53	***
51-110	0.00	0.43	102.06	***
111-230	0.61	0.37	32.97	***
231-470	0.51	0.12	192.03	***
over 470	0.00	0.02	1.24	n.s.
(ii) Altitude				
0- 60 m	0.00	0.18	30.41	***
61-120 m	0.00	0.29	54.72	***
121-180 m	0.01	0.24	40.42	***
181-240 m	0.06	0.16	10.33	**
241-300 m	0.12	0.13	0.04	n.s.
301-420 m	0.40	0.15	61.86	***
421-600 m	0.58	0.13	223.89	***
601-900 m	0.36	0.08	147.42	***
over 900 m	0.00	0.01	1.12	n.s.
(iii) Land Cover				
Bare Rock	0.05	0.02	0.74	n.s.
Woodland	0.14	0.15	0.01	n.s.
Heath	0.66	0.21	153.54	***
Scrub	0.68	0.36	57.17	***
Built-up	0.00	0.02	1.94	n.s.
Crops	0.06	0.64	200.27	***
Grassland	0.40	0.45	1.29	n.s.
Unclassified areas	0.10	0.02	24.53	***

probability (p):
n.s. not significant ; ** 0.99 ; *** 0.999

4.3 Errors

Errors can arise from four sources and are associated with input data, model coefficients, arithmetic combination of datasets, and output generalisation from 50 metre pixels (GIS resolution) to 1 km square (habitat suitability mapping resolution).

Each input dataset contains inherent errors deriving from cartographic representation at the map scale used (the degree ofgeneralisation, omission and exaggeration required), the classes represented in the map, and error in data recording. The importance of these errors depends on the nature of the phenomenon being mapped and also the type of analysis (Burrough, 1986). In particular, categorical mapping of continuous data leads to error both close to interpolated category boundaries, where mapping accuracy is important, and as distance from the category boundary increases, where the difference between the categorical and continuous value is greater. This is related to error from model coefficients.

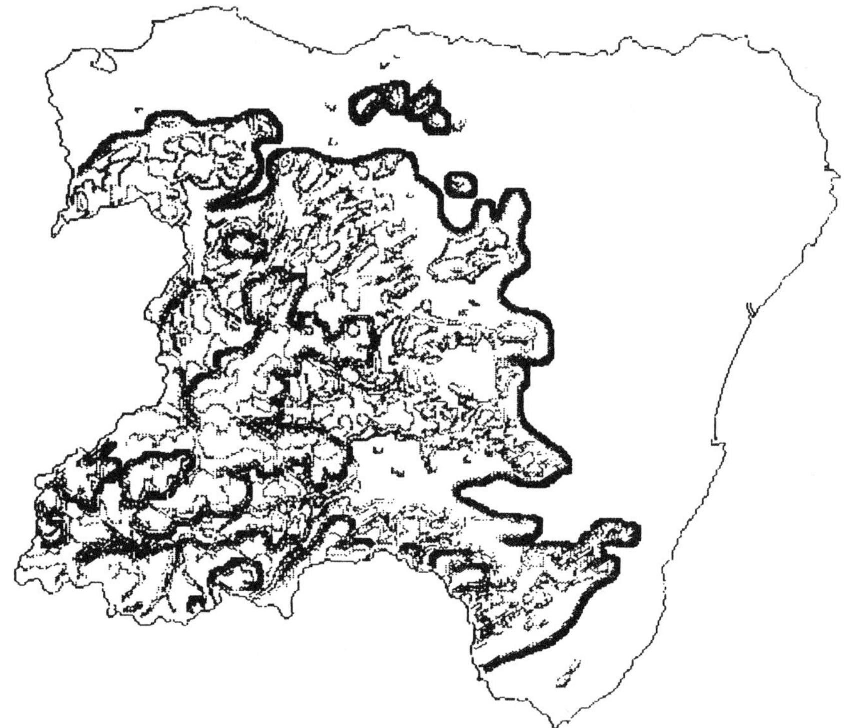

Fig. 2 Areas of potential greatest error (darker shades) on input data for altitude and accumulated frost.

Model coefficients have two sources of error related to (i) the accuracy of the estimate of the model coefficients and (ii) the spatial location of the unit being analysed in relation to the cartographic representation of map features (categorical representation of continuous data). Since categorical data are used for the analysis in this example, errors associated with model coefficients are considered greatest at the contour between adjacent map categories, both in relation to location accuracy and correct allocation of category.

The location of errors arising from input data and model coefficients can be indicated with a map showing distance from input data category boundaries (Fig. 2). The closer to the category boundary, the greater the error is likely to be.

Combining values with arithmetic functions compounds errors (Burroughs, 1986) and a procedure for quantifying this error when using model coefficients derived from either regression equations or kriging has been described by Heuvelink et al (1989). No direct estimate of this error is made in the present study although the location of areas where the error is potentially greatest is indicated by the map of error associated with input data (Fig. 2) and the analysis of error associated with changing resolution from 50 metre pixels to 1 km grid squares of the National Grid.

4.4 Mapping probabilities and errors

The probability of Red deer presence in each 1 km square is calculated as the mean of the probabilities for all pixels in the square (Fig. 3). This calculation, although itself introducing some error through map generalisation, introduces the possibility of indicating the potential errors derived from other sources.

The calculated probabilities vary within each 1 km square. Largely this is the result of variation in modelling results due to land cover variation but is also due to errors associated with the sources described. The within square variation can be used to provide a measure of confidence in calculated probabilities.

Each 1 km square contains 400 50m pixels; the standard deviation of probabilities for all pixels in the 1 km square is used to indicate the "uncertainty" associated with the probability estimate consequent on the errors described (Fig. 4).

This combination of calculated probability and "uncertainty" map outputs allows evaluation of the model procedure for predicting distribution and interpretation of the result as a habitat suitability map. Squares with extensive areas of suitable habitat have a high overall probability and also a high degree of confidence (low "uncertainty"). Equivalently, squares with extensive areas of unsuitable habitat have a low overall probability together with a high degree of confidence in the estimate (low "uncertainty"). Other squares have a combination of probability and "uncertainty" depending both on internal variability of habitat within squares and the accuracy and location of category boundaries in each of the original input datasets. This is one of the more important aspects of the "uncertainty" measure. It allows areas where input data limitations are important to modelling to be identified. Both maps are necessary therefore for full interpretation of the modelling result; the map of probabilities used alone would give a false impression of precision. Taken with the map of "uncertainty" and with the map showing areas of greatest potential error associated with input data (Fig. 2) the advantages and limitations of the probability map can be assessed.

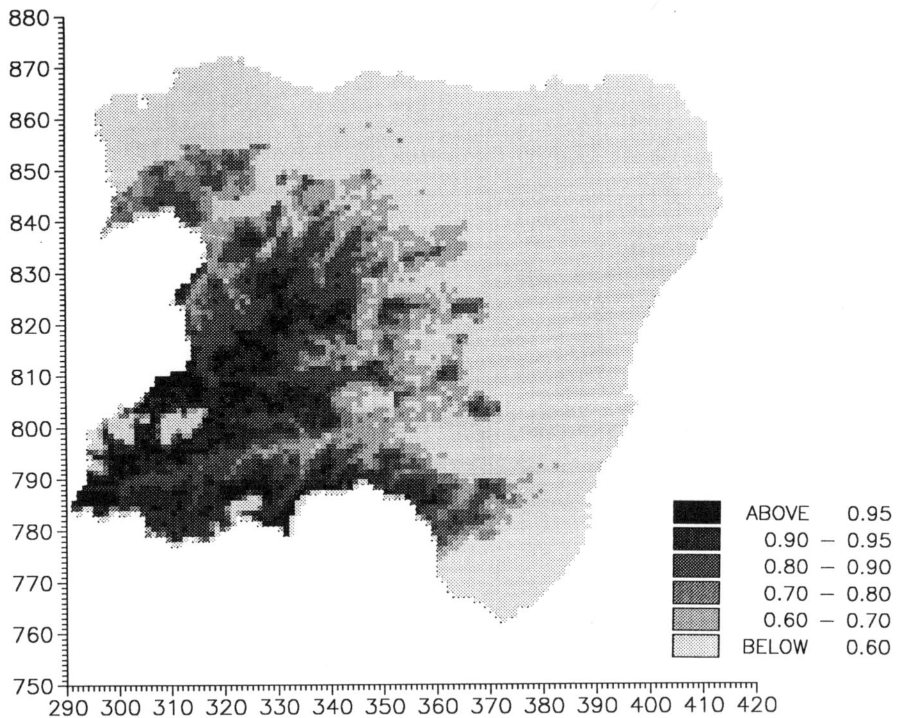

Fig. 3 Model output: the probability of Red deer occurrence in Grampian Region (in 1 kilometre grid squares). The map may also be considered to represent habitat suitability for Red deer during winter.

4.5 Testing the model output

The probabilities and uncertainty scores for 1 km squares in Moray District from which Red deer are recorded are shown in Table 2. In all but two squares the probability of occurrence is in excess of 0.5 and in the majority of squares the standard deviation of values within the square is low indicating the probability to be consistent across the square. Sixteen of the 37 squares have a probability of Red deer occurrence in excess of 0.76 and a standard deviation of less than 0.10. These figures suggest the model to be successful in predicting the distribution of Red deer in the Deer Management Group Area in Moray District.

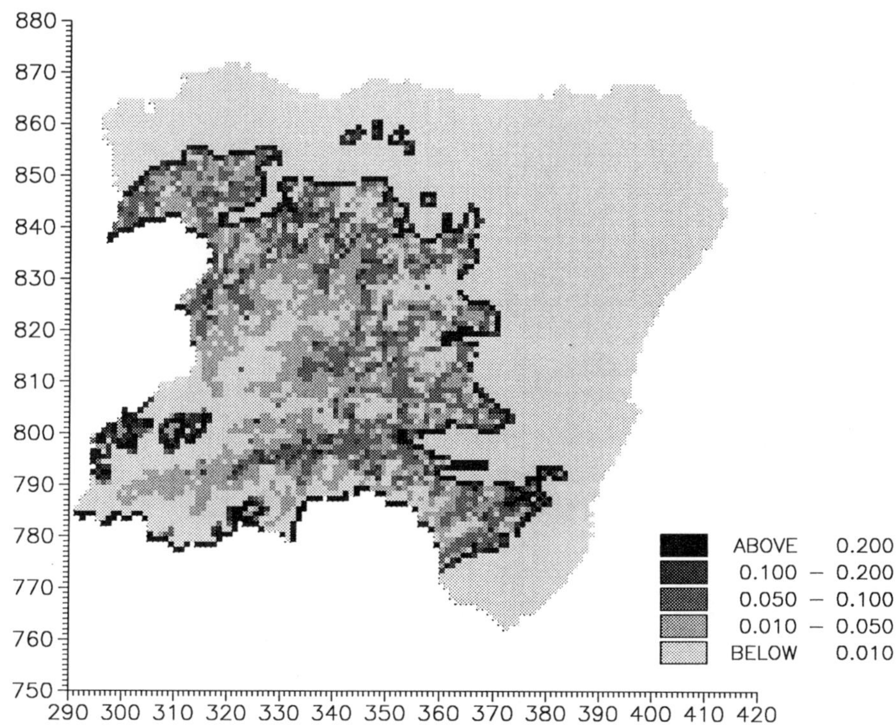

Fig. 4 Model output: a measure of the uncertainty associated with the probability of occurrence of Red deer in Grampian Region (Fig. 3). Values are the standard deviation of individual model results for pixels in each 1 kilometre grid square.

5. Applications of probability map of distribution

The overall success of the inductive spatial modelling procedure for modelling Red deer distribution (Table 2) is only an indication of the value of applying the method to habitat evaluation. For many applications it may be the quantitative map representation of habitat suitability (rather than distribution) which is relevant and this may be of value independent of any measure of success in distribution prediction. For example, an area may be classed as "suitable" even though the species is not recorded; equally an area may be classed as "unsuitable" although the species is present. Both of these forms of "misclassified" site may be of special interest in relation to both limitations of wildlife survey in extensive areas (where search is often inefficient, time-consuming and expensive) and for understanding of species ecology. Differences between prediction and recorded information also merit further attention in order to improve modelling.

Table 2: *Bayesian Model Habitat Suitability Probabilities and Uncertainty measure for sample of 1 kilometre squares with Red deer present in Moray District. Values are number of 1 kilometre squares (n=37).*

	Probability of Occurrence							
Uncertainty	0.00-0.12	0.13-0.25	0.26-0.37	0.38-0.50	0.51-0.62	0.63-0.75	0.76-0.87	0.88-0.99
High								
0.41-0.50								1
0.31-0.40						1		
0.21-0.30								5
0.11-0.20				2	1	2	3	2
0.00-0.10						4	7	9
Low								

Habitat suitability maps however, may find application in evaluating a variety of land use change or other scenarios or as an input to resource management programmes. For example, mapped habitat suitability allows prediction of areas which may become occupied or unoccupied if distribution of a species expands or contracts. Although such changes will have biological or environmental causes and require additional information for accurate prediction, a map of habitat suitability represented as a range of probabilities of occurrence offers an objective framework for evaluating scenario outcomes. Similarly, an objective assessment of habitat suitability provides a rational basis for management decisions incorporating impact on species habitat.

5.1 Inductive spatial modelling and wildlife habitat evaluation

Inductive spatial modelling based on Bayes' theorem, as described for Red deer in Grampian Region, is one of a number of methods which can be used to identify suitability of a site on the basis of its habitat characteristics. However, the Bayesian approach, as implemented here, does have a number of specific advantages, particularly when operated within the context of a Geographic Information System. The advantages are related both to the process by which the probability model is developed, and to the management tool which it provides. These can be summarised:-

(a) inductive spatial modelling provides a search method whereby hypotheses relating spatial aspects of wildlife population (distribution) to habitat characteristics can be generated;

(b) the inputs to the model and the hypotheses generated are derived through automated analysis of relationships between habitat characteristics and wildlife data;

(c) the method, as implemented, identifies threshold (critical) values where conditional probabilities for presence and absence associated with specific habitat characteristics can alter habitat suitability. This offers potential for changing input data to assess management impact on habitat suitability;

(d) the approach incorporates a procedure for assessing statistical significance of relationships between datasets;

(e) the combined use of Bayes Theorem and measures of statistical significance of model inputs allows a measure of uncertainty to be associated with the model output.

These advantages are supplemented by benefits which have more general application within Geographic Information Systems. The approach is relevant to a wide variety of spatial data and allows models to be developed which make full use of available data. With increasing accessibility to spatial environmental and wildlife data in digital form and development of both Geographic Information Systems and expert systems this method of modelling offers great potential in ecology and environmental management.

6. Acknowledgement

I should like to thank Mrs Jane Morrice who digitised much of the information in the Geographic Information System for Grampian Region and A.R. Sibbald, Dr. R.V. Birnie and Prof. T.J. Maxwell for comments on a draft of this paper.

7. References

Abler, R.F. (1987). The National Science Foundation National Center for Geographic Information and Analysis. International Journal of Geographical Information Systems, 1:303-326.

Agee, J.K., Stitt, S.C.F., Nyquist, M. and Root, R. (1989). A geographic analysis of historical Grizzly Bear sightings in the North Cascades. Photogrammetric Engineering and Remote Sensing, 55:1637-1642.

Aspinall, R.J. (1990). Inductive spatial modelling as a tool for wildlife habitat evaluation. Proceedings International Union of Forestry Research Organisations, Montreal 1990.

Birse, E.L. and Robertson, L. (1970). Assessment of Climatic Conditions in Scotland. 2. Based on exposure and Accumulated Frost. Map and Explanatory Pamphlet. The Macaulay Institute for Soil Research, Aberdeen.

Burrough, P.A. (1986). Principles of Geographical Information Systems for Land Resources Assessments. Clarendon, Oxford: 194pp.

Clutton-Brock, T.H., Guiness, F.E. and Albon, S.D. (1982). Red deer: behaviour and ecology of two sexes. Edinburgh University Press, Edinburgh.

Department of the Environment (1987). Handling Geographic Information. Report of the Committee of Enquiry chaired by Lord Chorley. HMSO, London.

Grubb, T.G. (1988). Pattern recognition - a simple model for evaluating wildlife habitat. USDA Forest Service Research Note RM-487. 5pp.

Heuvelink, G.B.M., Burrough, P.A. and Stein, A. (1989). Propagation of errors in spatial modelling with GIS. International Journal of Geographical Information Systems, 3:303-322.

Hubbard, N.K. and Wright, R. (1982). A semi-automated approach to land cover classification of Scotland from LANDSAT. In: Remote Sensing and the Atmosphere. Proceedings of the Eighth Annual Conference of the Remote Sensing Society, Liverpool, pp. 212-221.

Mitchell, B., Staines, B. and Welch, D. (1977). Ecology of Red Deer. ITE, Huntingdon. 74pp.

National Center for Geographic Information and Analysis (1989). The research plan of the National Center for Geographic Information and Analysis. International Journal of Geographical Information Systems, 3:117-136.

Openshaw, S. (1987). An automated geographical analysis system. Environment and Planning, A, 18:143.

Openshaw, S., Charlton, M., Wymer, C. and Craft, A. (1987). A Mark 1 Geographical Analysis Machine for the automated analysis of point data sets. International Journal of Geographical Information Systems, 1:335-358.

Red Deer Commission (1981). Red Deer Management. HMSO, Edinburgh: 100pp.

Scepan, J., Davis, F. and Blum, L.L. (1987). A Geographic Information System for managing California Condor habitat. GIS 87 Second Annual International Conference on GIS, San Francisco, California, Oct 26-30, 1987.

Shellard, H.C. (1959). Averages of accumulated temperature and standard deviation of monthly mean temperature over Britain, 1921-50. Prof. Notes Meteorological Office, London. No. 125.

Stenback, J. M., Travlos, C.B., Barrett, R.H., Congalton, R.G. (1987). Application of remotely sensed digital data and a GIS in evaluating deer habitat suitability of the Tehema Deer winter range. GIS 87 Second Annual International Conference on GIS, San Francisco, California, October 26-30, 1987.

Walker, A.D., Campbell, C.G.B., Heslop, R.E.F., Gauld, J.H., Laing, D., Shipley, B.M. and Wright, G.G. (1982). Eastern Scotland: Soils and Land Capability for Agriculture. Macaulay Institute for Soil Research, Aberdeen. 206pp.

Walker, P.A. and Moore, D.M. (1988). SIMPLE. An inductive modelling and mapping tool for spatially-oriented data. International Journal of Geographical Information Systems, 2:347-363.

Wright, G.G. and Morrice, J.G. (1988). Potato crop distribution and subdivision on soil type and potential water deficit: an integration of satellite imagery and environmental spatial database. International Journal of Remote Sensing, 9:683-699.

Computer modelling of water quality in coastal and inland ecosystems

D.E. Reeve, S.J. Phillips, C.R. Hoggart,
J. Osment

Sir William Halcrow & Partners Limited

1. Introduction

Waterbodies, especially those to which man has added excess nutrients, need careful management to achieve acceptable water quality conditions. Often, most notably when the body of water is enclosed or partially enclosed, high concentration of nutrients lead to an over-abundance of algae or macrophytes, a build-up of sediments and, in severe cases, the onset of anaerobic conditions following the death and decay of the particular plant species. The use of mathematical models capable of predicting the outcome of various operational strategies can help in the restoration of such waterbodies. Restoration may take many years and so the model chosen must be capable of long-term quality predictions in reasonable computer run times. In this case accurate spatial resolution is not normally required as the response of the waterbody as a whole is being sought.

However, mathematical models may also be used to describe water quality changes which occur over much shorter time scales, such as the decay of bacteria discharged from a sewage outfall or the effect of a shock load of organic matter from a pollution incident. In this case the modelling needs are very different, the requirement being for a two-dimensional plan model which can make accurate spatial predictions of the spread and decay of pollutants over periods of hours or days.

Each of the above types of mathematical model are described below by reference to two case studies. The Taw-Torridge Management Plan demonstrates the application of DAWN - a two-dimensional hydrodynamic and water quality model, which assessed various options for sewage treatments in a coastal area of Devon, UK. The Tunis North Lake Study describes a further application of DAWN and shows how it may be coupled to ZEBRA, a model capable of simulating the ecological interactions within, on this occasion, a nutrient-rich coastal lagoon. In this case both long and short term water quality simulations were required, with the hydrodynamic output from DAWN being used to provide input to ZEBRA.

2. DAWN

DAWN (Dispersion Advection With Nesting) is a two-dimensional computer program designed to model both hydrodynamics, and up to ten water quality determinands. Various interdependent water quality determinands such as BOD (biochemical oxygen demand), nitrogen, and dissolved oxygen are "built in" to the program, whilst other, independent parameters (such as coliforms, or metals) may be selected by the user as appropriate.

The hydrodynamic solution is obtained by simultaneous solution of the depth-averaged continuity and St. Venant equations using a central implicit finite difference scheme in conjunction with the "alternating direction" method of solution. The transport of water quality determinands is governed by the nonlinear depth averaged advection diffusion equation (Rinaldi et al, 1979). This equation is solved using a modified version of the explicit Eulerian-Lagrangian approach of Cheng et al (1984).

The program has been specifically designed to be easily implemented for different model applications. Its modular form ensures that the creation of a specific model for each application entails the modification of only those units of the program directly concerned with the generation of bathymetry and boundary conditions, which are necessarily unique to each application. For instance, the recently completed model of the Tunis North Lake was required to represent the one-way valves used to control the flow into and out of the lake. The model was therefore arranged so that at each outlet gate location, as the outside water level falls below the lake water level the gate is considered to open and discharge is permitted. Discharge is controlled by the outside water level, and continues until the outside level rises just sufficiently to prevent further outward flow, at which point the gate is considered to close. No further flow is then permitted until the outside water level again falls below the level inside the gate. The model representation of the inlet gate is identical to that representing the outlet gates, but acting in reverse.

For economic use of computer time, the program is designed to handle systems of nested grids, with each successively finer grid being driven by boundary conditions generated by the grid above.

3. ZEBRA

The water quality/ecological process model ZEBRA (Zoned Eutrophication and Biological Recirculation Algorithm), describes the response and feedback of plankton and macrophytes to nutrient loadings in both water and sediment layers. The set of coupled, nonlinear differential equations describing these interactions (see, for example: Orlob, 1983; Canale and

Auer, 1982; Golterman, 1980; and Evans and Parslow, 1985) are solved numerically.

ZEBRA can be run independently from DAWN and is most easily applied to situations in which the area of interest may be considered as a linear series of interlinked boxes. Biological and chemical conditions within each box are assumed to be uniform, though the boxes need not be of equal depth or volume.

Water quality within each box may also vary due to the exchange of fluid between adjacent boxes. Both forward and reverse flows across the common borders of adjacent boxes may be specified, and additional discharges and abstractions from each box included. The flow rates may be specified either from observations or from output of one- or two- dimensional hydrodynamic models, and may be fixed in time (initial conditions) or be in the form of time series (boundary conditions). For long term runs (of the order of several years) in regions of relatively small tidal range, the fluxes between boxes are calculated by averaging the flows computed by DAWN over many tidal cycles.

ZEBRA is also driven by water temperature, light intensity, salinity and nutrient loadings which may be specified either directly as time series or by assigning a particular functional form to the temporal variation. The efficiency of the solution algorithm is such that the PC-based version takes of the order of 1 hour CPU time to simulate the ecological changes over one year.

4. Case studies

DAWN is the fundamental component of the Taw-Torridge Water Quality Model. In November 1986 South West Water Authority, now South West Water Services Ltd (SWW) were required by the Secretary of State for the Environment to produce a comprehensive plan for water quality throughout the Torridge Estuary in North Devon. Since the Taw Estuary combines with the Torridge at Appledore, SWW decided that this water quality plan should consider both estuaries and additionally, Bideford Bay, into which the estuaries discharge.

As part of the work required to produce the plan, a comprehensive environmental survey of the area was undertaken and the data collected during this survey were used for the calibration and verification of the model.

The nesting capabilities of DAWN were most suitable for this application where greater resolution was required for the two estuaries than throughout Bideford Bay. The entire modelled area is covered by a coarse grid having a resolution of 180m and the estuaries are covered by a finer resolution grid of 60m.

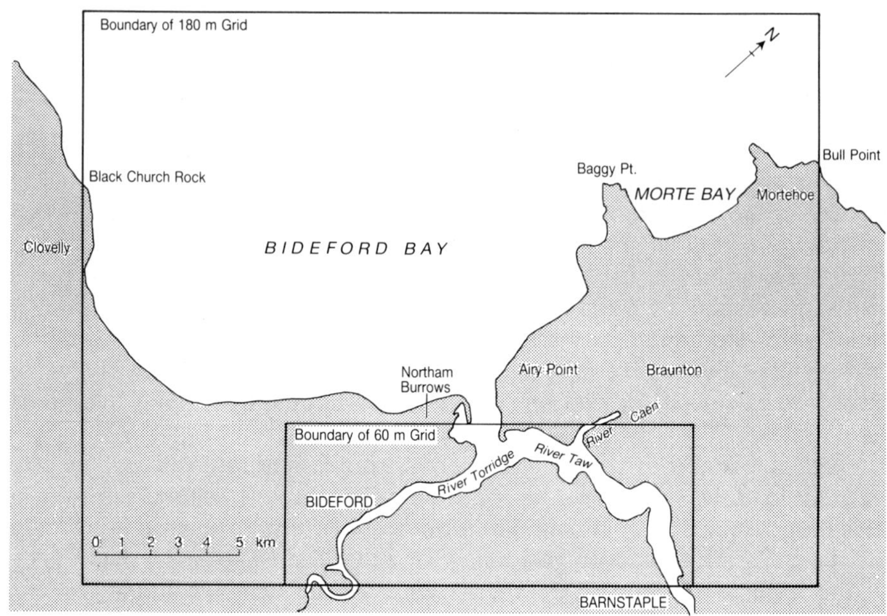

Fig. 1: Extent of the coarse and fine grids for the Taw-Torridge study.

The objective of the water quality modelling study was to use the model to assess water quality resulting from proposed sewage disposal options and thus act as a tool to aid decision making, particularly where no firm conclusions could be drawn from the consideration of other criteria. The basis for the assessment and comparison of the options were the environmental quality standards for the study area particularly those related to the EC directive regarding the quality of bathing waters. Water quality parameters investigated were total and faecal coliforms, total ammonia and dissolved oxygen.

A series of model simulations was carried out to investigate the water quality resulting from various sewage disposal options. After consideration of all other factors and the results of the modelling study, two disposal options were selected for further, more detailed investigation.

A second application of DAWN, this time in conjunction with ZEBRA, is the Tunis North Lake Water Quality and Ecological Model.

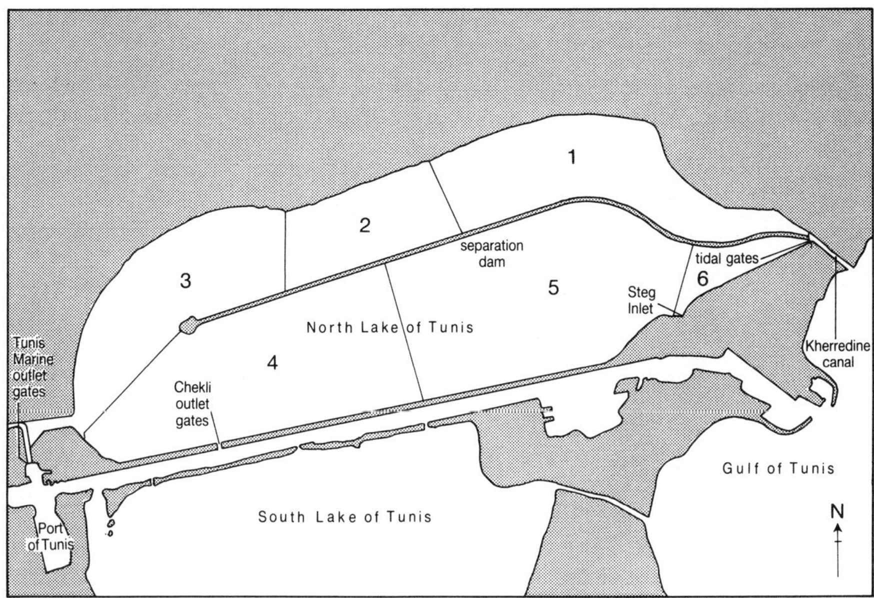

Fig. 2: Plan of Tunis North Lake showing the boundaries of the boxes used in ZEBRA.

The North Lake of Tunis is a shallow seawater lagoon on the north coast of Tunisia, east of the city of Tunis. The lake has been polluted over many decades due to effluent discharges from the city of Tunis and its surrounding communities. This has resulted in severe eutrophication and the growth of excessive quantities of macroalgae, particularly ulvae. The death and decay of these algae in conditions of high temperature, low wind and no rainfall has often led to anoxic conditions in the lake resulting in the rapid growth of sulphate reducing bacteria, evolution of hydrogen sulphide and the death of fish and other water life. These so-called "catastrophic situations" are characterised by red coloured water and obnoxious smells.

In order to restore the lake the Societe de Promotion du Lac de Tunis (SPLT) commissioned a Dutch consortium (the Lake Group) to design and construct a water circulation system to create a lake that was free from pollution and which would remain so.

Following completion of the construction stage of the restoration work SPLT commissioned Sir William Halcrow and Partners to supply a hydrodynamic and water quality model of the lake, together with the necessary computer hardware. The model will be used as a management

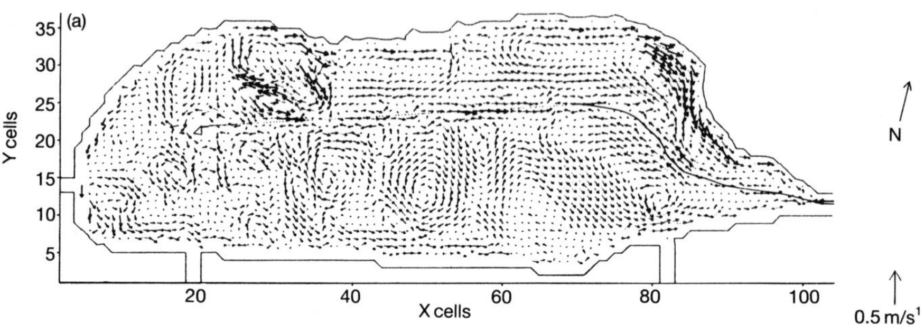

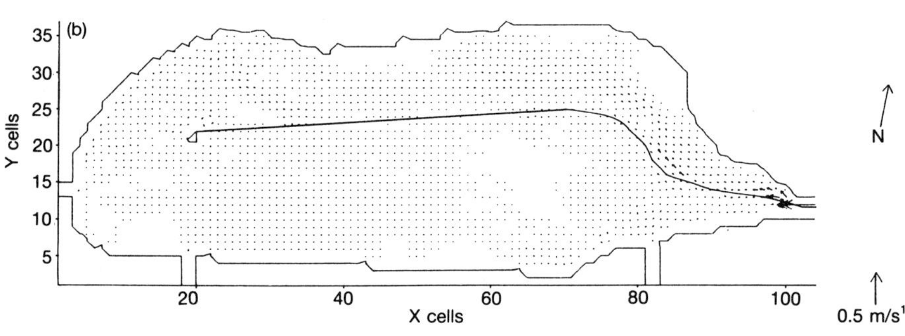

Fig. 3: Tidal currents throughout Tunis North Lake. (a) for 10 m.s^{-1} NW wind. (b) for zero wind.

tool to investigate both long-term water quality, taking into account the effects of algae, and short-term pollution incidents.

DAWN is used to simulate the hydrodynamics in the lake and short-term water quality scenarios whilst ZEBRA is used to simulate the long-term response and feedback of the aquatic biomass to nutrient concentrations in the water and sediment. The lake is represented as a series of 6 boxes in ZEBRA, as shown in Fig. 2; the forward and reverse flows across the box boundaries are obtained from DAWN which provides the connection between the two models.

Field data collected by the Lake Group were used to calibrate and verify the models. Fig. 3 shows currents averaged over a tidal cycle simulated with a 10 $m.s^{-1}$ wind from 315° and with zero wind. The complex flow and gyre patterns in Fig. 3a are typical of those produced when the wind blows from any direction. Note the lack of flow in the southern part of the lake (Fig. 3b) which is particularly prone to "catastrophic events" in calm conditions.

5. Hardware

For the Taw-Torridge Study the model was run on SWW's VAX 8530 and DEC 3100 workstations. The model operated at about one third of real time although this was dependent on the number of determinands being modelled and the grid resolution.

In the case of the Tunis North Lake Study, both models were installed on an IBM-compatible microcomputer for shipping as a complete, ready-to-run package to Tunisia. The machine, a Compaq 386/20e, was fitted with a 110mbyte hard disk and 8 mbytes of extra RAM memory. The model programs, written in Fortran, were compiled using an 80386 compiler able to access all of the additional memory. With a maths co-processor DAWN run-times, with a full suite of 13 determinands, were approximately one third real time.

6. Discussion

The two modelling studies described above have both been performed in coastal areas. However, the coupled model system also has wide application to inland waters. In the near future it is planned to apply similar models to rivers, inland lakes and reservoirs, especially where nutrient rich water is to be impounded. This will allow accurate predictions of the future trophic status of the retained water and enable suitable plans to be made to prevent water quality problems.

7. References

Canale, R.P. and Auer, M.T. (1982). Ecological Studies and Mathematical Modelling of Coastal, Estuarine and River Waters. J. Great Lakes Res., 8(1):112-125.

Cheng, R.T., Casulli, V. and Milford, S.N. (1984). Eulerian-Lagrangian Solution of the Convection Dispersion Equation in Natural Coordinates. Water Resources Research, 20(7):944-952.

Evans, G.T. and Parslow, J.S. (1985). A model of Annual Plankton Cycles. Biological Oceanography, 3(3):327-346.

Golterman, H. (1980). Quantifying the Eutrophication Process: Difficulties Caused, for Example, by Sediments. Prog. Wat. Tech., 12:63-80.

Orlob, G.T. (1983). Mathematical Modelling of Water Quality: Streams, Lakes and Reservoirs. Wiley: 518pp.

Rinaldi, S., Soncini-Sessa, R., Stelfest, H. and Tamura, H. (1979). Modelling and Control of River Quality. McGraw-Hill: 380pp.

Growth of a sand wave

S.J. Wakes, S. Hibberd, C. Baker

University of Nottingham

1. Background

There has been much speculation as to the nature of sand waves and why they occur in such places as deserts and river beds. Consequently there have been many papers on the subject including those by Kennedy (1969), Englund and Fredsoe (1982) and it has also been discussed at the Euromech Colloquia 156 (1982) and 192 (1985). When a steady flow of unidirectional fluid moves over a sedimentary bed, the sediment is disturbed and a pattern of bed forms is generated. These are either initiated from a plane bed or they are modified or migrate if in existence already. Typically two forms exist in subcritical flow; ripples with a wavelength dependant on grain size; and dunes with a wavelength dependant on fluid depth; the two forms may coexist and also be of comparable sizes. An experimental study of these can be found in Yalin (1972).

The motion of entrained sand is usually divided into either bed load or suspended load. In the bed load transportational mode the sediment grain moves by either rolling or sliding, continuous contact with the bed; or by saltation, regular jumps returning back to the bed. The grains go into suspended mode if the local turbulence is strong enough for the grain to be entrained into the main flow. The dynamics of the grain is thus a balance between mixing and settlement. There have been many, highly complex computer models for sediment movement, an example as given by Van Rijn (1985).

Most work in this field has been in two areas; the initial growth, i.e. finding preferential wavelengths; and the final steady state solution. Little work has been done on the growth or evolution of such bed forms.

2. Kinematic approach

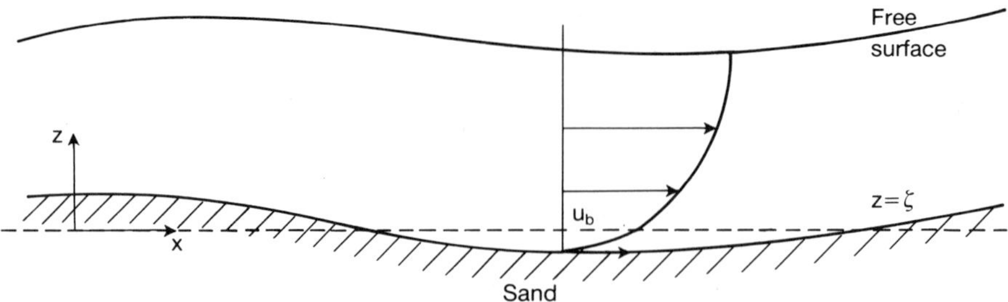

Fig. 1: Definition sketch for a 2D flow with a changing bed profile.

For a two-dimensional flow with a changing bed profile, as in Fig. 1, the overlying flow is linked to the sediment transport upon the bed by the Polya-Exner equation

$$(1-n)\frac{\partial \xi}{\partial t} + \frac{\partial q_b}{\partial x} = 0 \tag{2.1}$$

where q_b is the horizontal flux of sediment transported along the bed, and n the porosity of the bed. Equation (2.1) has been investigated by many authors, in the most detail by Van Rijn (1985), and it arises from a mass conservation argument. The sediment flux q_b depends upon the overlying flow and sand bed interaction and can be considered as driven by the excess shear stress τ_s above that required for sediment movement. Therefore one model is

$$q_b = q_b(\tau_s), \qquad where \quad \tau_s = \tau_s(\xi) \tag{2.2}$$

This equation has been extensively investigated experimentally (e.g. see Yalin, 1972) whilst $\tau_s = \tau_s(\xi)$ is found by solving the problem of turbulent flow over non-plane topography. This is not an easy problem and would have to be solved numerically. The inclusion of (2.2) into (2.1) gives

$$(1-n)\frac{\partial \xi}{\partial t} + q_b'(\xi)\frac{\partial \xi}{\partial x} = 0 \tag{2.3}$$

which has kinematic wave solutions. Examples of similar approaches are those associated with flood waves or traffic flows (Whitham, 1974). This equation is used to study established wave forms or for predicting growth rates of waves of a given wave number. It has been used by Fredsoe (1974) to study the steady migration of bed forms, but gives limited agreement with observations (Allen, 1982).

Papers studying the instability of an initially plane bed have all relied upon the solving of the final equations by numerical methods. Many of them employ an intricate fluid flow model over the sediment bed with a simplistic sediment transport equation as the link between them. No matter how accurate or complex the computer model it will always be the Polya-Exner equation upon which the inherent correctness of the model depends. The solution is not to strive to improve computer accuracy but to improve the sediment transport equation. This is demonstrated by looking at Smith (1970) who used the kinematic approach with an eddy viscosity model to conclude that all bed form wavelengths are unstable which is contrary to experimental data. Richards (1980) used a more sophisticated turbulence model but had to introduce a constant β, attributed to local slope effects, into q_b of the form

$$q_b = q_b\,(\tau_s - \beta\frac{\partial\xi}{\partial x}) \tag{2.4}$$

in order to isolate two wavelengths that were the most unstable. Sumer and Bakioglu (1984) complete a mathematically similar type of analysis with comparable results. Recently reported work by De Jong (1989) has suggested that internal waves may play an important role in selecting the bed form wavelengths. It showed that a concentration profile in a one layer fluid had considerable effect and that most of the changes take place for long waves. It has been seen through extensive experimental and observational evidence that sand particles at the bed surface become engulfed within the dunes leading to cross-bedding, e.g. see Siever (1988). This throws doubt upon the use of the kinematic boundary condition on the bed as used by some of the aforementioned papers. The other conditions most commonly used are that of zero flux of water through the bed and a tangential condition involving a slip velocity u_b. It is the use of a singular bed load layer that causes such confusion in deciding upon bottom boundary conditions.

3. Two layer model

In view of the shortcomings of the kinematic models discussed above, an approach that models the bed load region as a discrete layer of thin, but finite thickness, was decided upon to imitate reality as closely as possible. This layer is then sediment laden and within it sediment is transferred between the flow and bed by turbulent erosion and natural deposition, or stored within this region of varying width. Lyn (1988), using laser-doppler anemometry, confirms that such a layer exists and is not simply confined to a few sand grain diameters in thickness. This paper also showed that the presence of the mixture in a fluid effects the velocity of the mixture only near the bed, thus adding strength to the argument for this model. Two layer models have been suggested before by Shirasuna, McTigue (1981), Raudkivi (1983) and Armanini and di Silvio (1988), but little analysis exists in this context. Analogous problems where thin sedimentary layers are important include those in the theory of turbidity currents and avalanches. A thin boundary layer also has a crucial effect on waves in shallowing water, Mahony and Pritchard (1980).

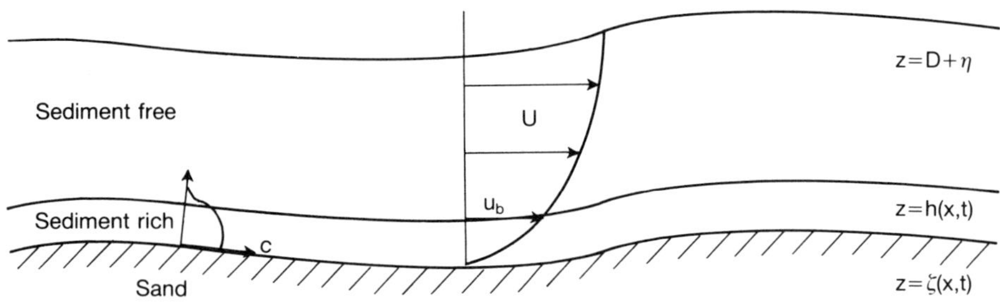

Fig. 2: Definition sketch for two layer model.

The simplified flow situation is shown in Fig. 2, with the sand bed denoted by $z = \xi(x,t)$ and a further interface $z = h(x,t)$, with the thickness of the layer being $\delta = h-\xi$. The typical thickness is δ_0, with $\delta_0 \ll D$, where D is the typical fluid depth. Sediment is assumed to be confined mainly to this thin layer and has a velocity profile matching the outer flow with value u_b. The interface boundary condition will satisfy a kinematic condition when the upper extent of the bed load layer may be given by taking a zero flux of sediment. The conservation of sediment mass equation reduces to

$$(1-n)\frac{\partial\xi}{\partial t} + \frac{\partial q_b}{\partial x} = -\frac{\partial(C\delta)}{\partial t} + f_{nb} \tag{3.1}$$

depth averaged across the thin layer; where C is the depth averaged sediment concentration, and f_{nb} is the normal sediment flux at the bed. The

Polya-Exner equation (2.1) is retrieved under the conditions $C\delta$ and f_{nb} are negligible; that is, the total sediment load/unit length is negligible. The form of q_b can be found from consideration of the sediment mixture momentum equations within this thin layer. A shallow layer approximation leads to substantial simplification although eventually the equations will have to be solved numerically. The equations reduce to

$$\frac{\partial q_b}{\partial t} + q_b \frac{\partial q_b}{\partial x} = -\frac{1}{\rho}\frac{\partial p_I}{\partial x} - \frac{Rg}{2}\frac{\partial (C\delta)}{\partial x} - \tau \qquad (3.2)$$

where: $\tau = -(\upsilon u_z)$, υ being the fluid viscosity; p_I is the pressure at the interface; and $R_g=((\rho_s-\rho_a)/\rho_a)g$, with ρ_s and ρ_a being the densities of the sediment and fluid respectively and g being the acceleration due to gravity. The terms on the right-hand side of (3.2) are the driving forces due to the interface pressure, the movement of sediment up and down slopes on the bed, and due to the difference in shear of the flow at the interface and on the bed, respectively. The left-hand side is the response of the fluid sediment mixture to these forces.

4. Discussion

It can be seen that this two layer model gives the more intricate sediment transport equation required and so when solved there will be more confidence in the solutions for the entire fluid, and so the numerical accuracy can be improved with the confidence that the model can support it.

5. References

Allen, J.R.L. (1982). Sedimentary Structures. Elsevier.

Armanini, A. and di Silvio, G. (1988). A one-dimensional model for the Transport of a sediment mixture in non-equilibrium conditions. Journal of Hydraulic Research, 26:275-291.

Englund, F. and Fredsoe, J. (1982). Sediment ripples and dunes. Annual Review of Fluid Mechanics, 14:13-37.

Fredsoe, J. (1974). On the development of dunes in erodible channels. Journal of Fluid Mechanics, 64:1-16.

De Jong, B. (1982). On the formation of dunes in open channel flow on an initially flattened erodible bed. Euromech, 156:119-126.

De Jong, B. (1989). Bed waves generated by internal waves in alluvial channels. ASCE Hydraulics, 115:801-817.

Kennedy, J.F. (1969). The formation of sediment ripples, dunes and antidunes. Annual Review of Fluid Mechanics, 1:147-168.

Lyn, D.A. (1988). A similarity approach to turbulent sediment laden flow in open channels. Journal of Fluid Mechanics, 193:1-26.

McTigue, D.F. (1981). Mixture theory for suspended sediment transport. ASCE Hydraulics, 107:659-673.

Mahony, J.J. and Pritchard, W.G. (1980). Wave reflection from beaches. Journal of Fluid Mechanics, 101:809-832.

Raudkivi, A.J. (1983). Thoughts on ripples and dunes. Journal of Hydraulic Research, 21:315-321.

Richards, K.J. (1980). The formation of ripples and dunes on an erodible bed. Journal of Fluid Mechanics, 99:597-618.

Van Rijn, L.C. (1985). Mathematical models for sediment concentration profiles in steady flow. Euromech, 192.

Shirasuna, T. Formation of sand waves. Proc 15th Congr Int Ass Hydr Res, p.107-114.

Smith, J.D. (1970). Stability of a sand bed subjected to a shear flow of low froude number. Journal of Geophysical Research, 75:5929-5940.

Sumer, B.M. and Bakioglu, M. (1984). On the formation of ripples on an erodible bed. Journal of Fluid Mechanics, 144:177-190.

Whitham, G.B. (1974). Linear and Nonlinear Waves. Wiley, New York.

Yalin, M.S. (1972). Mechanics of Sediment Transport. Pergamon Press.

SLOPES: the shoreline and orthogonal process emulation system

J. Hardisty

University of Hull

1. Introduction

Early attempts at predicting the two- and three-dimensional form of wave dominated nearshore seabed bathymetries concentrated on statistical correlations between the beach gradient and some measure of the incident waves and were later developed to include the full wave controlled orthogonal profile (cf King, 1972, for a review of the empirical models). Models based upon a better understanding of the processes began with Bagnold (1963), and have evolved into either numerical models based upon computer simulations (e.g. Davidson-Arnott, 1981; Horikawa, 1988) which "run" the process equations over a given initial bathymetry, or analytical models which attempt to solve the process equations directly for an equilibrium seabed gradient (e.g. Bowen, 1980; Bailard and Inman, 1981; Hardisty, 1986, 1990). Hardisty (1990) noted that although the analytical approach offers an understanding of the beach response through the use of state-space representations of system stability, the equations can presently be solved only for small sub-sections of the system and the predictive capability is therefore rather limited. Alternatively, the numerical approach promises a powerful predictive capability, but the computer programs are becoming so complex that it is difficult to understand the response of the beach system to changing inputs. Anderson and Sambles (1988) allude to the same problem and argue in favour of a new, more open modelling architecture in order to recover the ability to explain geomorphological responses. In the present paper, this suggestion is addressed through the construction and operation of a numerical model which was specifically designed to enable an understanding of the system through an analysis of the sensitivity of the geomorphological response to either inputs or processes. In addition Anderson and Sambles (1988) regret that the complex coding and machine specificity involved in existing models of hydrological and hillslope systems limits the transportability of the result and often prevents the incorporation of new ideas or algorithms without difficult re-coding. The present paper also addresses this problem by demonstrating the use of a common and commercially available spreadsheet "Microsoft

Excel" as a modelling environment. The spreadsheet operates on a microcomputer, is easily modified if additional processes are to be included, or existing algorithms changed, and incorporates a useful set of graphical routines which enable the output of results.

The present paper therefore has three specific aims. These are firstly to demonstrate the use of a microcomputer-based spreadsheet for geomorphological modelling; secondly to present a model which attempts to facilitate sensitivity analyses of the geomorphological response of the system and thirdly to use the model to examine the roles of the different hydrodynamic and sediment dynamic processes in the beach system.

2. Theory

The model has been called the Shoreline and Orthogonal Process Emulation System and will be referred to by the acronym SLOPES. SLOPES (Appendices I and II) operates by taking the deep water wave length and wave period, shoaling the waves over a given pre-existing orthogonal bathymetry and calculating the resulting nearbed flows. These represent the hydrodynamic processes and are detailed below. The resulting nearbed flows drive onshore and offshore sediment transport. These represent the sediment dynamic processes and are detailed below. Finally net sediment transport for each wave is used, by continuity, to change the seabed profile and hence to influence the consequent iterations of the model. These represent the morphodynamic processes and are detailed below. In the following description, parameters which remain constant and are input into the model through the control panel are termed "independent", whilst parameters which are calculated separately for each cell are termed "local".

2.1 Hydrodynamic processes

The theoretical basis for the hydrodynamic processes which effect a monochromatic wave train is well established and the following is referenced to Komar (1976) although corresponding equations can be found in any number of text books (e.g. Sleath, 1984; Dyer, 1986; Horikawa, 1988 etc). The deep water wave length and celerity are calculated from the independent wave period (Equations I.1 and I.2 in Appendix I; Komar, 1976, p37). The sediment bulk density is calculated from the independent sediment density and the independent sediment concentration (Eq.I.3). The elevation of the still water level with respect to an absolute datum is calculated from the independent amplitudes and phase differences of the solar and lunar semi-diurnal tidal components (Eq.I.1 and Doodson and Wartburg, 1941). The pre-existing profile of the seabed is calculated from an exponential power law which has been found to hold for American beaches (Eq.I.5 and I.6; Bowen, 1980).

The local wave length is calculated for the local water depth using an iterative, numerical solution of the dispersion equation (I.7 and Komar, 1976 p41), and the local wave celerity is calculated from the local wave length and the wave period (Eq.I.8 and Komar, 1976, p46). The local value of substitution variable "n" in the wave power relationship is calculated from the local water depth and the wave number (Eq.I.9 and Komar, 1976, p46). The local wave height is calculated from the deep water value through three groups of processes. Firstly, first order wave shoaling theory is utilised (Eq.I.10 and Komar, 1976, p104). Secondly allowance is made for the loss in wave energy due to seabed friction (Eq.I.12 and Horikawa, 1988, p76). Thirdly wave height changes due to refraction changes are calculated by assuming simple shore-parallel, but not necessarily equally spaced submarine contours and calculating the local angle of incidence (Eq.I.12, and Komar, 1976, p110) and hence the local wave height change due to refraction (Eq.I.13, and Komar, 1976, p104) from the deep water and local values of the wave celerity and from the deep water, independent value of the angle of incidence. The monochromatic wave train is assumed to break when the wave height exceeds a function of the water depth (Eq.I.14, and Dyer, 1986, p297), and thereafter the wave height is taken to be given by the same function. The local values of the surf scaling parameter and the Ursell number are calculated (Eq.I.15 and I.16).

These processes permit an estimation to be made of the local values of the wave height and length and are then used to generate local seabed currents. The nearbed oscillatory currents and drift current are calculated from second order theory (Eq.I.18 and Eq.I.19 and Komar, p 49) which is attenuated by an empirical coefficient (Hardisty, 1989).

2.2 Sediment dynamic processes

The theoretical basis for the transport of bedload and suspended load beneath waves on the seabed is far less well developed than are the wave equations. The choice of process functions is based largely on the Bagnold type of energetics model, and on some recent results for the important effect of bedslope on the sediment transport rate.

The instantaneous bedload rate is calculated from the excess stress function (Eq. I.20, Hardisty, 1983) and in which both the calibration coefficient, k_b, and the threshold velocity, u_{cr}, have been modified for the slope inclusive effects (Eq.I.21 and Eq.I.22, Whitehouse and Hardisty, 1988). The horizontal bed value of the calibration coefficient is taken from a review of marine applications of this particular transport function (Eq.I.23, Hardisty and Hamilton, 1984) and the horizontal bed value of the threshold velocity is also taken from a review of a range of laboratory and field data and is dependent upon the sediment grain size (Eq.I.24 and Eq.I.25, Komar and Miller, 1973). The net bedload sediment transport in each cell and for each wave is then calculated by integrating the instantaneous transport rates over the wave cycle (Eq.I.26).

In an earlier paper (Hardisty, 1990) attempted to utilise the energetics approach to relate the suspended load transport rate to the cube of the nearbed current. However, recent work by others, and the author's own observations using low light underwater television cameras on English Channel beaches suggests that this approach is unlikely to be successful, but that the nearbed suspension concentration and the bedload concentration are closely related. The two processes in fact appear to merge when finer sediments and accelerating flow fields are considered. Instead therefore a Rousian description of the suspended concentration profile is related to a reference concentration close to the bed, and this reference concentration is related to the bedload concentration. This approach appears to have a better physical basis and is analogous to the analysis reported by Vincent et al (1982). Specifically the peak bedload transport rate is calculated (Eq.I.27), and used to determine a reference concentration at a height z_a=1cm above the bed (Eq.I.28 and Eq.I.29). The Rouse concentration profile (Eq.I.30 and I.31, Vincent et al, 1982) is then calculated by assuming that the bedload concentration forms the reference concentration at a fixed height above the seabed. The corresponding, residual current profile is taken from second order theory (Eq.I.32, Komar, 1976 p49) and the product of the concentration and current profiles is integrated over the bottom 100cm of the water column and again through the wave cycle to obtain the net suspended load transport per wave in each cell (Eqs. I.33 and I.34).

Finally the net transports in each cell are used to erode or deposit sediment weights, and hence volumes and hence generate vertical height changes in accordance with the mass continuity equation (Eq.I.36). The results of each model run depend on initial boundary conditions and iteration sequences defined in the model's control panel as described below.

2.3 Computer implementation

Traditionally, environmental models are written in one of the higher level programming languages such as BASIC, Fortran or Pascal and sometimes operate on a microcomputer, but more commonly on a mainframe, and occasionally a supercomputer is required. There are many respects in which a linear, high level language is appropriate for the equally linear process paths by which the operation of the geomorphological system is envisioned.

There are however many limitations, not the least of which were identified by Sambles and Anderson (1988) as being the difficulties associated with either, on the one hand, understanding the results of a complex simulation in terms of the operative processes or, on the other, of revising the model and transporting it between different users operating within different computer environments.

In order to address some of these problems, it was decided that SLOPES would be implemented on a microcomputer-based spreadsheet. The choice between the two industry standard packages Lotus 123 (for MS-DOS) machines and Excel (for both MS-DOS and Apple Macintosh machines) was made in favour of the latter because it appeared to offer greater flexibility and graphical capabilities and because the School of Earth Resources in the University of Hull had developed a Macintosh network for undergraduate use. It was anticipated that this route might allow the teaching of environmental modelling to students without the usual prerequisite of computer programming courses. In the event this has proven to be possible, although SLOPES is rather more sophisticated than other, more typical, undergraduate models.

Microsoft Excel (Microsoft Corporation, 1986) is a spreadsheet consisting of up to 256 columns and 16,384 rows of cells. The columns are labelled from the left beginning with A and continuing through to Z, then AA to AZ and so on to IV. The rows are numbered from the top so that the top left hand cell is A1 and so on. Like all spreadsheets, each cell in Excel can contain numbers, strings or functions, and it is the ability to define the relationships between cells in numerical terms which allows the spreadsheet to be used for computer modelling. The present implementation of SLOPES operates within Excel on a Macintosh SE with 1mbyte of RAM and a 20mbytes hard disk. A Radius A3 display screen is used.

SLOPES was coded into columns A through to FE of the spreadsheet, and in its present implementation covers the first thirty nine rows. The model is two-dimensional and represents the vertical plane along the orthogonal path from the waters edge out to a particular seawards limit. Presently the maximum water depth is set at 9.25m, giving thirty seven model cells. Effectively the model shoals the deep water wave from row thirty nine to row three in the spreadsheet, whilst each of the processes operate from left to right in the different columns. Thus column O calculates the local wave length, columns P to AI calculate the local wave height, columns AP to CC calculate the local seabed currents and columns CE to EE calculate the net mass transports per wave.

The model will run a single iteration (equivalent to the passage of one wave from deep water to the shoreline) in about 90s which is approximately the same timescale as for the prototype processes. However, repeated iterations with the same, or different, incident wave conditions are controlled from a separate "command Macro" which is coded and uses a system recorder to duplicate keyboard functions.

3. Operating procedure

The model is run from the "Control Panel" (Fig. 1) which occupies columns G and H in the spreadsheet and allows the various processes to be, quite literally switched "on" and "off". In a first series of experiments referred to below the various process functions were individually introduced into the model in order to investigate their separate effect on the orthogonal profile response. For these experiments the model was run for a single iteration and operation is then initiated from the menu bar. Repeated iterations are controlled from a programmable "macro" which is called from within the spreadsheet and which controls the step length and number of iterations. Control of the experiment is then via the macro rather than the menu, and the model iterates automatically, and changes and saves the orthogonal profile evolution results.

The nearshore orthogonal profiles at Spurn Head on the English coast of the North Sea are close to the exponential form described above, and it is therefore not unreasonable to utilise wave data from that site for these simulations. The Institute of Oceanographic Sciences have analysed a twelve month wave record from the Dowsing Light vessel some 20 kms offshore from the site, and found that, typically, deep water wave heights are 1-2m with a mean annual zero crossing period of about 5-8 seconds. Operating the model at the lower end of this range produced slowly varying bathymetries, and therefore the experiments referred to in this paper have been generated with T=8s and H_∞=2m. These were taken as input parameters for the process sensitivity analysis. In addition to the input parameters, the SLOPES model contains a series of coefficients in the hydrodynamic and sediment dynamic equations, and these are shown by Column D in Fig. 1.

It is most useful to represent the first-order orthogonal profile, and the response of that profile to time and to changing inputs through time in terms of a few characteristics. SLOPES is a geomorphological model, and therefore the geomorphological variables are under consideration. The results were therefore characterised by the response function, $\partial z/\partial t$, which describes the system response to change, and by the evolving seabed profile itself. Each of the simulations represented the response function and profile after ninety six prototype hours of operation for a range of increasingly-complicated process functions.

	A	B	C	D	E	F
1						
2	SLOPES		CONSTANTS		units	
3						
4	Program Number WM_003		Wave Period	8.0	seconds	
5	Version Number 89.00		Deep Water wave length	99.8	m	
6			Deep water wave direction	15.0	degrees	
7			Deep water wave celerity	12.4800	m/s	
8			tidal phase	0.0000	°	
9			Deep water wave height	2.0000	m	
10			Wave friction factor	0.0100		
11			Attenuation coefficient	0.7500		
12	Software and Packaging is © Unico Geosystems Ltd, 1990		sediment grain diameter	0.1000	cm	
13			tan phi	0.6000		
14	Unico Geosystems Ltd		sediment density	2.6500	gm/cm^3	
15	School of Earth Resources		sediment concentration	0.6300		
16	University of Hull, HU6 7RX, U.K.		sediment bulk density	1.6695	gm/cm^3	
17			tidal height,	0.0000	m	
18	Telephone 0482 465374		density of seawater	1.0250	gm/cm^3	
19	Fax:		M2	2.2320	m	
20			S2	0.7880	m	
21			tidal phase	-4.0000	°	
22			tidal height	0.0000	m	
23						
24			bedload k	2.82E-07	gm cm-4 s-2	
25			fluid viscosity	0.010	cgs	
26						
27			Time steps	24.000	hrs	
28			RJSW m	7.000		
29						
30						
31						
32						
33						
34						
35						
36						
37						
38						
39						

	G	H	I	J	K	L	M	N	O	P	Q	R	S	T
1	CONTROL PANEL		h	distance	gradient	grain diam.	friction		W'length	kh	h/L∞		cosh(kh)	sinh(kh)
2			m	m		cm	factor		m					
3	DEFAULT : constant height	off	0.250	6	0.0417	0.10	0.0100		1.57	0.9992	0.0025		1.5429	1.1750
4	Shoaling height	on	0.500	17	0.0227	0.10	0.0100		3.14	0.9995	0.0050		1.5432	1.1754
5	Frictional loss	on	0.750	32	0.0167	0.10	0.0100		4.71	0.9999	0.0075		1.5437	1.1760
6	Refraction loss	on	1.000	49	0.0147	0.10	0.0100		6.28	1.0005	0.0100		1.5444	1.1769
7	Breaking loss	on	1.250	68	0.0132	0.10	0.0100		7.84	1.0012	0.0125		1.5453	1.1781
8			1.500	89	0.0119	0.10	0.0100		9.40	1.0021	0.0150		1.5463	1.1795
9			1.750	113	0.0104	0.10	0.0100		10.96	1.0032	0.0175		1.5476	1.1811
10			2.000	138	0.0100	0.10	0.0100		12.51	1.0044	0.0200		1.5491	1.1830
11	DEFAULT : first-order flow	off	2.250	164	0.0096	0.10	0.0100		14.06	1.0058	0.0225		1.5507	1.1852
12	Second-order flow	on	2.500	192	0.0089	0.10	0.0100		15.59	1.0074	0.0250		1.5526	1.1876
13	Second-order drift	on	2.750	222	0.0083	0.10	0.0100		17.12	1.0091	0.0275		1.5546	1.1903
14			3.000	253	0.0081	0.10	0.0100		18.64	1.0110	0.0300		1.5569	1.1932
15			3.250	285	0.0078	0.10	0.0100		20.16	1.0131	0.0326		1.5593	1.1964
16	DEFAULT : Integral bedload transport	off	3.500	319	0.0074	0.10	0.0100		21.66	1.0153	0.0351		1.5620	1.1999
17	Suspended load transport	on	3.750	353	0.0074	0.10	0.0100		23.15	1.0176	0.0376		1.5648	1.2036
18	Threshold inclusive transport	on	4.000	389	0.0069	0.10	0.0100		24.64	1.0202	0.0401		1.5679	1.2076
19	Slope inclusive transport	on	4.250	426	0.0068	0.10	0.0100		26.11	1.0229	0.0426		1.5712	1.2118
20			4.500	465	0.0064	0.10	0.0100		27.57	1.0257	0.0451		1.5746	1.2163
21			4.750	504	0.0064	0.10	0.0100		29.01	1.0287	0.0476		1.5783	1.2211
22	N.B. Defaults reset automatically		5.000	544	0.0063	0.10	0.0100		30.44	1.0319	0.0501		1.5822	1.2261
23			5.250	585	0.0061	0.10	0.0100		31.86	1.0353	0.0526		1.5863	1.2314
24			5.500	628	0.0058	0.10	0.0100		33.27	1.0387	0.0551		1.5906	1.2369
25			5.750	671	0.0058	0.10	0.0100		34.66	1.0424	0.0576		1.5951	1.2427
26			6.000	715	0.0057	0.10	0.0100		36.03	1.0462	0.0601		1.5999	1.2488
27			6.250	760	0.0056	0.10	0.0100		37.39	1.0502	0.0626		1.6048	1.2552
28			6.500	807	0.0053	0.10	0.0100		38.74	1.0543	0.0651		1.6100	1.2618
29			6.750	853	0.0054	0.10	0.0100		40.07	1.0586	0.0676		1.6154	1.2687
30			7.000	901	0.0052	0.10	0.0100		41.38	1.0630	0.0701		1.6211	1.2759
31			7.250	950	0.0051	0.10	0.0100		42.67	1.0676	0.0726		1.6269	1.2833
32			7.500	1000	0.0050	0.10	0.0100		43.95	1.0723	0.0751		1.6330	1.2910
33			7.750	1050	0.0050	0.10	0.0100		45.21	1.0772	0.0776		1.6393	1.2990
34			8.000	1101	0.0049	0.10	0.0100		46.45	1.0822	0.0801		1.6459	1.3073
35			8.250	1153	0.0048	0.10	0.0100		47.67	1.0874	0.0826		1.6527	1.3158
36			8.500	1206	0.0047	0.10	0.0100		48.88	1.0927	0.0851		1.6597	1.3246
37			8.750	1260	0.0046	0.10	0.0100		50.06	1.0982	0.0876		1.6670	1.3337
38			9.000	1314	0.0046	0.10	0.0100		51.23	1.1038	0.0901		1.6745	1.3431
39			9.250	1369	0.0045	0.10	0.0100		52.38	1.1096	0.0926		1.6823	1.3528

Fig. 1: Opening columns of the model as shown on the screen display.

4. Results

The sensitivity of the model to various process elements was investigated by initially running a "minimum process solution" in which first order wave theory was utilised without any of the shoaling transformations and with horizontal bedload transport functions. The result was nearbed flows which varied by less than 15% from deep water to the shore. The maxima, for example, range from 45.5 $cm.s^{-1}$ in 9.25m of water to 50.1 $cm.s^{-1}$ in 0.25m of water. Since the flows were symmetrical and there were no bedslope effects in the transport function, then there was no net bedload transport and hence no geomorphic response. There was, however, bedload transport and with $u_{cr}=0$ the rate varied with the cube of the flow speed, increasing from 0.0233 $gm.cm^{-1}.s^{-1}$ in deep water to 0.0355 $gm.cm^{-1}.s^{-1}$ at the shoreline.

Solutions were then obtained for the systems response to a series of increasingly sophisticated process elements. In general, it was found that all of the processes detailed above lead to net onshore transport, except for the effect of second order drift currents on the suspended load concentration profile and for the offshore transport generated by gravitational effects. The monochromatic description of the breaking waves also led, almost inevitably, to break-point bar development and a shorewards flattening of the profile. Typically, as shown in Fig. 2, the total load increases shorewards towards the breakers with the suspended component, in this simulation, contributing more than half the of the total transport. The spatial change in the transport rate increases at an increasing rate in a shorewards direction, and the inclusion of the suspended load increases $\partial I/\partial x$ by almost an order of magnitude. Shorewards of the breakers, rates are much lower. However, there is evidence of net seawards sediment transport of suspended load which tends to balance the onshore bedload transport and to result in negligible geomorphic response in shallow water. The behaviour of the response function, $\partial z/\partial t$, through time is shown at two locations in Fig. 2c, where, with increasing bed elevation at the bar crest, $\partial z/\partial t$ is positive but $\partial^2 z/\partial t^2$ is negative, indicative of stability. The rates are relatively constant enabling an estimation to be made of the relaxation time of the system. Given that $\partial^2 z/\partial t^2$ is 44mm/24hr over 72 hours (approximately 0.025 mm/hr/hr), then a value for $\partial z/\partial t$ of 438 mm/24hr (18.25 mm/hr) will reduce to zero in 730 hours. The relaxation time of the system at the bar crest in this simulation is therefore more than four weeks. Conversely, in the region seawards of the bar where $\partial z/\partial t$ is negative, $\partial^2 z/\partial t^2$ is positive which is again indicative of stability.

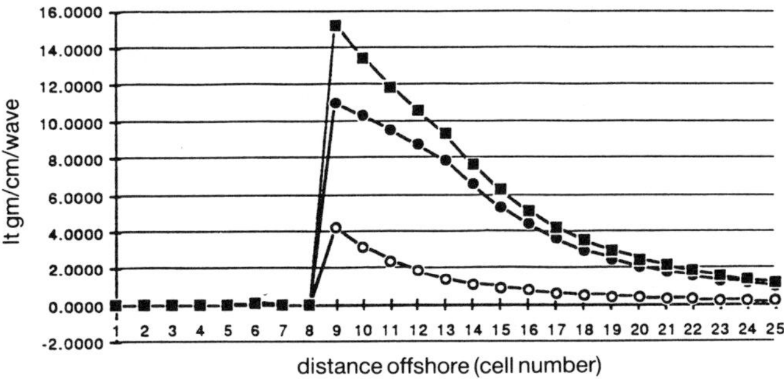

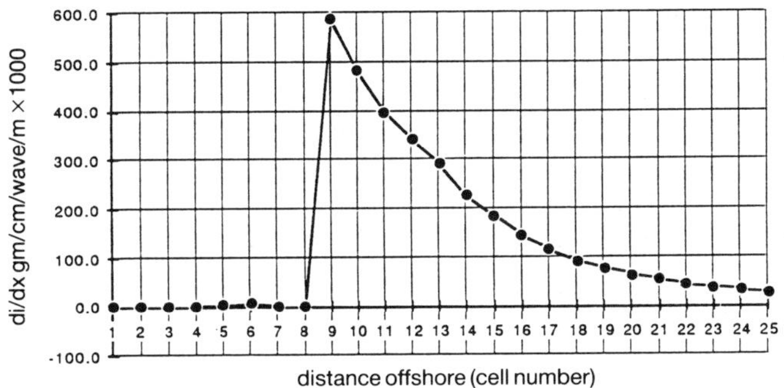

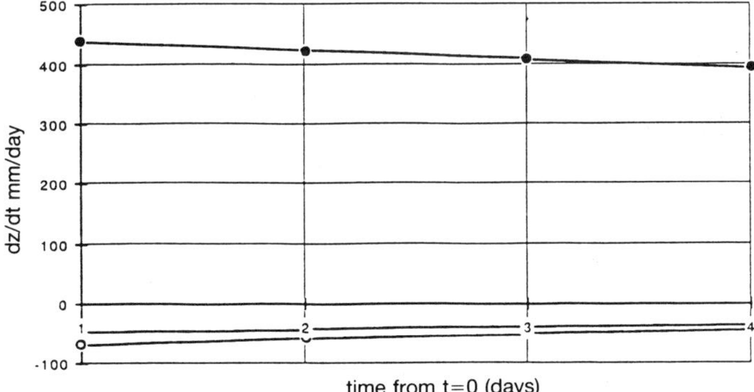

Fig. 2: (a) Bedload (open diamonds), suspended load (closed diamonds) and total load transport rates (closed squares) at the seawards boundary of each cell in the total load solution; (b) $\partial I/\partial x$; and (c) the response function plotted through time for the bar crest (x=138m and h=1.56, 1.4, 0.73 and 0.34m: closed circles) and offshore from the bar (x=164m and h=2.32, 2.38, 2.42, and 2.47m: open circles).

The system is therefore stable, and responds by more quickly establishing an equilibrium when the slope-inclusive effects (Fig. 3) are represented in the model, though the absolute values depend upon the initial bathymetry. It is none-the-less apparent that the full SLOPES model is generating a system response which is of the same order of magnitude as the prototype values.

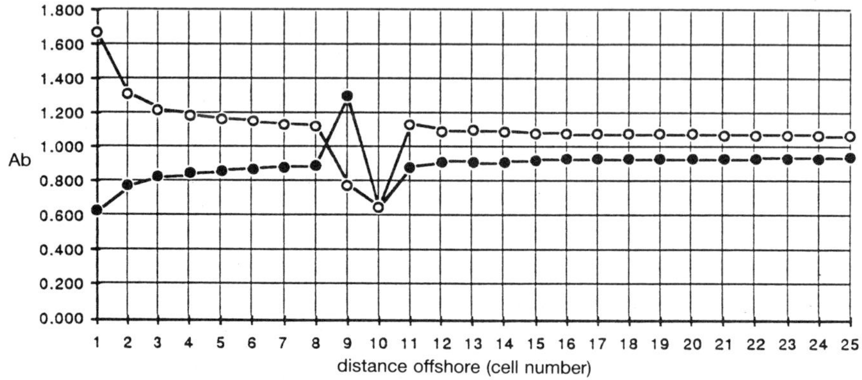

Fig. 3: The bedload slope correction factor, A_b, which the model applies to the shorewards transport (solid circles) and to the seawards transport (open circles) after 72 hours in the slope-inclusive solution.

5. References

Anderson, M.G. and Sambles, K.M. (1988). A review of the bases of geomorphological modelling. In: Modelling Geomorphological Systems, M.G. Anderson, editor. Wiley, Chichester, pp.1-32.

Bagnold, R.A. (1963). Mechanics of marine sedimentation. In: The Sea, Vol.3, Ideas and Observations, M.N. Hill, editor. Interscience, New York, pp.507-526.

Bailard, J.A. and Inman, D.L. (1981). An energetics model for a plane sloping beach: local transport. J. Geophys. Res, 86:2045-2053.

Bowen, A.J. (1980). Simple models of nearshore sedimentation, beach profiles and longshore bars. In: The Coastline of Canada, S.B. McCann, editor. Geol. Surv. of Canada, pp.1-11.

Davidson-Arnott, R.G.D. (1981). Computer simulation of nearshore bar formation. Earth Surface Processes and Landforms, 6:23-34.

Doodson, A.T. and Warburg, H.D. (1941). Admiralty Manual of Tides. HMSO, London, 270pp.

Dyer, K.R. (1986). Coastal and Estuarine Sediment Dynamics. Wiley, Chichester. 342pp.

Hardisty, J. (1983). An assessment and calibration of formulations for Bagnold's bedload equation. J. Sedim. Pet., 53:1007-1010.

Hardisty, J. (1986). A morphodynamic model for beach gradients. Earth Surface Processes and Landforms. 11:327-333.

Hardisty, J. (1989). Morphodynamic and experimental assessment of wave theories in intermediate water depths. Earth Surface Processes and Landforms, 14:107-118.

Hardisty, J. (1990). A note on suspension transport in the beach gradient model. Earth Surface Processes and Landforms, 15:91-96.

Hardisty, J. and Hamilton, D. (1984). Measurements of sediment transport on the sea bed southwest of England. Geo-Marine Letters, 4:19-23.

Horikawa, K. (1988). Nearshore Dynamics and Coastal Processes. University of Tokyo Press, Tokyo. 522pp.

King, C.A.M. (1972). Beaches and Coasts, 2nd Ed. Arnold, London. 570pp.

Komar, P.D. (1976). Beach Processes and Sedimentation. Prentice-Hall, New Jersey. 429pp.

Komar, P.D. and Miller, M. (1973). The threshold of sediment movement under oscillatory waves. J. Sedim. Pet., 43:1101-1110.

Microsoft Corporation, 1986. Microsoft Excel User's Guide. 364pp

Sleath, J.F.A., 1984. Sea Bed Mechanics. Wiley, New York. 335pp.

Vincent, C.E., Young, R.A. and Swift, D.J.P. (1982). The relationship between bedload and suspended sand transport on the Inner Shelf, Long Island, New York. J. Geophys. Res., 87:4163-4170.

Whitehouse, R.S.J. and Hardisty, J. (1988). Experimental assessment of two theories for the effect of bedslope on the threshold of bedload transport. Marine Geology, 79:135-139.

Appendix I. Formulae utilised in the SLOPES model.

$$L_\infty = 1.56\, T^2 \quad \text{I.1}$$

$$C_\infty = L_\infty / T \quad \text{I.2}$$

$$\rho_b = \rho_s\, C \quad \text{I.3}$$

$$z_t = Z_o + M_2 \sin(28.9841 t^\circ) + S_2 \sin(30.0000 t^\circ + p^\circ) \quad \text{I.4}$$

$$x = (13.33h)^{1.5} \quad \text{I.5}$$

$$\tan\beta = 0.05 x^{-0.33} \quad \text{I.6}$$

$$L = \frac{gT^2}{2\pi} \tanh\left(\frac{2\pi h}{L}\right) \quad \text{I.7}$$

$$C = \frac{L}{T} \quad \text{I.8}$$

$$n = \frac{1}{2}\left(1 + \frac{2kh}{\sinh(kh)}\right) \quad \text{I.9}$$

$$H_s = H_\infty \sqrt{\frac{1}{2n}\frac{C_\infty}{C}} \quad \text{I.10}$$

$$\Delta H_f = \int_{h=L_\infty/4}^{h} \frac{4\, f_e\, k^2\, H^2}{3\pi \sinh(kh)(\sinh 2kh + 2kh)}\, dx \quad \text{I.11}$$

$$\alpha = \sin^{-1}\left(\frac{\sin\alpha_\infty\, C}{C_\infty}\right) \quad \text{I.12}$$

$$\Delta H_r = H_\infty \sqrt{\frac{C_\infty}{2nC}\frac{\cos\alpha_\infty}{\cos\alpha}} \quad \text{I.13}$$

$$H_b = 0.72\, h\, (1 + 6.4 \tan\beta) \quad \text{I.14}$$

$$\varepsilon = \frac{4\pi^2 H_b}{gT^2 \tan^2\beta} \quad \text{I.15}$$

$$Ur = \frac{HL^2}{h^3} \quad \text{I.16}$$

$$u(t) = A_a \frac{\pi H}{T \sinh(kh)} \cos(\omega t) \quad \text{I.17}$$

$$u(t) = A_a\left(\frac{\pi H}{T\sinh(kh)}\cos(\omega t) + \frac{3}{4}\left(\frac{\pi H}{L}\right)^2 C\frac{1}{\sinh^4(kh)}\cos(2\omega t)\right) \qquad \text{I.18}$$

$$U = \frac{5}{4}\left(\frac{\pi H}{L}\right)^2 C\frac{1}{\sinh^2(kh)} \qquad \text{I.19}$$

$$i_b(t) = A_b(t)\, k_b \left(u(t)^2 - B_b(t)^2\, u_{cr}^2\right) u(t) \qquad \text{I.20}$$

$$A_b = \frac{\tan\phi}{\cos\beta\left(\tan\phi + \frac{u(t)\tan\beta}{|u(t)|}\right)} \qquad \text{I.21}$$

$$B_b = \sqrt{\frac{\tan\phi + \frac{u(t)\tan\beta}{|u(t)|}}{\tan\phi}\cos\beta} \qquad \text{I.22}$$

$$k_b = 0.282 \times 10^{-6} \qquad \text{I.23}$$

$$u_{cr} = 25.5\left(\frac{DH}{\sinh(kh)}\right)^{0.25} \quad \text{forD<0.05 cm} \qquad \text{I.24}$$

$$u_{cr} = D^{0.375}\left(\frac{H}{\sinh(kh)}\right)^{0.125} \quad \text{for D>0.05 cm} \qquad \text{I.25}$$

$$I_b = \int_{t=0}^{t=T} i_b(t)\; dt \qquad \text{I.26}$$

$$i_{max} = A_{bmax}\, k_b\, (u_{max}^2 - B_{bmax}^2\, u_{cr}^2)\, u_{max} \qquad \text{I.27}$$

$$C_a = 330\,\frac{i_{max}}{\rho_s\, u} \qquad \text{I.28}$$

$$z_a = 1 \qquad \text{I.29}$$

$$C_z = C_a\left(1 - A_c \ln\frac{z}{z_a}\right) \qquad \text{I.30}$$

$$A_c = 0.22 \qquad \text{I.31}$$

$$U_z = \frac{1}{4}\left(\frac{\pi H}{T}\right)\left(\frac{\pi H}{L}\right)\frac{1}{\sinh^2(kh)}$$

$$\{2\cosh\left(2kh\left(\frac{z}{h}-1\right)\right) + 3 + 2kh\left(3\frac{z^2}{h^2} + 4\frac{z}{h} + 1\right)\sinh 2kh$$

$$+ 3\left(\frac{\sinh 2kh}{kh} + \frac{3}{2}\right)\left(\frac{z^2}{h^2} - 1\right)\} \qquad \text{I.32}$$

$$\overline{i_s} = \frac{1}{T}\int_{t=0}^{t=T}\int_{z=0}^{z=1} U_z\, C_z\; dz\; dt \qquad \text{I.33}$$

$$I_s = T\,\overline{i_s} \qquad \text{I.34}$$

$$\frac{\partial z}{\partial t} = \frac{1}{\rho_b}\left(\frac{\partial i}{\partial t} + \frac{1}{U}\frac{\partial i}{\partial x}\right) \qquad \text{I.35}$$

Appendix II. Symbols.

A_a	attenuation coefficient for seabed currents	
A_b	slope coefficient for bedload rate parameter	
b	depth of bedload layer	m
B_b	slope coefficient for threshold velocity	
C	sediment concentration on the bed	
C_a	sediment concentration at height z_a above bed	
C_z	sediment concentration at height z above bed	
C_∞	deep water wave celerity	m s^{-1}
D	sediment grain diameter	m
f_w	wave friction factor	
ΔH_f	change in wave height due to friction	m
ΔH_r	change in wave height due to refraction	m
H	local wave height	m
H_b	breaking wave height	m
H_s	shoaled wave height	m
H_∞	deep water wave height	m
$i_b(t)$	instantaneous bedload rate	kg m^{-1} s^{-1}
i_s	suspended load transport rate	kg m^{-1} s^{-1}
I_s	net suspended load transport per wave	kg m^{-1}
L	local wave length	m
L_∞	deep water wave length	m
h	water depth	m
k	wave number (=2π/L)	
k_b	bedload rate parameter	kg m^{-4} s^{-2}
n	shoaling abreviation	
M_2	amplitude of lunar semi-diurnal tide	m
S_2	amplitude of solar semi-diurnal tide	m
ℓ	tidal phase	°
T	Wave period	s
u_{cr}	threshold velocity	m s^{-1}
u(t)	seabed current	m s^{-1}
U_r	Ursell number	
U_z	Velocity of second order drift current	m s^{-1}
x	distance offshore	m
z	depth below sea surface (positive upwards)	m
z_t	sea surface elevation	m
Z_o	datum height	m
α	local wave orthogonal direction	°
α_∞	deep water wave orthogonal direction	°
β	seabed gradient	
ε	surf scaling parameter	
ϕ	angle of internal friction of sediment	
ρ_b	sediment bulk density	kg m^{-3}
ρ_s	sediment density	kg m^{-3}
ω	wave radian frequency (=2π/T)	
max	local, maximum values of periodic parameters	

Index